Springer Undergraduate Texts in Mathematics and Technology

Springer Undergraduate Texts in Mathematics and Technology (SUMAT) publishes textbooks aimed primarily at the undergraduate. Each text is designed principally for students who are considering careers either in the mathematical sciences or in technology-based areas such as engineering, finance, information technology and computer science, bioscience and medicine, optimization or industry. Texts aim to be accessible introductions to a wide range of core mathematical disciplines and their practical, real-world applications; and are fashioned both for course use and for independent study.

More information about this series at http://www.springer.com/series/7438

Tomas Sauer

Continued Fractions
and Signal Processing

 Springer

Tomas Sauer
Universität Passau
Passau, Germany

ISSN 1867-5506 ISSN 1867-5514 (electronic)
Springer Undergraduate Texts in Mathematics and Technology
ISBN 978-3-030-84362-5 ISBN 978-3-030-84360-1 (eBook)
https://doi.org/10.1007/978-3-030-84360-1

Mathematics Subject Classification: 11A55, 40A15, 65D32

This Springer imprint is published by the registered company Springer Nature Switzerland AG
The registered company address is: Gewerbestrasse 11, 6330 Cham, Switzerland

Dedicated to the memory of my thesis advisor Hubert Berens who was a "Doktorvater" in the best meaning of the word.

Preface

Fractions are strange animals. Nowadays, everyone is or at least should be aware of what $\frac{3}{4}$ and $\frac{6}{8}$ stand for, and that these two expressions actually mean the same rational number, although they do not represent the same meter in dance music. For the ancient Egyptians on the other hand, these fractions were not really what they wanted; they preferred to deal with pure reciprocals or *unit fractions* like $\frac{1}{5}$ or $\frac{1}{9}$. The German word for that is *Stammbruch*; Google's automatic translation *Stem break* is a hint that artificial intelligence still has the potential for improvement. In Egyptian mathematics, fractions were actually written as sums of unit fractions and there even existed tables for such decompositions [53]. In this spirit, we (nowadays) consider $\frac{a}{b}$ with $a > b$ and write this fraction as

$$\frac{a}{b} = a_0 + \frac{c}{b} = a_0 + \frac{1}{\frac{b}{c}},$$

where the fraction appearing in the denominator of the unit fraction on the right-hand side is again similar to the type $\frac{a}{b}$; the numerator is larger than the denominator. This idea defines a recursive process and allows us to write each fraction formally as

$$\frac{a}{b} = a_0 + \frac{1}{[a_1; a_2, \ldots]} =: [a_0; a_1, a_2, \ldots].$$

This is the definition of a *continued fraction* . It turns out that any rational number, i.e., any fraction, has *exactly* one finite representation of the form $[a_0; a_1, \ldots, a_n]$ for some n, at least if trivial ambiguities are excluded. Therefore, the embarrassment of multiple representations such as $\frac{3}{4} = \frac{6}{8}$ is avoided here. Moreover, any real number can be expressed, again uniquely, by an *infinite* continued fraction expansion $[a_0; a_1, a_2, \ldots]$; most prominently known is the famous *golden ratio*

$$\frac{1+\sqrt{5}}{2} = [1; 1, 1, \ldots],$$

which we will identify as the most irrational number precisely due to the fact that it has the simplest possible continued fraction expansion. According to [61] who in turn refers to [3], this was the way the ancient Greeks introduced real numbers once they realized that $\sqrt{2} = [1; 2, 2, \ldots]$ is not rational, that is, $\sqrt{2}$ cannot be written as a fraction p/q with $p \in \mathbb{Z}$ and $q \in \mathbb{N}$.

Even after these very simple observations it should not come as a big surprise any more that continued fractions play a fundamental role in Number Theory and even in music. This will be the content of the first major chapter of this book. But once one deals with rational numbers and starts to think beyond numbers, one may be tempted to also consider more general rational objects, in particular rational functions in the sense of quotients of polynomials. Continued fractions with polynomials in them were an important topic in the eighteenth and nineteenth centuries with major contributions by Bernoulli, Euler, Gauss and many more. Their contribution will occupy the major part of this book and will relate to concepts like quadrature, moment problems and sparse recovery which all appear, not quite accidentally in connection with Signal Processing, which will make these classical mathematical tools relevant for modern applications as well. Not that they have ever been irrelevant, just to record that.

The last topic is to describe polynomials which have zeros that are restricted to certain locations, either outside the unit circle or in the left half-plane. The latter, the so-called *Hurwitz polynomials* have a really nice characterization in terms of continued fractions.

Prerequisites and Recommended Reading

This book emerged from an introductory lecture on continued fractions, and still is supposed to act as elaborated lecture notes that view continued fractions from a Signal Processing perspective and focus on their role in numerical applications. To that end, it starts with an introductory chapter that summarizes the main concepts to be considered and outlines what will happen in the chapters to follow where results are worked out together with all the necessary details of sometimes a technical nature. The presentation aims at being mostly self-contained and should not rely on too much previous knowledge, except for the standard basic knowledge in Analysis and Linear Algebra. Some more complex detailed aspects or material that is not central to the main theme and therefore did not make it into the presentation, but are provided in the exercises so the reader can elaborate on them. The problems should be structured enough to provide sufficient information on the intermediate steps, thereby helping to work out these additional facts and complete the theory.

Some more specific remarks on the requirements needed for the individual chapters might be helpful for using this book for lectures or seminars. Of course Chap. 1 only surveys things to come and requires only curiosity from the readers' side.

Chapter 2 is almost completely elementary and requires not much more than the skill to manipulate fractions, to follow a "proof by induction" and to compute a derivative of polynomials or exponential functions once in a while. The price to be paid for that is Sect. 2.6 which deals with algebraic numbers and only remains very superficial. Here, it may be useful to have a look at [45] where, for example, also the transcendence of π and e are proved in part even by using continued fractions. For a deeper connection between math and music that could make a seminar of its own, the first choice is still [5], but also [82] and [69] are highly recommended.

Chapter 3 makes use of some basic concepts of algebra, especially Euclidean rings. Any standard literature on algebraic structures is more than sufficient here and some acquaintance with these concepts is not really necessary, only helpful. Due to the algorithmic nature of the chapter, [31] is a good choice for further reading in case some things are not clear.

Chapter 4 is mostly driven by explaining "Gauss' quadrature", i.e., the original approach by Gauss for developing what is nowadays known as Gaussian quadrature. The quadrature problem is treated in all classics in Numerical Analysis, for example [54]. For further background, I would recommend [33] which was written by an expert in quadrature and orthogonal polynomials. Although things become a bit more intricate and tricky in that chapter, everything is still mostly elementary and should be understandable from the presentation.

Chapter 5 uses the Dirac distribution which is popular in the engineering literature and on Wikipedia pages where it often is used like a function which frequently leads to incorrect proofs. Basic facts on distributions are pointed out in [114] in a very condensed way; a more substantial presentation to understand generalized functions is to view functions as functionals as provided in [34], but it is really not necessary here. What is important is an awareness that distributions are beautiful but treacherous. Again, the rest of the chapter is fairly elementary; infinite matrices and operators on sequence spaces work in a straightforward extension of the finite case. Some basic knowledge, however, in Functional Analysis, see, for example [63], could be useful.

Chapter 6 introduces basic concepts of mathematical signal processing, still based on the classics like [44]. Useful additional reading, especially on allpass and wavelets would be [106]; and of course [71]. This additional seconding will be helpful to obtain a more complete image of the underlying theory than what is briefly surveyed in Chap. 6; but the knowledge will be especially useful for working out seminar talks based on this chapter.

Chapter 7 uses a fundamental concept in Function Theory, namely the so-called *argument principle* which will be recalled in Theorem 7.4. It is a consequence of the residue theorem, hence it is located at the end of a one-semester lecture on Function Theory, and can be found, e.g., in [25, 50, 100]. Its main application is identity (7.11) and if one is willing to "believe" this identity, one can restrict the needed Complex Analysis to elementary manipulations of complex numbers that should be known from calculus/analysis. The other concepts in this chapter are provided and nicely introduced in the first chapters of [27].

Literature on Continued Fractions

The literature on continued fractions is not overly huge, but definitely existent. Besides **the** classic, Perron's two volumes [78, 79], one has to mention Khinchin [60] as the other extreme; while Perron is truly encyclopedic, covers a lot of issues, he is not so easy to read (even besides the fact that the books are in German), Khinchin's book is thin, but a very well-written introduction to continued fractions. Even if some material has been added meanwhile and some of the proofs were modified, it still may be recognized as the guideline for Chap. 2. Another classic book where basics on continued fractions can be found, of course in the context of Number Theory, is the one by Hardy and Wright [45]. The special application to music from Sect. 2.7 is covered in almost all books on Mathematical Music Theory, from the highly recommended and mathematically substantial masterpiece by Benson [5] to [23, 82]. Their connection is also mentioned more from the perspective of Computer Algebra in [31] and [61]. Apologies to all I forgot to mention here. Applications in Analysis similar to what we consider in later chapters were already treated in [107] which is even older than Perron's work; and in [68]. Connections to orthogonal polynomials can be found in [16]. The statistical context is presented in greater detail in [10].

There are also quite recent books on the issue: Hensley [48] covers Number Theory, Ergodic Theory and complex numbers, and a very careful overview especially about number-theoretic aspects is also provided in [64]; geometric aspects and even geometry-based extensions to the multivariate case are pointed out in [57]. The role of continued fractions in Number Theory is once again considered in quite some depth in [9] which combines classical with more modern aspects. Finally, a more educational approach from a somewhat recreational perspective is provided in [101].

Some Personal Remarks

My first personal encounter with continued fractions happened by accident and was totally unintended as I look back. Working as a teaching assistant at the University of Erlangen, I was responsible for the exercises in the Numerical Analysis class of my boss and thesis supervisor Hubert Berens. When he arrived to teach Gaussian quadrature, he told me that he had read, probably in Chihara's book [16], that Gauss himself had not used orthogonal polynomials (which were unknown at this time as a formal concept anyway, so why should he?) but that he did something with continued fractions instead. His simple order was *Make some exercise out of it*. Unfortunately, nothing of sufficient detail could be found in modern literature, so the only possibility was to grab Gauss' collected works [112] from the library and to translate paper [32] from there. The proof is really a masterpiece of clarity and obscurity: one can follow any single step made there, but in the end one is clueless as to *why* he did things this way. Nevertheless, there is the explicit reference to Bernoulli's work on continued fractions [6] on which the magic of the continued fractions is based.

After these exercises continued fractions became dormant for about 15 years, until, then at the University of Giessen, I encountered Khinchin's booklet [60] on continued fractions due to his also very nice paper [59] on an axiomatic intro-duction of *entropy*; so I decided to make good use of the freedom of research and teaching by offering a lecture on continued fractions. The notes for this lecture are still the core of this book, and while working on them I finally understood what the idea was behind Gauss' approach and that this way of treating the problem was actually totally natural and almost straightforward: just approximate the generating function of the moment sequence with as many terms as possible—all this usage of mathematical concepts and methods that were quite standard at that time. That was the first time when the idea of turning this into a book somewhat came up; there was even a proposal and a report, but then for several reasons things once more were postponed for quite some time.

Finally, now at the University of Passau, internationalization made it necessary to translate the notes into English; having done research on Prony's problem meanwhile widened the field of applications and allowed me to address some points of the almost forgotten report from the past which all of a sudden made good sense to me. In the end, classical concepts and modern application merged continued fractions and Signal Processing into a homogeneous and nice mathematical theory. When finally the time came for writing up and polishing things, yet another unexpected event affected the book: the corona virus crisis in the spring of 2020 that forced the cancellation of long planned trips, conferences and workshops and give the book much more time and attention than anticipated. For example, Sect. 7.7 was completely worked out from scratch during this period. In the end, even bad things always also have some good effects, which is encouraging.

Passau, Germany   Tomas Sauer
2021

Acknowledgements

First, my thanks go to Hubert Berens, not only for accidentally getting me addicted to continued fractions, but especially for patiently teaching me how to do mathematics.

The very first version of this lecture was greatly improved during several discussions with and suggestions by H. Michael Möller that came up in the course of our collaboration.

In finishing the book I especially want to thank Dörte Rüweler who spent a lot of time in holding the exercise sessions for the lecture and provided me with several of the exercises that can now be found in the problem section. Benedikt Diederichs did a very careful reading of the notes, pointing out some mistakes and making valuable suggestions to improve content and presentation. I am lucky to have such people in my group. In addition, my Bachelor Student Markus Georgi found various typos and mistakes while preparing his thesis.

Ann Kostant did a great job in checking and improving the language for the book which I really appreciated. It has been a privilege to work with her.

And, last but not least, I want to thank my wife Astrid for love, patience and support over all the years.

Contents

Chapter 1
Continued Fractions and What Can Be Done with Them

*That's the reason they're called lessons,[…] because they lessen
from day to day.*

L. *Carroll*, Alice's Adventures in Wonderland

1.1 The First and Fundamental Definition

In its simplest form, a *continued fraction* is a fraction, i.e., a ratio of integer numbers, with a *denominator* that is written in the same style as a continued fraction. This informal introduction is, however, a self-referential and recursive definition, so that we better make clear immediately what we are really talking about.

Definition 1.1 For integers $a_0, \ldots, a_n \in \mathbb{Z}$, the associated *continued fraction* is the rational number

$$[a_0; a_1, \ldots, a_n] = a_0 + \cfrac{1}{a_1 + \cfrac{1}{a_2 + \cfrac{1}{\ddots + \cfrac{1}{a_{n-1} + \cfrac{1}{a_n}}}}} \qquad (1.1)$$

Still, this "dot notation" is neither exact nor does it give rise to a really well-defined object. We just have to look at the cases where $a_n = 0$ or $a_{n-1} = -1/a_n$ to get an idea of what can go wrong. In both cases we would divide by zero which is not really welcome in mathematics. Clearly, this can and will be easily overcome by choosing $a_k \in \mathbb{N}$ as *positive integers*, and only allow the somewhat special a_0 to take arbitrary

© The Author(s), under exclusive license to Springer Nature Switzerland AG 2021
T. Sauer, *Continued Fractions and Signal Processing*, Springer Undergraduate Texts
in Mathematics and Technology, https://doi.org/10.1007/978-3-030-84360-1_1

values in $\mathbb{Z}$. This is, in fact, the direction that we will pursue later, especially since it will provide further advantages.

A simple recursive definition of the continued fraction results from a closer inspection of (1.1), since it reveals that the denominator of the continued fraction is a continued fraction again, namely $[a_1; a_2, \ldots, a_n]$. In this way we obtain the recursive formula

$$[a_0; a_1] := a_0 + \frac{1}{a_1}, \qquad [a_0; a_1, \ldots, a_n] := a_0 + \frac{1}{[a_1; a_2, \ldots, a_n]}, \qquad n \in \mathbb{N}.$$
$$(1.2)$$

Definition (1.2) already shows us what would happen in the degenerate cases mentioned above. Assume for example that $[a_k; a_{k+1}, \ldots, a_n] = 0$. Then in an imprecisely written, formally incorrect way that is for illustrative purposes only,[1] we obtain the following "conclusion"

$$[a_{k-1}; a_k, \ldots, a_n] = a_{k-1} + \frac{1}{[a_k; a_{k+1}, \ldots, a_n]} = a_{k-1} + \frac{1}{0} = \infty,$$

$$[a_{k-2}; a_{k-1}, \ldots, a_n] = a_{k-2} + \frac{1}{[a_{k-1}; a_k, \ldots, a_n]} = a_{k-2} + \frac{1}{\infty} = a_{k-2};$$

and as long as we do not encounter the additional degeneracy $a_{k-2} = 0$, everything proceeds normally in the recursion. Hence division by zero is not such a sacrilege in this case, at least as long it does not happen too often. Nevertheless, it is even better to avoid all the trouble by choosing $a_0 \in \mathbb{Z}$ and $a_j \in \mathbb{N}$. This will be the concept that we will follow in the sequel, as well as for a lot of other good reasons beyond avoiding ambiguities.

Remark 1.1 *(Notation)* To make it clear: we are using the notation $\mathbb{N} := \{1, 2, \ldots\}$ and $\mathbb{N}_0 := \mathbb{N} \cup \{0\}$ for positive and nonnegative integers, respectively.

Moreover, the recursive identity (1.2) immediately allows us to develop "normal" fractions into continued fractions, as a very simple example shows. Let us consider $\frac{13}{7}$ and note that

$$\frac{13}{7} = 1 + \frac{6}{7} = 1 + \frac{1}{7/6} \qquad \text{and} \qquad \frac{7}{6} = 1 + \frac{1}{6} = [1; 6],$$

hence

$$\frac{13}{7} = 1 + \frac{1}{[1; 6]} = [1; 1, 6],$$

which already illustrates the simple pattern behind the continued fraction expansion of rational numbers.

On the other hand, since all operations involved in (1.1) are additions and divisions, we do not even have to restrict continued fractions to integer coefficients. Indeed,

---

[1] Please do not quote these formulas, and definitely not with a reference to this book.

we could in the same fashion even define *rational* continued fractions with the form $[r_0; r_1, \ldots, r_n]$ and with $r_j \in \mathbb{Q} \backslash \{0\}$. A simple and immediate formula is

$$[a_0; a_1, \ldots, a_k, \ldots, a_n] = a_0 + \cfrac{1}{a_1 + \cfrac{1}{\ddots + \boxed{a_k + \cfrac{1}{\ddots\, a_{n-1} + \cfrac{1}{a_n}}}}}$$

$$= a_0 + \cfrac{1}{a_1 + \cfrac{1}{\ddots\, a_{k-1} + \cfrac{1}{[a_k; a_{k+1}, \ldots, a_n]}}}$$

$$= \left[a_0; a_1, \ldots, a_{k-1}, [a_k; a_{k+1}, \ldots, a_n] \right]$$

$$= \left[a_0; a_1, \ldots, a_{k-1}, r_k \right],$$

which makes use of the *remainder* $r_k := \left[a_k; a_{k+1}, \ldots, a_n \right]$. As long as $a_j \in \mathbb{Z} \backslash \{0\}$ or even $r_j \in \mathbb{Q} \backslash \{0\}$, the continued fraction is a rational number which can be shown for example with a simple induction over the number of parameters in the formula (1.2). All that is needed is the fact that rational numbers form a *field*, and thus are closed under addition and taking reciprocals.

Every *finite* sequence $a_0, \ldots, a_n$ of numbers is the initial sequence of an *infinite* sequence $a = (a_j : j \in \mathbb{N}_0)$ which also enables us to consider an *infinite continued fraction* of the form

$$[a_0; a_1, \ldots] = a_0 + \cfrac{1}{a_1 + \cfrac{1}{a_2 + \cdots}}.$$

This way of writing the continued fraction is nice as a formal expression, but what is the value of such an infinite object? In principle this is clear: it is the *limit* of the continued fractions associated with the finite initial segments, i.e.,

$$[a_0; a_1, \ldots] = \lim_{n \to \infty} [a_0; a_1, \ldots, a_n],$$

but we have to understand when this limit really *exists*, that means when such an infinite continuous fraction converges. The criterion for that will be not only simple but also very handy and elegant. And it even works for continued fractions with rational coefficients.

Theorem 1.1 (Convergence criterion for continued fractions) *For $r_j \in \mathbb{Q}$, $r_j > 0$, $j \in \mathbb{N}$, the continued fraction $[r_0; r_1, \ldots]$ is convergent if and only if*

$$\sum_{j=0}^{\infty} r_j = \infty. \tag{1.3}$$

(1.3) *is trivially satisfied in the special case that* $r_j = a_j \in \mathbb{N}$.

This tells us that infinite continued fractions will be well behaved provided that $a_1, a_2, \ldots$ are chosen as *positive integers*. Since in this case the continued fraction $[a_1; a_2, \ldots]$ is positive, we allow $a_0 \in \mathbb{Z}$ to be capable of representing negative numbers as well. And indeed this approach gives us "everything".

Theorem 1.2 *Any* real number $x \in \mathbb{R}$ *can be written as a continued fraction* $[a_0; a_1, \ldots]$ *with* $a_0 \in \mathbb{Z}$ *and* $a_j \in \mathbb{N}_0$, $j \in \mathbb{N}$, *and this continued fraction is a* finite continued fraction *if and only if* x *is a* rational number.

Moreover, we will find out that the continued fraction expansion of a real number is *unique*, except for a little bit of an ambiguity that occurs for rational numbers only.

Remark 1.2 The statement in Theorem 1.2 is by far more elegant than considering the decimal expansion of a real number where the number is rational if and only if the expansion is *finite* or *periodic*.

According to [61, S. 359], which in turn refers to [3], the ancient Greek mathematicians used continued fractions for a first definition of a concept resembling real numbers; in general this occured after their discovery of the existence of *irrational numbers*.[2] They did not use "normal" fractions and in particular they did not use the infinite decimal expansions which are common nowadays.

This was a really good choice, since we will see that with the same effort that was measured in the number of digits used, continued fractions give a much better approximation for an irrational number than do fractions with an arbitrary numerator and denominator or with decimal expansions.

To approach this topic in a quantitative fashion, we consider the *approximation* of real numbers by rational numbers. It will turn out that the *best approximation* from $\mathbb{Q}$ to an irrational number, among all rational numbers with a certain maximal denominator, is a continued fraction. The best approximation has to be defined properly of course, as any irrational number can be approximated arbitrarily well with rational numbers; the standard case is a truncated or rounded *decimal expansion* as in $\pi \approx 3.1416$.

Taking into account the simple observation that any rational number can be written as $x = k + \frac{p}{q}$, $k \in \mathbb{Z}$, $0 \le p < q$, with the fairly irrelevant integer part k, the complexity of a rational number p/q can be measured by the size of its denominator. Therefore it is reasonable to ask which functions $\varphi : \mathbb{N} \to \mathbb{R}$ can describe the order

---

[2] Those who do not yet know the story about the Pythagoreans, their somewhat religious and rational view of the world and harmony as fraction with a small numerator and denominator, are recommended to find out about it. As popularized science this can be found, for example in [51, 85].

of approximation of x by rational numbers p/q in terms of its denominator q; this means that we have an expression for the error

$$\left| x - \frac{p}{q} \right| \leq \varphi(q). \tag{1.4}$$

Moreover, we may ask what is the fastest possible rate of decay φ. If p/q is a truncated or rounded decimal expansion, we have that $q = 10^{-n}$ for some n and the error is bounded by $\frac{1}{2} \times 10^{-n} = \frac{1}{2q} =: \varphi(q)$. Can we do better? The answer is yes and we will find out that the best possible rate is $\varphi(q) = \frac{1}{\sqrt{5}q^2}$, and any similar best approximant has to be a convergent of the continued fraction expansion of x:

$$x = [a_0; a_1, \dots], \quad \text{and} \quad \frac{p}{q} = \lfloor a_0; a_1, \dots, a_n \rfloor$$

for some $n \in \mathbb{N}$. This will be shown in Theorem 2.12.

This theory has the nice side effect since it will tell us that the real number that is hardest to approximate using rational numbers is the *golden ratio* $\frac{1+\sqrt{5}}{2}$, and the poor approximation rate will be a consequence of its particularly simple representation $[1; 1, 1, \dots]$ as a continued fraction. Indeed the simpler the continued fraction expansion of a number is, the harder it is to approximate with rational numbers in a way that will be specified in Sect. 2.4. This has implications in number theory as it allows us to distinguish between an *algebraic number*, i.e., a zero of a polynomial with integer coefficients, and a *transcendental* number, i.e., a real number that does not have this property. In fact, a classical result by Liouville states that for any algebraic number a that is the zero of a polynomial of degree n, one has

$$\left| a - \frac{p}{q} \right| > \frac{C}{q^n}$$

for some constant $C > 0$, so that $\varphi(q)$ cannot be chosen to be of a better order than q^{-n}. This was significantly improved by the Thue–Siegel–Roth Theorem from which we can conclude that for any $\varepsilon > 0$,

$$\left| a - \frac{p}{q} \right| > \frac{C(\varepsilon)}{q^{2+\varepsilon}};$$

hence the approximation $\varphi(q)$ cannot be of a better order than q^{-2} for algebraic numbers. In contrast to that rather slow order of approximation, transcendental numbers can be constructed with an arbitrary approximation rate φ, for example as with the famous Liouville number that we will meet in Corollary 2.6. This explains why continued fractions play an important role in number theory which will, however, only be touched on in this book.

Moreover continued fractions also answer the question as to why the musical octave consists of 12 semitones and how many tones the next reasonable scale should have—besides the pentatonic ones and our 12 semitone octave. Of course the number can only be 53.

1.2 Continued Fractions with Polynomials in Them

Continued fractions can be built from various objects. We already observed this in the case of nonzero integers $\mathbb{Z} \setminus \{0\}$ and rational numbers $\mathbb{Q}$. It will, however, turn out that most of the concept works whenever we can add and multiply objects, in other words over any *ring*. To ensure the existence of continued fractions for any rational object, however, *Euclidean* rings will be the preferred choice and we will discover that division with a remainder plays a crucial role when computing a continued fraction representation.

To illustrate this, we do a naïve continued fraction expansion of a rational function as in our previous $\frac{13}{7}$ example.

Example 1.1 *(Rational continued fraction)* We have

$$\frac{x^4 + 2x^2 + 1}{x^3 + x + 1} = x + \frac{x^2 - x + 1}{x^3 + x + 1} = x + \frac{1}{\dfrac{x^3 + x + 1}{x^2 - x + 1}} = x + \frac{1}{x + 1 + \dfrac{x}{x^2 - x + 1}}$$

$$= x + \frac{1}{x + 1 + \dfrac{1}{x - 1 + \frac{1}{x}}} = [x; x + 1, x - 1, x],$$

where each component of the continued fraction is now a polynomial of degree 1.

A short reflection on what we really did in the computations above tells us that we actually performed division with a remainder in each step, also known as *Euclidean division*. Therefore we will also consider continued fractions of (univariate) polynomials, which are expressions such as

$$[p_0; p_1, \ldots, p_n], \qquad p_j \in \Pi = \mathbb{K}[x]$$

for a suitable field $\mathbb{K}$. Such a finite continued fraction will give a *rational function*

$$[p_0(x); p_1(x), \ldots, p_n(x)] = \frac{f(x)}{g(x)}, \qquad f, g \in \Pi, \tag{1.5}$$

while their limits for $n \to \infty$ will be more special objects. This will be our first interesting question in that context: what should be the analogy of real numbers in the sense of a class of numbers that can be written as infinite continued fractions or more precisely, as limits of initial segments of infinite continued fractions. For such limits we will encounter formal *Laurent series*, i.e., expressions of the form

$$\lambda(x) = \sum_{j=0}^{\infty} \lambda_j \, x^{-j}, \qquad \lambda_j \in \mathbb{K}, \tag{1.6}$$

just like we could imagine real numbers as an integer plus an infinite expansion with respect to a given base. Nevertheless besides a lot of formal similarities, the situation also changes significantly, since no longer is any Laurent series the limit of initial parts of an infinite continued fraction. This is classical theory, and fundamental results in this direction already date back to D. Bernoulli in 1775 and to Euler who related series and continued fractions. One formula to be derived from this theory is, for example, Euler's continued fraction expansion for the number e as

$$e = 1 - \cfrac{1}{1 - \cfrac{1}{3 - \cfrac{2}{4 - \cfrac{3}{\ddots}}}}$$

The fascinating point of this expansion is that it is not obtained by working on the transcendental number e or its series representation directly, but it is derived via a continued fraction expansion for the power series of the function e^x, and then just specifying x to 1.

As Example 1.1 already indicates, each of the p_j is normally an affine or constant polynomial, in other words a polynomial of degree at most 1. Also in this case, we will pursue some kind of *Approximation Theory* in trying to approximate a given function, represented by a power series in some best possible way with a rational object which will be by convergents of the associated continued fraction. Here "best possible" means that as many terms as possible coincide in the series and the approximation. To be more precise, the fundamental question is whether one can approximate a sequence as in (1.6) with a continued fraction, i.e., if we can write

$$\lambda(x) = [p_0; p_1, p_2, \dots], \qquad p_j \in \Pi_1 \backslash \Pi_0, \tag{1.7}$$

with the additional requirement that each p_j be a polynomial of degree *exactly* one. We will find out that the initial segments of the continued fraction, provided that it converges, have the property that

$$\lambda(x) - [p_0; p_1, \dots, p_n] = O\left(x^{-2n-1}\right) \tag{1.8}$$

which yields the analogy of the approximation rate $\varphi(q) = O(q^{-2})$ in (1.4). The converse does not hold true in general, but the existence of a representation in the sense of (1.6) and then (1.7) can be characterized and leads us to yet another nice mathematical theory. The main tools are the *Hankel matrices*

$$\Lambda_n := \left[\lambda_{j+k-1} : j, k = 1, \dots, n\right] = \begin{bmatrix} \lambda_1 & \lambda_2 & \dots & \lambda_n \\ \lambda_2 & \lambda_3 & \ddots & \vdots \\ \vdots & \ddots & \ddots & \lambda_{2n-2} \\ \lambda_n & \dots & \lambda_{2n-2} & \lambda_{2n-1} \end{bmatrix} \in \mathbb{R}^{n \times n}, \qquad (1.9)$$

defined for $n \in \mathbb{N}$, and they provide a very simple characterization, see Theorem 4.4.

Theorem 1.3 (Limits of continued fractions) *A Laurent series* λ *satisfies* (1.7) *if and only if* $\det \Lambda_n \neq 0$, $n \in \mathbb{N}$.

An infinite continued fraction that has a given Laurent series as its limit will be called *associated* to the series or to the underlying sequence of coefficients in that series. Associated continued fractions with *affine* components provide an optimal rate of convergence in the sense of (1.8), and connect among others to the theory of moments of square positive linear functionals.

The idea of meeting as many terms as possible in a series by a rational function connects continued fractions to a more general concept of approximating the power series by rational functions, known as *Padé approximation*, a wide area by itself. And indeed, we will identify convergents of continued fractions as particular elements of the so-called *Padé table* in Sect. 3.6.

1.3 Continued Fractions and Moments

Continued fractions with especially simple coefficients in (1.5) as already mentioned are those where each p_j is an *affine polynomial* of the form $p_j(x) = \alpha_j x + \beta_j$. As indicated in Example 1.1, this is even the generic case. Moreover these are the associated continued fractions for a given Laurent series.

Provided that $\alpha_j < 0$, these continued fractions will have a close relationship with orthogonal polynomials, polynomial sequences $f_j \in \Pi$, $j \in \mathbb{N}_0$, with the property that

$$\langle f_j, f_k \rangle = c_j \, \delta_{j,k}, \qquad c_j > 0, \qquad j, k \in \mathbb{N}_0,$$

where $\langle \cdot, \cdot \rangle$ denotes an *inner product*. In fact orthogonal polynomials can even be characterized and parameterized using continued fractions. A result in this direction shows that orthogonal polynomials and certain finite continued fractions are equivalent.

Theorem 1.4 *For each sequence* f_j, $j \in \mathbb{N}_0$, *of orthogonal polynomials with alternating leading coefficients, there exist coefficients* $\alpha_j < 0$ *and* β_j, $j \in \mathbb{N}_0$, *such that*

$$\left[0; \alpha_1 x + \beta_1, \dots, \alpha_j x + \beta_j\right] = \frac{g_j(x)}{f_j(x)},$$

and vice versa.

Eventually this theory allows us to construct orthogonal polynomials and even quadrature formulas using continued fractions. This connection is classical, and actually this was the way Gauss originally constructed what is nowadays known as the *Gaussian quadrature formula*. We will revisit and hopefully finally understand the original approach from [32], and the very natural idea behind it. The approach relies on considering an inner product of the form

$$\langle f, g \rangle = \int_{\mathbb{R}} f(x) \, g(x) \, w(x) \, dx, \qquad w \geq 0,$$

with, for convenience, a compactly supported continuous *weight function* $w : \mathbb{R} \to \mathbb{R}$ that describes an *integral* $f \mapsto I(f) := \langle f, 1 \rangle$. The integral can be described in terms of the Laurent series or as a *generating function* for the *moment sequence*

$$\mu(x) = \sum_{j=1}^{\infty} \mu_{j-1} x^{-j}, \qquad \mu_j = \langle (\cdot)^j, 1 \rangle = \int_{\mathbb{R}} x^j \, w(x) \, dx, \qquad j \in \mathbb{N}_0, \qquad (1.10)$$

and then to approximate this sequence, continued fractions are the method of choice. Once again this connects to Hankel matrices. Setting $\lambda_j = \mu_{j-1}$, $j \in \mathbb{N}$, the inner product of two polynomials $f(x) = f_0 + \cdots + f_n x^n$ and $g(x) = g_0 + \cdots g_n x^n$ of the degree at most n can be written in terms of the Hankel matrix Λ_n as

$$\langle f, g \rangle = f^T \Lambda_{n+1} g = [f_0, \ldots, f_n] \begin{bmatrix} \mu_0 & \cdots & \mu_n \\ \vdots & \ddots & \vdots \\ \mu_n & \cdots & \mu_{2n} \end{bmatrix} \begin{bmatrix} g_0 \\ \vdots \\ g_n \end{bmatrix};$$

it becomes clear, at least intuitively, that once more Hankel matrices must be of fundamental importance. In particular, if the integral is *definite*, i.e., if $\langle f, f \rangle > 0$, $f \neq 0$, then all matrices Λ_n are *positive definite* giving det $\Lambda_n > 0$; furthermore there always exists a continued fraction expansion for the Laurent polynomial defined by the moment sequence with affine coefficients whose leading terms all satisfy $\alpha_j < 0$.

The denominators of the convergents are then the orthogonal polynomials with respect to the inner product, and the fact that their zeros are all real and simple follows directly from the fact that they belong to a continued fraction expansion for a sequence λ with det $\Lambda > 0$.

Having all these facts at hand, *Gaussian quadrature* simply becomes a combination of these observations with some basic aspects of polynomial interpolation, which Gauss originally did in a quite ingenious fashion. We will present this approach in Sect. 4.1.

1.4 Continued Fractions and Prony

A seemingly unrelated problem is *Prony's problem* that is concerned with the recovery of a function written in Prony's original style,

$$f(x) = \sum_{j=1}^{s} f_j \, \rho_j^x, \qquad \rho_j \in \mathbb{C} \backslash \{0\}, \qquad f_j \neq 0, \tag{1.11}$$

from samples $f(0), \ldots, f(N)$ of f at certain integers. The intuition underlying (1.11) is that the number s for terms in the sum is comparatively small; in Prony's original case for analyzing the vaporization of alcohol, it was $s = 3, 4$, hence the representation is assumed to be *sparse* and efficient, the latter being ensured by the requirement that $f_j \neq 0$. This means that we are only interested in components that really add to the sum in (1.11). The connection to continued fractions is once more a Hankel matrix.

As already shown by Prony in 1795, the bases ρ_j of the exponential parts are obtained as the zeros of the polynomial $q \in \Pi_s$, $q(x) = q_0 + \cdots + q_s x^s$, in which the coefficients are a nontrivial solution of the homogeneous problem

$$\begin{bmatrix} f(0) & \ldots & f(s) \\ \vdots & \ddots & \vdots \\ f(s) & \ldots & f(2s) \end{bmatrix} \begin{bmatrix} q_0 \\ \vdots \\ q_s \end{bmatrix} = 0.$$

Once the ρ_j are found—and this is the nonlinear part of the problem—the coefficients f_j in (1.11) can be obtained by solving the Vandermonde linear system

$$\begin{bmatrix} 1 & \ldots & 1 \\ \rho_1 & \ldots & \rho_s \\ \vdots & \ddots & \vdots \\ \rho_1^{s-1} & \ldots & \rho_s^{s-1} \end{bmatrix} \begin{bmatrix} f_1 \\ \vdots \\ f_s \end{bmatrix} = \begin{bmatrix} f(0) \\ \vdots \\ f(s-1) \end{bmatrix};$$

this is the easier part of the story except of course when one really wants to compute things, and the conditioning of the matrix gains relevance.

The theory derived for Prony's problem also allows us to investigate *Hankel operators*, i.e., infinite matrices of the form

$$M := \begin{bmatrix} \mu_0 & \mu_1 & \ldots \\ \mu_2 & \mu_2 & \ldots \\ \vdots & \vdots & \ddots \end{bmatrix},$$

and we will encounter *Kronecker's Theorem* that characterizes when such an operator has finite rank.

Theorem 1.5 (Kronecker's Theorem) *The operator M has finite rank if and only if $\mu(x)$ is a rational function.*

Hankel operators of finite rank are encoded by the finite information of the rational representation $\mu(x) = p(x)/q(x)$ and can be written as a finite continued fraction expansion as well. This expansion gives a recurrence for the denominators of the convergents and the polynomial $q \in \Pi_n$ itself has the property that

$$M_n q = 0, \qquad M_n := \begin{bmatrix} \mu_0 & \cdots & \mu_n \\ \vdots & \ddots & \vdots \\ \mu_n & \cdots & \mu_{2n} \end{bmatrix}. \tag{1.12}$$

Now (1.12) implies in turn that the sequence μ must be obtained by sampling a sequence of the form (1.11); once more the ρ_j are the zeros of the denominator polynomial q.

In the case of moment sequences for a definite inner product, as in the preceding section, even more is true. Since all the matrices M_n are positive definite, and hence nonsingular, the natural question arises as to whether it is possible to extend M_n, i.e., the initial moments $\mu_0, \ldots, \mu_{2n}$ in such a way that the resulting infinite moment sequence defines a Hankel operator of rank $n + 1$. This is what is called a *flat extension* of the truncated moment sequence $\mu_0, \ldots, \mu_{2n}$ of order n. And flat extensions always exist and have a meaning.

Theorem 1.6 (Flat extension) *Any truncated moment sequence of order n for a definite inner product admits a (positive) flat extension that corresponds to a Gaussian quadrature formula.*

Flat extension means that rank $M_{n+1} = $ rank $M_n = n + 1$, and therefore there exists a vector $q = (q_0, \ldots, q_n)$, unique up to normalization, such that $M_{n+1}q = 0$ and the zeros $\zeta_1, \ldots, \zeta_{n+1}$ of the associated polynomial $q(x) = q_0 + \cdots + q_{n+1}x^{n+1}$ define a discrete measure

$$\sum_{j=1}^{n+1} f_j \delta_{\zeta_j}, \qquad f_j > 0,$$

that is precisely the Gaussian quadrature formula of degree $n + 1$ for the inner product. Moreover the weights f_j are exactly the coefficients in the Prony representation (1.11) of the function f, which is sampled at the integers to yield the moments $\mu_k = f(k)$. These weights f_j are positive if and only if the underlying measure is positive.

With different techniques, flat extensions can also be constructed for truncated sequences that do not necessarily come from moments of a definite inner product.

Theorem 1.7 *Whenever M_n is nonsingular, there exists a one parameter family of flat extensions for the truncated sequence.*

1.5   Digital Signal Processing

We will also be interested in yet another modern application of mathematics where surprising or not, continued fractions will play a crucial role. This is *Digital Signal Processing*, and more precisely the construction of rational digital filters.

A *digital filter* is the fundamental operator in signal processing; it consists of convolving a given discrete signal, represented as an infinite sequence with the so-called *impulse response* of the filter which is yet another signal by itself. For practical purposes and to ensure that the convolution is well defined, the standard assumption in the beginning is that the impulse response must be a finitely supported signal. Any finitely supported signal $f = (f_k : k \in \mathbb{Z})$ defines a *Laurent polynomial*

$$f(z) = \sum_{k \in \mathbb{Z}} f_k z^{-k},$$

called the *z transform* of f. This adds a useful algebraic flavor to Signal Processing as Laurent polynomials can be multiplied, factorized, and further mistreated in various ways.

Rational filters provide an extension to filters with an *infinite* impulse response that still only depend on finite information, i.e., they are based on finitely many parameters, and can be realized technically. They are characterized by the fact that their impulse response can be written as a *rational function*

$$f(z) = \frac{p(z)}{q(z)}, \qquad p, q \in \mathbb{C}[z]. \tag{1.13}$$

In contrast to a filter with a finite impulse response whose z transform is just a Laurent polynomial, rational filters have a serious issue with stability, which means that after a while the feedback loop of the filter may take over and can get so overexcited that the filter can hardly be controlled any more.

This (unwanted) behavior can be characterized by the fact that the rational function (1.13) has poles *inside* the complex unit disc. For that, a filter is called *stable* provided that all its zeros lie *outside*[3] the unit disc. This brings us to yet another interesting mathematical problem: how can we characterize the fact that a polynomial has such a zero structure, of course without factorizing the polynomial or explicitly computing the zeros in some other way.

The same behavior also appears in the context of difference equations with constant coefficients. This is a well-known connection; also a lot of signal processing can also be formulated in terms of *System Theory* and therefore in terms of difference equations. So it is no surprise that once again stability is characterized by the fact

---

[3] The inside/outside issue varies through the literature. This is due to the fact that one might either consider $q(z)$ or $q(z^{-1})$ for the denominator, since in the Laurent setting, the numerator and the denominator of the filter can be multiplied by any power of z simultaneously without changing the rational function.

that all zeros lie inside the unit circle, even if this is obtained by different methods there.

1.6 Zeros of Polynomials and Hurwitz

The main message of stability is the following: to be reasonable, the denominator of a rational filter or the (Laurent) polynomial associated with a difference equation must not have poles inside the unit disc. Hence it is reasonable to look for criteria for the location of zeros or conditions that ensure that the localization of zeros is as wanted. A simple criterion with a short and elegant proof is the following.

Theorem 1.8 (Eneström–Kakeya) *If the coefficients of the polynomial $p(z) = p_0 + p_1 z + \cdots + p_n z^n$ satisfy $p_0 > \cdots > p_n > 0$, then p has no zero in the unit disc.*

With the rational linear transformation $z = \frac{w+1}{w-1}$, the "bad" zeros inside the unit disc are shifted into the right half-plane,

$$\mathbb{H}_+ := \{z = x + iy \in \mathbb{C} : x > 0\},$$

and the "good" polynomials are now the ones that have all their zeros in the left half-plane. Such a polynomial with real coefficients and all its zeros in $\mathbb{H}_-$ is called a *Hurwitz polynomial*.

For Hurwitz polynomials, there exist a rich and fascinating theory. Much of that is based on decomposing the polynomial f into $f(z) = g(z^2) + zh(z^2)$, where the polynomial g collects the coefficients of f with an even index and the polynomial h those with an odd index, respectively. Being a Hurwitz polynomial can now be described in terms of continued fractions.

Theorem 1.9 (The Stieltjes Theorem) *The polynomial f is a Hurwitz polynomial if and only if the ratio of its parts has a continued fraction expansion*

$$\frac{h(x)}{g(x)} = [c_0; d_1 x, c_1, \ldots, d_n x, c_m], \tag{1.14}$$

where $c_0 \geq 0$ and $c_j, d_j > 0$, $j = 1, \ldots, m$.

This formula allows us to parameterize or enumerate *all* possible Hurwitz polynomials by listing all continued fractions of the form (1.14) with positive c_j, d_j. Moreover, f being a Hurwitz polynomial is also equivalent to g, h being a so-called *positive pair* which means that they have interlacing real and negative zeros. It should come as no surprise that again this is also connected to Hankel matrices. Expanding h/g as

$$\frac{h(x)}{g(x)} = c_0 + \sum_{j=1}^{\infty} \mu_j x^{-j}$$

and defining the *Markov numbers*

$$\gamma_j = (-1)^{j+1}\mu_j, \qquad j \in \mathbb{N},$$

then there is the following characterization of Hurwitz polynomials to be proved once more by continued fraction techniques.

Theorem 1.10 *The polynomial f with parts $g, h \in \Pi_m$ is a Hurwitz polynomial if and only if the two Hankel matrices*

$$\begin{bmatrix} \gamma_1 & \cdots & \gamma_m \\ \vdots & \ddots & \vdots \\ \gamma_m & \cdots & \gamma_{2m-1} \end{bmatrix}, \quad \begin{bmatrix} \gamma_2 & \cdots & \gamma_{m+1} \\ \vdots & \ddots & \vdots \\ \gamma_{m+1} & \cdots & \gamma_{2m} \end{bmatrix}$$

are strictly positive definite.

This theorem could also be formulated in terms of the coefficients μ_j directly, but it would then result in certain principal minors having an alternating sign, while the above formulation with a single positive definite matrix is simply more elegant.

It even turns out that the above matrices are not only positive definite but even *totally positive*, which means that all their minors are positive as well. This is eventually turned into a characterization by the total positivity of all minors of a size up to m of the *infinite* Hankel operator. In other words: whether *infinite* Hankel operators of some finite rank are associated to a Hurwitz polynomial or not is determined by all minors of a certain maximal size. Of course this size is nothing but the rank of a Hankel operator.

1.7   And What Else?

It is clear that the issues presented in this lecture are only a fraction of the multitude of aspects of the theory of continued fractions that could be continued and extended on and on. For example, one can find in [60] some measure theory of continued fractions, i.e., how are they distributed on the real line. And already the two volumes of Perron's book [78, 79] contain a lot that is not even mentioned here, from number theoretical concepts like Pell's equation, to the question, for example under which conditions a continued fraction, seen as a power series, converges to an *analytic function*. The literature is much richer than these two classics—just search, find out and enjoy.

Chapter 2
Continued Fractions of Real Numbers

And now I must stop saying what I am not writing about, because
there's nothing so special about that; every story one chooses to
tell is a kind of censorship, it prevents the telling of other tales ...
S. Rushdie, Shame

2.1 Convergents and Continuants

Our first step in the direction of understanding continued fractions consists of having a closer look at the expression $[a_0; a_1, \ldots, a_n]$ and its meaning. This leads us to the most fundamental notion in the theory of continued fractions, which is still well defined even for *rational* coefficients of the continued fraction.

Definition 2.1 Given numbers $a_j \in \mathbb{Q}$, $j = 0, 1, \ldots$, the nth *convergent* of the infinite continued fraction $[a_0; a_1, \ldots]$ is defined as the finite continued fraction $[a_0; a_1, \ldots, a_n]$.

First, note that the nth convergent of a continued fraction can always be written as the quotient of two polynomials in the variables $a_0, \ldots, a_n$:

$$[a_0; a_1, \ldots, a_n] = \frac{p_n(a_0, \ldots, a_n)}{q_n(a_0, \ldots, a_n)}. \qquad (2.1)$$

This is trivially true for $n = 0$, as we then only have the constant polynomial r_0, and in general (2.1) follows inductively from the definition (1.2):

T. Sauer, *Continued Fractions and Signal Processing*, Springer Undergraduate Texts in Mathematics and Technology, https://doi.org/10.1007/978-3-030-84360-1_2

$$\begin{aligned}
\left[a_0; a_1, \ldots, a_{n+1}\right] &= a_0 + \cfrac{1}{\left[a_1; a_2, \ldots, a_{n+1}\right]} = a_0 + \frac{q_n\left(a_1, \ldots, a_{n+1}\right)}{p_n\left(a_1, \ldots, a_{n+1}\right)} \\
&= \frac{a_0\, p_n\left(a_1, \ldots, a_{n+1}\right) + q_n\left(a_1, \ldots, a_{n+1}\right)}{p_n\left(a_1, \ldots, a_{n-1}\right)}.
\end{aligned}$$

This representation even gives a recursive way to obtain p_{n+1} and q_{n+1} explicitly as

$$\begin{aligned}
p_{n+1}\left(a_0, \ldots, a_{n+1}\right) &= a_0\, p_n\left(a_1, \ldots, a_{n+1}\right) + q_n\left(a_1, \ldots, a_{n+1}\right), \\
q_{n+1}\left(a_0, \ldots, a_{n+1}\right) &= p_n\left(a_1, \ldots, a_{n+1}\right).
\end{aligned} \tag{2.2}$$

Since $\left[a_0; a_1, \ldots, a_{n+1}\right] = \dfrac{p_{n+1}\left(a_0, \ldots, a_{n+1}\right)}{q_{n+1}\left(a_0, \ldots, a_{n+1}\right)}$, the second identity in (2.2) yields

$$q_n\left(a_0, \ldots, a_n\right) = p_{n-1}\left(a_1, \ldots, a_n\right), \tag{2.3}$$

and allows us to conclude that

$$\left[a_0; a_1, \ldots, a_n\right] = \frac{p_n\left(a_0, \ldots, a_n\right)}{p_{n-1}\left(a_1, \ldots, a_n\right)}. \tag{2.4}$$

Moreover from the first identity in (2.2), we have the recurrence relation

$$p_{n+1}\left(a_0, \ldots, a_{n+1}\right) = a_0\, p_n\left(a_1, \ldots, a_{n+1}\right) + p_{n-1}\left(a_2, \ldots, a_{n+1}\right), \tag{2.5}$$

initialized by

$$p_{-2} := 0, \quad p_{-1} := 1, \tag{2.6}$$

for the numerator as well.

Definition 2.2 The polynomials $p_n\left(x_0, \ldots, x_n\right) : \mathbb{Q}^{n+1} \to \mathbb{Q}$ are called *continuants*.

Remark 2.1 Continuants have been considered, if not introduced already by Euler.

Let us consider some first examples:

$$\begin{aligned}
\left[a_0; \right] &= a_0, \\
\left[a_0; a_1\right] &= a_0 + \frac{1}{a_1} = \frac{a_0 a_1 + 1}{a_1}, \\
\left[a_0; a_1, a_2\right] &= a_0 + \frac{1}{\left[a_1; a_2\right]} = a_0 + \frac{a_2}{a_1 a_2 + 1} = \frac{a_0 a_1 a_2 + a_0 + a_2}{a_1 a_2 + 1}, \\
\left[a_0; a_1, a_2, a_3\right] &= a_0 + \frac{a_2 a_3 + 1}{a_1 a_2 a_3 + a_1 + a_3} = \frac{a_0 a_1 a_2 a_3 + a_0 a_1 + a_0 a_3 + a_2 a_3 + 1}{a_1 a_2 a_3 + a_1 + a_3},
\end{aligned}$$

which all looks nicely symmetric in the variables, although with increasing complexity.

The next result is an explicit recurrence relation for the numerator and the denominator of the convergent.

Theorem 2.1 (Fundamental recurrence relation) *For $k \geq 1$, the kth convergent can be written with numerator and denominator satisfying the following* recurrence relation:

$$p_k = a_k p_{k-1} + p_{k-2}, \qquad p_{-1} = 1, \quad p_0 = a_0$$
$$q_k = a_k q_{k-1} + q_{k-2}, \qquad q_{-1} = 0, \quad q_0 = 1. \qquad (2.7)$$

Proof The case $k = 1$ has been computed explicitly in the above examples. To advance the induction hypothesis from k to $k + 1$, we use the canonical representation

$$[a_1; a_2, \ldots, a_{k+1}] =: \frac{\widetilde{p}_k}{\widetilde{q}_k}$$

of the "shifted" continued fraction, and obtain by the definition of continued fractions that

$$\frac{p_{k+1}}{q_{k+1}} = a_0 + \frac{1}{[a_1; a_2, \ldots, a_{k+1}]} = a_0 + \frac{\widetilde{q}_k}{\widetilde{p}_k} = \frac{\widetilde{p}_k a_0 + \widetilde{q}_k}{\widetilde{p}_k}.$$

Using the induction hypothesis (2.7) for $\widetilde{p}_k$ and $\widetilde{q}_k$ and taking into account the shift of the indices there, we get that we can choose p_{k+1} and q_{k+1} to be

$$p_{k+1} = a_0 \left(a_{k+1} \widetilde{p}_{k-1} + \widetilde{p}_{k-2} \right) + \left(a_{k+1} \widetilde{q}_{k-1} + \widetilde{q}_{k-2} \right)$$
$$= a_{k+1} \left(a_0 \widetilde{p}_{k-1} + \widetilde{q}_{k-1} \right) + \left(a_0 \widetilde{p}_{k-2} + \widetilde{q}_{k-2} \right) = a_{k+1} p_k + p_{k-1},$$
$$q_{k+1} = a_{k+1} \widetilde{p}_{k-1} + \widetilde{p}_{k-2} = a_{k+1} q_k + q_{k-1},$$

which completes the induction. $\qquad \square$

It is well known that the representation of a fraction as a quotient of integers is not unique; just consider $\frac{1}{2} = \frac{2}{4} = \frac{3}{6} = \ldots$, and only the normal form where the numerator and the denominator are *coprime* is unique. The same holds true for the representation of a convergent, which we make unique given the above recurrence. It will turn out later that under additional assumptions this representation with integer parameters is even *irreducible*, but at the moment we take it as it is and use the following definition.

Definition 2.3 The values defined in (2.7) are called the numerator and denominator in the *canonical representation* of the kth convergent

$$[a_0; a_1, \ldots, a_k] = \frac{p_k}{q_k}$$

of a continued fraction with arguments $a_j \in \mathbb{Q}, \ j \in \mathbb{N}_0$.

Corollary 2.1 *For $k \geq 0$, we have that*

$$q_k \, p_{k-1} - p_k \, q_{k-1} = (-1)^k, \tag{2.8}$$

or

$$\frac{p_{k-1}}{q_{k-1}} - \frac{p_k}{q_k} = \frac{(-1)^k}{q_{k-1} \, q_k}, \tag{2.9}$$

respectively.

Proof We multiply the first line of the recurrence (2.7) by $-q_{k-1}$ and the second one by p_{k-1} to get

$$
\begin{aligned}
q_k \, p_{k-1} - p_k \, q_{k-1} &= -a_k p_{k-1} q_{k-1} - q_{k-1} \, p_{k-2} + a_k p_{k-1} q_{k-1} + q_{k-2} p_{k-1} \\
&= -(q_{k-1} \, p_{k-2} - q_{k-2} p_{k-1}) = \cdots = (-1)^k \, (q_0 \, p_{-1} - q_{-1} \, p_0) \\
&= (-1)^k,
\end{aligned}
$$

which is (2.8). If we divide this identity by $q_{k-1} \, q_k$, we end up with (2.9). $\square$

And there is one more cute formula.

Theorem 2.2 *For $k \geq 2$, one has*

$$p_k \, q_{k-2} - q_k \, p_{k-2} = (-1)^k a_k \quad or \quad \frac{p_k}{q_k} - \frac{p_{k-2}}{q_{k-2}} = \frac{(-1)^k a_k}{q_{k-2} \, q_k}, \tag{2.10}$$

respectively.

Proof The proof is not particularly surprising: we multiply the two lines of (2.7) by q_{k-2} and $-p_{k-2}$, respectively, add the expressions, and end up with

$$q_k \, p_{k-2} - p_k \, q_{k-2} = a_k \, (p_{k-1} q_{k-2} - q_{k-1} \, p_{k-2}) = -a_k (-1)^{k-1} = (-1)^k a_k$$

due to (2.8). $\square$

This apparently innocent theorem already provides information about the convergence of convergents for infinite continued fractions, at least in the case that $a_j \in \mathbb{Q}_+$, $j \in \mathbb{N}$, where $\mathbb{Q}_+$ stands for the set of all nonnegative rational numbers.

Corollary 2.2 *If $a_j \in \mathbb{Q}_+$, $j \in \mathbb{N}$, then the sequence of convergents of an even order, $[a_0; a_1, \ldots, a_{2k}]$, is monotonically increasing, and the convergents of an odd order, $[a_0; a_1, \ldots, a_{2k+1}]$ decrease monotonically. Moreover,*

$$\inf_{k \in \mathbb{N}} [a_0; a_1, \ldots, a_{2k-1}] \geq \sup_{k \in \mathbb{N}} [a_0; a_1, \ldots, a_{2k}]. \tag{2.11}$$

Proof A view of the recurrence (2.7) shows that $q_k > 0$, $k \in \mathbb{N}$, and that it is even monotonically increasing as long as all a_j are strictly positive.[1] Then (2.10) yields that

$$\frac{p_{2k}}{q_{2k}} - \frac{p_{2(k-1)}}{q_{2(k-1)}} = \frac{(-1)^{2k} a_{2k}}{q_{2(k-1)} q_{2k}} > 0, \tag{2.12}$$

or

$$\frac{p_{2k+1}}{q_{2k+1}} - \frac{p_{2k-1}}{q_{2k-1}} = \frac{(-1)^{2k+1} a_{2k+1}}{q_{2k-1} q_{2k+1}} < 0, \tag{2.13}$$

respectively. Next, we show that any convergent of an even order is smaller than any convergent of an odd order. To that end, let $m, m' \in \mathbb{N}$ and $\ell \geq \max\{m, m'\}$. From (2.9) with $k = 2\ell + 1$, it follows that

$$\frac{p_{2\ell}}{q_{2\ell}} = \frac{p_{2\ell+1}}{q_{2\ell+1}} + \frac{(-1)^{2\ell+1}}{q_{2\ell} q_{2\ell+1}} < \frac{p_{2\ell+1}}{q_{2\ell+1}},$$

and the already proven monotonicity properties (2.12) and (2.13) of convergents yield

$$\frac{p_{2m}}{q_{2m}} < \frac{p_{2\ell}}{q_{2\ell}} < \frac{p_{2\ell+1}}{q_{2\ell+1}} < \frac{p_{2m'+1}}{q_{2m'+1}},$$

as claimed. From this, (2.11) is immediate. $\qquad\square$

Let us make it clear what Corollary 2.2 means. Even-order convergents form a monotonically *increasing* sequence; odd order convergents on the other hand form a monotonically *decreasing* sequence. Moreover, the decreasing sequence is bounded from below and thus has to converge. In the same way, the increasing sequence of odd order convergents, being bounded from above must converge as well, which enables us to conclude the following fact.

Corollary 2.3 *The sequence of convergents* $[a_0; a_1, \ldots, a_k]$, $k \in \mathbb{N}$, *has at most two accumulation points, namely*

$$\lim_{k \to \infty} [a_0; a_1, \ldots, a_{2k}] \quad and \quad \lim_{k \to \infty} [a_0; a_1, \ldots, a_{2k+1}],$$

and converges if and only if equality holds in (2.11).

Moreover, this enclosing type of convergence is also welcome, since at any finite step it gives us an upper and a lower estimate for the limit – provided the limit exists, of course.

We close this section by extending our toolbox with two more formulas for continued fractions and their convergents.

---

[1] And even some zeros among the a_j would not hurt as soon as we once reached a positive value.

Proposition 2.1 *For* $1 \leq k \leq n$, *we have that*

$$[a_0; a_1, \ldots, a_n] = \frac{p_{k-1} r_k + p_{k-2}}{q_{k-1} r_k + q_{k-2}}, \qquad r_k := [a_k; a_{k+1}, \ldots, a_n], \qquad (2.14)$$

as well as[2]

$$\frac{q_k}{q_{k-1}} = [a_k; a_{k-1}, \ldots, a_1]. \qquad (2.15)$$

Proof From the recurrence (1.2) for the definition of continued fractions, it follows that

$$[a_{k-1}; a_k, \ldots, a_n] = a_{k-1} + \frac{1}{[a_k; a_{k+1}, \ldots, a_n]} = a_{k-1} + \frac{1}{r_k} = [a_{k-1}; r_k],$$

$$[a_{k-2}; a_{k-1}, \ldots, a_n] = a_{k-2} + \frac{1}{[a_{k-1}; r_k]} = [a_{k-2}; a_{k-1}, r_k],$$

$$\vdots$$

$$[a_0; a_1, \ldots, a_n] = [a_0; a_1, \ldots, a_{k-1}, r_k].$$

If p_{k-1}, q_{k-1} are the numerator and denominator of the $(k-1)$-convergent and p_k, q_k are the components of the kth convergent of $[a_0; a_1, \ldots, a_{k-1}, r_k]$, then (2.7) yields that

$$[a_0; a_1, \ldots, a_n] = [a_0; a_1, \ldots, a_{k-1}, r_k] = \frac{p_k}{q_k} = \frac{r_k p_{k-1} + p_{k-2}}{r_k q_{k-1} + q_{k-2}},$$

which is precisely (2.14).

Formula (2.15) will be proved by induction on k. Since the continued fraction only starts at a_1, the case $k = 1$ takes the form

$$[a_1;] = a_1 = \frac{q_1}{q_0} = q_1 = p_0 = a_1.$$

Having verified (2.15) for some $k \geq 1$, we simply place the induction hypothesis (2.15) into (2.7) and get

$$q_{k+1} = a_{k+1} q_k + q_{k-1} = q_k \left(a_{k+1} + \frac{q_{k-1}}{q_k} \right) = q_k \left(a_{k+1} + \frac{1}{[a_k; a_{k-1}, \ldots, a_1]} \right)$$

$$= q_k [a_{k+1}; a_k, \ldots, a_1],$$

which is exactly what we wanted.   □

---

[2] Note that here the order of the components in the continued fraction expansion is reversed; they occur in reverse order.

2.1.1 Problems

2.1 Prove the symmetry property

$$p_n(x_0, \ldots, x_n) = p_n(x_n, \ldots, x_0)$$

of continuants; see [61, S. 357].

2.2 Two fractions $\frac{p}{q}$ and $\frac{p'}{q'}$ are called *Farey neighbors*, if

$$\left| \frac{p}{q} - \frac{p'}{q'} \right| = \frac{1}{qq'}.$$

Prove that every fraction

$$\frac{p}{q} < \frac{a}{b} < \frac{p'}{q'}$$

between two Farey neighbors has a denominator $b \geq q + q'$.

2.3 Show that every fraction between the kth convergent and the $(k+1)$-convergent of a continued fraction has the denominator at least $q_k + q_{k+1}$.

2.4 Prove the formulas

$$[0; 0, a_1, a_2, \ldots, a_n] = a_1 + [0; a_2, \ldots, a_n]$$
$$1 - [0; a_1, a_2, \ldots, a_n] = [0; 1, a_1 - 1, a_2, \ldots, a_n]$$

that hold for $n \geq 1$.

2.5 Show that

$$[a_0; a_1, \cdots, a_n] = \frac{\det \begin{bmatrix} a_0 & 1 & & & \\ -1 & a_1 & 1 & & \\ & \ddots & \ddots & \ddots & \\ & & -1 & a_{n-1} & 1 \\ & & & -1 & a_n \end{bmatrix}}{\det \begin{bmatrix} a_0 & 1 & & & \\ -1 & a_1 & 1 & & \\ & \ddots & \ddots & \ddots & \\ & & -1 & a_{n-2} & 1 \\ & & & -1 & a_{n-1} \end{bmatrix}}.$$

Hint: look at continuants.

2.2 Infinite Continued Fractions and Their Convergence

In this section we consider the situation of an *infinite continued fraction* of the form $[a_0; a_1, \ldots]$ and its convergence. To that end, we will assume that

$$a_j > 0, \qquad j = 1, 2, \ldots. \tag{2.16}$$

We still do not (yet) assume that the coefficients are integers, but we will give the motivation as to why it is really a good choice to select them as integers. Indeed, an inspection of the proofs will show that in principle everything works and all identities and properties are also valid for $a_1, a_2, \ldots \in \mathbb{Q}_+$. However, we will show in the next section that continued fractions with integer entries are "sufficient" anyway, and we can make our lives significantly easier by not enforcing an ultimate generality, especially since we will then even get convergence for free.

Our goal here is to collect information about the *convergence* of infinite continued fractions and in particular to prove Theorem 1.1. We start with some preliminary remarks that will clarify the real meaning of convergence.

Definition 2.4 The infinite continued fraction $[a_0; a_1, \ldots]$ is called *convergent* if the limit

$$[a_0; a_1, \ldots] := \lim_{n \to \infty} [a_0; a_1, \ldots, a_n]$$

exists and is finite. Otherwise the continued fraction is called *divergent*.

Remark 2.2 From now on, we will no longer emphasize that the infinite continued fraction is infinite and just speak of a continued fraction. The "$\ldots$" notation should speak for itself.

Proposition 2.2 *If the continued fraction $a = [a_0; a_1, \ldots]$ converges, then also all the remainders are infinite continued fractions $r_k = \big[a_k; a_{k+1}, \ldots\big]$ that converge. Conversely, if at least one r_k converges, then so does a and hence all r_k.*

Proof We choose any $k, n \in \mathbb{N}$ and consider the nth convergent

$$r_{k,n} := \frac{p'_n}{q'_n} = \big[a_k; a_{k+1}, \ldots, a_{k+n}\big]$$

of the remainder r_k. Using (2.14), we get that

$$\frac{p_{k+n}}{q_{k+n}} = \big[a_0; a_1, \ldots, a_{k+n}\big] = \big[a_0; a_1, \ldots, a_{k-1}, r_{k,n}\big] = \frac{p_{k-1} r_{k,n} + p_{k-2}}{q_{k-1} r_{k,n} + q_{k-2}}. \tag{2.17}$$

Solving this rational equation for $r_{k,n}$ yields

$$r_{k,n} = \frac{p_{k-2}\,q_{k+n} - q_{k-2}\,p_{k+n}}{q_{k-1}\,p_{k+n} - p_{k-1}\,q_{k+n}} = \frac{p_{k-2} - q_{k-2}\,\dfrac{p_{k+n}}{q_{k+n}}}{q_{k-1}\,\dfrac{p_{k+n}}{q_{k+n}} - p_{k-1}},$$

and therefore, due to the convergence of $[a_0; a_1, \ldots]$ to a,

$$r_k := \lim_{n\to\infty} r_{k,n} = \frac{p_{k-2} - q_{k-2}\,a}{q_{k-1}\,a - p_{k-1}}.$$

If the limit of the denominator were zero and therefore the sequence $r_{k,n}$, $n \in \mathbb{N}_0$, is divergent, we only have to look at the values $r_{k,2n+1}$ to see that we now face a contradiction; by Corollary 2.2 the values $r_{k,2n+1}$, $n \in \mathbb{N}$, would form a *monotonically decreasing* sequence that diverges to $+\infty$.

For the converse, assume that the limit $r_{k,n} \to r_k$ for $n \to \infty$ exists; then we have

$$\lim_{n\to\infty}[a_0; a_1, \ldots] = \frac{p_{k-1}\,\lim\limits_{n\to\infty} r_{k,n} + p_{k-2}}{q_{k-1}\,\lim\limits_{n\to\infty} r_{k,n} + q_{k-2}} = \frac{p_{k-1}\,r_k + p_{k-2}}{q_{k-1}\,r_k + q_{k-2}} =: a$$

and the continued fraction converges which implies by the first part of the proof that *all* remainders converge. $\qquad\square$

Next, we will get a *quantitative* approximation about convergence which will turn out to be one of the central results in continued fraction theory with plenty of consequences.

Theorem 2.3 *If $a = [a_0; a_1, \ldots]$ is convergent, then for any $k > 0$, we have the estimate*

$$\left| a - \frac{p_k}{q_k} \right| < \frac{1}{q_k\,q_{k+1}}. \tag{2.18}$$

Proof The strikingly short and simple proof relies on the *monotonic convergence* of convergents: if k is even, then by Corollary 2.2

$$\frac{p_k}{q_k} < a < \frac{p_{k+1}}{q_{k+1}},$$

and (2.9) yields that

$$0 < a - \frac{p_k}{q_k} < \frac{p_{k+1}}{q_{k+1}} - \frac{p_k}{q_k} = \frac{1}{q_k\,q_{k+1}},$$

whereas for odd k, the estimate

$$0 > a - \frac{p_k}{q_k} > \frac{p_{k+1}}{q_{k+1}} - \frac{p_k}{q_k} = -\frac{1}{q_k\,q_{k+1}}$$

holds. Together this gives (2.18). $\qquad\square$

Now we have already provided all the tools needed to prove our criterion for convergence. Let us first formulate it once more.

Theorem 2.4 (Theorem 1.1 on p. 3) *For any choice of $a_0 \in \mathbb{Q}$, $a_j \in \mathbb{Q}_+$, $j \in \mathbb{N}$, the infinite continued fraction $[a_0; a_1, \dots]$ converges if and only if*

$$\sum_{j=0}^{\infty} a_j = \infty. \tag{2.19}$$

Since (2.19) trivially holds true whenever $a_j \geq 1$, we can immediately state the following consequence for Theorem 2.4.

Corollary 2.4 *Any continued fraction $[a_0; a_1, \dots]$ with $a_0 \in \mathbb{Z}$ and $a_j \in \mathbb{N}$, $j \in \mathbb{N}$, converges.*

Proof (of Theorem 2.4) By Corollary 2.2, we have to show that the sequences of the even and odd convergents have *the same* limit since we already know that individually they converge. If the convergents are faithful to their name and converge, the identity (2.9) implies that $(q_k q_{k-1})^{-1}$ converges to zero for $k \to \infty$, which by (2.18) is necessary for convergence. In other words the infinite continued fraction converges if and only if

$$\lim_{k \to \infty} q_k q_{k+1} = \infty. \tag{2.20}$$

Let us now assume that the sequence in (2.19) converges. This means that $a_k \to 0$ for $k \to \infty$, and there exists $k_0 \in \mathbb{N}$ such that $a_k < 1$ for $k \geq k_0$. The recurrence (2.7) for q_k tells us that these values have to be positive for $k \geq 1$, and consequently that

$$q_k = a_k q_{k-1} + q_{k-2} > q_{k-2}$$

holds. Hence either $q_{k-1} \leq q_{k-2}$ and therefore $q_{k-1} < q_k$, or $q_{k-1} > q_{k-2}$. Let us assume that $k \geq k_0$. In the first case another application of (2.7) yields that

$$q_k < a_k q_k + q_{k-2} \quad \Rightarrow \quad q_k < \frac{q_{k-2}}{1 - a_k},$$

while in the second case we have that

$$q_k < (1 + a_k) q_{k-1} = \frac{1 - a_k^2}{1 - a_k} q_{k-1} < \frac{q_{k-1}}{1 - a_k}.$$

Since exactly one of these cases has to happen, there exists $\ell \in \{k-1, k-2\}$ such that

$$q_k < \frac{q_\ell}{1 - a_k}, \qquad k \geq k_0.$$

If $\ell \geq k_0$, we can repeat the argument and obtain that

$$q_k < \frac{q_m}{(1 - a_k)(1 - a_\ell)}$$

for some $m \in \{k - 2, k - 3, k - 4\}$ and that eventually[3]

$$q_k < \frac{q_{\ell_m}}{(1 - a_k)(1 - a_{\ell_1}) \cdots (1 - a_{\ell_{m-1}})}, \qquad \ell_m < k_0 \leq \ell_{m-1}, \qquad (2.21)$$

where $\ell_j \in \{k - j, \ldots, k - 2j\}$. Since the series in (2.19) converges by assumption, the same also holds true for the infinite product[4]

$$0 < \lambda := \prod_{j=k_0}^{\infty} (1 - a_j) \leq \prod_{j=0}^{m-1} (1 - a_{\ell_j}), \qquad \ell_0 = k. \qquad (2.22)$$

Setting $Q := \max \{q_j : j < k_0\}$, we deduce from (2.21) that $q_k < Q/\lambda$ for $k \geq k_0$ and the sequence $q_k q_{k+1}$ is bounded by

$$q_k q_{k+1} \leq \frac{Q^2}{\lambda^2}, \qquad k \geq k_0,$$

hence cannot diverge. Since this divergence was necessary for the convergence of the continued fraction, however, (2.19) is also a necessary condition for convergence.

For the converse, we suppose that the series diverges and therefore satisfies (2.19). Since we still have $q_k > q_{k-2}, k \geq 2$, we define $q := \min \{q_0, q_1\}$ and find that $q_k > q$ for any $k \geq 2$. Once more, we use the recurrence relation, this time to get the estimate

$$q_k \geq a_k q + q_{k-2} \geq (a_k + a_{k-2}) q + q_{k-4} \geq \cdots,$$

from which

$$q_{2k+\epsilon} \geq q_\epsilon + q \sum_{j=1}^{k} a_{2j+\epsilon} \qquad \epsilon \in \{0, 1\},$$

and so we have that

$$q_{2k} + q_{2k+1} \geq q_0 + q_1 + q \sum_{j=2}^{2k+1} a_j \qquad \Rightarrow \qquad q_k + q_{k+1} > q \sum_{j=0}^{k+1} a_j$$

---

[3] After iterating to the "bitter end".

[4] To quote Khinchin [60]: "...the infinite product [...], as we know, converges: that is, it has positive value ..." To be complete and self-contained, we give a proof of this folklore result in Lemma 2.1 later.

follows. This in turn implies that

$$\max\{q_k, q_{k+1}\} \geq \frac{q}{2} \sum_{j=0}^{k+1} a_j,$$

and we can use the above estimate for the larger of these values, and $q_k > q$ or $q_{k+1} > q$, respectively, for the smaller one to conclude that

$$q_k\, q_{k+1} > \frac{q^2}{2} \sum_{j=0}^{k+1} a_j \to \infty, \quad k \to \infty,$$

which yields convergence.   □

To complete the proof and to be self-contained, we recall some folklore result which is useful in various situations. And its proof is somewhat cute anyway.

Lemma 2.1 *For $a_j \in [0, 1)$, $j \in \mathbb{N}$, the infinite product*

$$\prod_{j=1}^{\infty} (1 - a_j)$$

*has a **positive** limit and is thus convergent if and only if the infinite series*

$$\sum_{j=1}^{\infty} a_j$$

converges.

Remark 2.3 An infinite product is called *convergent* if the limit of the partial products exists and is not zero. An infinite product with limit zero is still called divergent.

Proof Since $a_j \in [0, 1)$, the *partial products* $(1 - a_1) \cdots (1 - a_n)$, $n \in \mathbb{N}$, form a monotonically decreasing sequence of positive numbers, so the limit

$$0 \leq \lambda = \prod_{j=1}^{\infty} (1 - a_j) = \lim_{n \to \infty} \prod_{j=1}^{n} (1 - a_j)$$

has to exist, and the only question is whether it is zero or not. It is easy to see that $\lambda = 0$ if a_j is not converging to zero; if that were the case, we would have infinitely many factors smaller than $1 - \varepsilon$ for some $\varepsilon > 0$, and their infinite product would already be zero. Therefore the only interesting case is the one in which $a_j \to 0$ for $j \to \infty$ where the proof is based on the estimate[5]

---

[5] Figure 2.1 shows that this is satisfied on an even larger region than $[0, \frac{1}{2}\log 2]$, but (2.23) is sufficient for the proof.

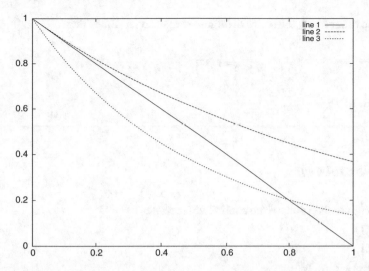

Fig. 2.1 The three functions from estimate (2.23) which also holds for values of x larger than $\frac{1}{2}\log 2 \approx 0.35$

$$e^{-2x} \leq 1 - x \leq e^{-x}, \qquad 0 \leq x \leq \frac{1}{2}\log 2. \tag{2.23}$$

Indeed at $x = 0$, all three expressions have value 1 and their derivatives satisfy

$$-2e^{-2x} \leq -1 \leq -e^{-x}, \qquad 0 < x < \frac{1}{2}\log 2,$$

so that a simple Taylor argument of order zero with an integral remainder verifies (2.23). If $a_j \to 0$, then there exists some n_0 such that $a_j < \frac{1}{2}\log 2$, $j \geq n_0$, and we get

$$\prod_{j=n_0}^{\infty} \left(1 - a_j\right) \geq \prod_{j=n_0}^{\infty} e^{-2a_j} = \exp\left(-2\sum_{j=n_0}^{\infty} a_j\right) \tag{2.24}$$

as well as

$$\prod_{j=n_0}^{\infty} \left(1 - a_j\right) \leq \prod_{j=n_0}^{\infty} e^{-a_j} = \exp\left(-\sum_{j=n_0}^{\infty} a_j\right). \tag{2.25}$$

If the series converges, then so does the subsequence starting at n_0, say to a limit a, and (2.24) yields that

$$\lambda \geq e^a \prod_{j=0}^{n_0-1} \left(1 - a_j\right) > 0.$$

If, on the other hand, the series diverges, we get from (2.25) that

$$\lambda \le e^{-\infty} \prod_{j=0}^{n_0-1} \left(1 - a_j\right) = 0,$$

as claimed. □

2.2.1 Problems

2.6 Compute the value of the continued fractions

1. $\left[0; \overline{3}\right]$,
2. $\left[0; \overline{4}\right]$,
3. $\left[0; 1, \overline{3}\right]$.

2.7 Show that

$$\prod_{p \in \mathbb{P}} \frac{p-1}{p} = \infty$$

where $\mathbb{P}$ stands for the set of *prime numbers*.

2.3 Continued Fractions with Integer Coefficients

Having realized in Corollary 2.4 that continued fractions with positive integer coefficients behave nicely and always converge, and that negative numbers can be compensated by a_0 alone, we will next convince ourselves that this class of continued fractions is fully *sufficient* for studying real and rational numbers. Since in this case we get convergence for free, continued fractions with integer coefficients give us an easy and direct access to real numbers, provided we can indeed represent any real number in this way. This can be seen as an analogy to a *digit expansion* with *base B*

$$x = \sum_{j=1}^{\infty} x_j \, B^{-j}, \qquad x_j \in \{0, \ldots, B-1\},$$

which also converges since the partial sums are bounded:

$$\sum_{j=n}^{\infty} x_j \, B^{-j} \le B^{1-n}, \qquad n \in \mathbb{N}. \tag{2.26}$$

In this case we do not have to worry about convergence issues as well.

Theorem 2.5 *Any nonnegative rational number $x = \frac{p}{q}$ can be represented by a finite continued fraction with positive integer coefficients.*

Proof We assume that p/q is the *normalized form* of the fraction, that is $\gcd(p, q) = 1$ and $p \geq 0, q > 0$; otherwise we could just divide by $\gcd(p, q)$ and multiply both by -1 if needed. Next we define, as in the Euclidean algorithm, cf. [31], a_0 and r by the process of *division with remainder* where the *remainder r* of p modulo q is obtained as

$$p = a_0 q + r, \qquad 0 \leq r < q.$$

If $r = 0$, then we have the simple form $x = \frac{p}{q} = a_0 = [a_0;]$, otherwise we get

$$\frac{p}{q} = \frac{a_0 q + r}{q} = a_0 + \frac{r}{q} = a_0 + \frac{1}{\frac{q}{r}} = \left[a_0; \frac{q}{r}\right]. \tag{2.27}$$

Now we do induction on the numerator q with the trivial initial case $q = 1$. Then since $r < q$ we get by the induction hypothesis that

$$\frac{q}{r} = [a_1; a_2, \ldots, a_k], \qquad a_j \in \mathbb{N},$$

which we substitute in (2.27) to get

$$\frac{p}{q} = a_0 + \frac{1}{[a_1; a_2, \ldots, a_k]} = [a_0; a_1, \ldots, a_k],$$

which is an expression of a finite continued fraction. Conversely, that any finite continued fraction defines a rational number has been mentioned several times and lies in the nature of the definition in Eq. (1.1). $\qquad\square$

From the proof, we get the following estimate for the *length* of a continued fraction.

Definition 2.5 The *length* of a continued fraction $[a_0; a_1, \ldots, a_k]$ is the number k of its coefficients and ∞ if the continued fraction is infinite.

Corollary 2.5 *If the fraction $\frac{p}{q}$ can be written as $\frac{p}{q} = [a_0; a_1, \ldots, a_k]$, then the length k of the continued fraction is $\leq q$.*

Remark 2.4 *(Continued fractions with positive integer components):*

1. Formally, Theorem 2.5 holds only for *nonnegative* rational numbers $x \in \mathbb{Q}_+$, but it is easily extended to $\mathbb{Q}$. Indeed, we only have to set

$$a_0 := \lfloor x \rfloor := \max\{k \in \mathbb{Z} : k \leq x\}, \qquad \text{hence,} \qquad r_0 := x - a_0 \in [0, 1),$$

and then proceed by determining the other components by the rule

$$a_j = \left\lfloor \frac{1}{r_{j-1}} \right\rfloor \in \mathbb{N}, \qquad r_j = r_{j-1} - \frac{1}{a_j}, \qquad j \in \mathbb{N}, \qquad (2.28)$$

which gives

$$r_{j-1} = a_j + r_j = \left[a_j; \frac{1}{r_j} \right],$$

so that the iteration (2.28) determines the coefficients of

$$x = [a_0; a_1, \ldots, a_k], \qquad a_0 \in \mathbb{Z}, \quad a_j \in \mathbb{N}, \ j \geq 1,$$

where Theorem 2.5 ensures that the expansion is finite and the iteration terminates after finitely many steps.

2. The above procedure also gives a way to define a *normal form* for the continued fraction expansion of any given rational number, and except perhaps for a_0, this normal form always consists of *positive* integers only.
3. The same process when applied to a *real number* will eventually lead to a convergent *continued fraction expansion* of that number. We will see that this expansion will even enable us to do number theory and that the expansions of rational, algebraic, and transcendental numbers can be easily distinguished.

The recurrence (2.7) for the canonical representation of the kth *convergent* shows us that in the representation according to Theorem 2.5, the numerator p_k is always an integer and the denominator is even a positive integer. What is not clear yet is whether this representation is already optimal, i.e., *normalized*, or if the two can have a nontrivial common divisor. The answer to that question is "irreducible" and the proof is strikingly simple.

Theorem 2.6 *The canonical representation* $\frac{p_k}{q_k}$ *of the kth convergent is* irreducible.

Proof Any common divisor of p_k and q_k would also divide the expression

$$q_k \, p_{k-1} - p_k \, q_{k-1} = (-1)^k$$

from (2.8) and therefore it can only be ± 1.   □

The *recurrence relation* for the convergents immediately implies that the denominators of the canonical representations satisfy

$$q_k = a_k \, q_{k-1} + q_{k-2} > a_k \, q_{k-1} \geq q_{k-1},$$

and that obviously the growth is related to the coefficients of the continued fractions: the larger the values a_k, the faster the denominators grow. But in any case the growth rate is at least exponential, namely

$$q_k \geq 2^{(k-1)/2}, \qquad k \geq 1. \qquad (2.29)$$

This is yet another consequence of the recurrence relation, which yields together
with the monotonic growth of the q_k that

$$q_k > (a_k + 1)\, q_{k-2} \geq 2\, q_{k-2},$$

from which (2.29) follows[6] with the initial conditions $q_0 = 1$ and $q_1 = a_0 \geq 1$.

Definition 2.6 (*Intermediate fractions*) For $k \geq 2$, any of the fractions

$$x_{j,k} := \frac{p_{k-2} + j\, p_{k-1}}{q_{k-2} + j\, q_{k-1}}, \qquad j = 0, \ldots, a_k,$$

is called an *intermediate fraction* between the $(k-2)$- and kth convergent of the
continued fraction.

The name *intermediate fraction* is easily explained: setting $j = 0$, we get the
canonical representation of the $(k-2)$-convergent, while the other extreme case
$j = a_k$ gives the kth convergent of the continued fraction. This is once more a direct
consequence of the recurrence relation (2.7).

Proposition 2.3 *For even k, the intermediate fractions form a monotonically* increasing *sequence; for odd k, a monotonically* decreasing *one.*

Proof For $j \geq 0$, we consider the difference

$$\frac{(j+1)\, p_{k-1} + p_{k-2}}{(j+1)\, q_{k-1} + q_{k-2}} - \frac{j\, p_{k-1} + p_{k-2}}{j\, q_{k-1} + q_{k-2}}$$

$$= \frac{(j+1)p_{k-1}q_{k-2} + jp_{k-2}q_{k-1} - jp_{k-1}q_{k-2} - (j+1)p_{k-2}q_{k-1}}{((j+1)q_{k-1} + q_{k-2})\,(jq_{k-1} + q_{k-2})}$$

$$= \frac{p_{k-1}q_{k-2} - q_{k-1}p_{k-2}}{((j+1)q_{k-1} + q_{k-2})\,(jq_{k-1} + q_{k-2})} = \frac{(-1)^k}{((j+1)q_{k-1} + q_{k-2})\,(jq_{k-1} + q_{k-2})},$$

which is positive for even k and negative for odd k. $\square$

The next concept is a seemingly strange but is nevertheless a meaningful way of
adding two fractions.

Definition 2.7 The *mediant* between the fractions a/b and c/d is defined as

$$\frac{a}{b} \oplus \frac{c}{d} := \frac{a+c}{b+d}. \tag{2.30}$$

---

[6] Working this out as a formally correct and complete induction makes a nice and worthwhile
exercise. Just a hint to the reader.

Remark 2.5 Definition 2.7 is a nice example that even "forbidden" mathematical operations, like a too naïve addition of fractions, can be meaningful when considered properly in the right context. Another example for that is the *Hadamard product* of two matrices, cf. [52]. Note however that the value of the mediant depends on the *fraction*, i.e., the pair formed by numerator and denominator, and not just on the rational number that they represent:

$$\frac{2}{4} \oplus \frac{1}{3} = \frac{3}{7} \neq \frac{2}{5} = \frac{1}{2} \oplus \frac{1}{3}.$$

In this terminology, an *intermediate fraction* is the mediant between two successive convergents of two consecutive fractions; more precisely, the kth intermediate fraction is the mediant between the kth and the $(k-1)$-convergent.

As we know from Proposition 2.3, the value of the mediant and therefore of the intermediate fraction depends on the *representation* of the fraction itself. Precisely, the intermediate fractions are mediants of

$$\frac{p_{k-2}}{q_{k-2}} \quad \text{and} \quad \frac{j\,p_{k-1}}{j\,q_{k-1}} = \frac{p_{k-1}}{q_{k-1}},$$

and are monotonically increasing or decreasing with respect to j depending on k, and more precisely on the parity of k. We can also view it differently and state it in the following formal way.

Proposition 2.4 *The jth intermediate fraction is the mediant between the $(j-1)$-intermediate fraction and the $(k-1)$-convergent, i.e.,*

$$\frac{j\,p_{k-1} + p_{k-2}}{j\,q_{k-1} + q_{k-2}} = \frac{(j-1)\,p_{k-1} + p_{k-2}}{(j-1)\,q_{k-1} + q_{k-2}} \oplus \frac{p_{k-1}}{q_{k-1}}.$$

In general, the value of the mediant always lies between the values of the two fractions; more precisely,

$$b, d > 0, \quad \frac{a}{b} < \frac{c}{d} \quad \Rightarrow \quad \frac{a}{b} < \frac{a+c}{b+d} < \frac{c}{d}. \tag{2.31}$$

The assumption $a/b < c/d$ or equivalently $bc - ad > 0$ is not a restriction as long as the two underlying rational numbers[7] are not equal. The inequalities in (2.31) now follow from the observation that

$$\frac{a+c}{b+d} - \frac{a}{b} = \frac{ab + bc - ab - ad}{(b+d)b} = \frac{bc - ad}{b^2 + bd} > 0$$

---

[7] Even if this may get boring to the reader due to excessive repetition of the argument, one still has to distinguish between the *fraction*, i.e., the pair of numerator and denominator and the *rational number* represented by the fraction, as this makes a difference for mediants as we already saw above. And $\frac{1}{2}$ and $\frac{2}{4}$ are different fractions representing the same rational number.

and

$$\frac{c}{d} - \frac{a+c}{b+d} = \frac{bc + cd - ad - cd}{d(b+d)} = \frac{bc - ad}{bd + d^2} > 0.$$

Hence any intermediate fraction is enclosed by two successive convergents. To that end, consider the sequence of potential intermediate fractions b_j defined by

$$b_j := \frac{j\, p_{k-1} + p_{k-2}}{j\, q_{k-1} + q_{k-2}} = b_{j-1} \oplus \frac{p_{k-1}}{q_{k-1}}, \qquad b_0 := \frac{p_{k-2}}{q_{k-2}}. \tag{2.32}$$

Being identified as a mediant in (2.32), b_j lies between b_{j-1} and p_{k-1}/q_{k-1}. Since the kth convergent is just b_{a_k} and since the limit $a = [a_0; a_1, \dots]$ of an infinite continued fraction is enclosed by the $(k-1)$- and kth convergent, we always find the limit between b_1 and p_{k-1}. On the other hand,

$$b_1 = \frac{p_{k-1} + p_{k-2}}{q_{k-1} + q_{k-2}} = \frac{p_{k-1}}{q_{k-1}} \oplus \frac{p_{k-2}}{q_{k-2}}$$

is the mediant between the $(k-2)$- and $(k-1)$-convergent, and thus lies between these two convergents.

Before this gets too confusing, we illustrate the situation for even k:

$$\frac{p_{k-2}}{q_{k-2}} = b_0 < b_1 = \frac{p_{k-2}}{q_{k-2}} \oplus \frac{p_{k-1}}{q_{k-1}} < \dots < b_{a_k} = \frac{p_k}{q_k} < a < \frac{p_{k-1}}{q_{k-1}}, \tag{2.33}$$

for odd k, all the inequality signs simply have to be reversed. If we replace k by $k+2$ in (2.33), we conclude that for any even k the relation

$$\frac{p_k}{q_k} < \frac{p_k}{q_k} \oplus \frac{p_{k+1}}{q_{k+1}} < a < \frac{p_{k+1}}{q_{k+1}} \tag{2.34}$$

holds, while for odd k, we again have the same with reversed inequality signs. This simple observation has a very interesting immediate consequence for the *quality of approximation* of continued fractions.

Theorem 2.7 *For* $a = [a_0; a_1, \dots]$ *and* $k \geq 0$, *we have that*

$$\frac{1}{q_k(q_{k+1} + q_k)} < \left| a - \frac{p_k}{q_k} \right| < \frac{1}{q_k\, q_{k+1}}. \tag{2.35}$$

This theorem tells us that the upper estimate for the *convergence rate* of continued fractions is practically optimal. Since the denominators q_k of the convergents are monotonically increasing, that is, $q_{k+1} > q_k$ and therefore also $q_{k+1} + q_k < 2q_{k+1}$, we get that

$$\frac{1}{q_k(q_{k+1} + q_k)} > \frac{1}{2\, q_k\, q_{k+1}},$$

which gives us the slightly coarser but more illustrating enclosure

$$\frac{1}{2\, q_k\, q_{k+1}} < \left| a - \frac{p_k}{q_k} \right| < \frac{1}{q_k\, q_{k+1}}. \tag{2.36}$$

Since the q_k grow at least like $2^{k/2}$, the factor 2 in (2.36) is more or less irrelevant, and we can say that the kth convergents converge like 2^{-k}. In other words, any convergent determines approximately at least one *binary digit* of the fraction.

Proof *(of Theorem 2.7)* The upper estimate in (2.35) is precisely Theorem 2.3; for the lower estimates we have a closer look at the mediants; indeed, (2.34) says that the mediant $p_k/q_k \oplus p_{k+1}/q_{k+1}$ of the kth and $(k+1)$-convergent lies between the kth convergent and the value a of the continued fraction, hence the mediant is closer to a than the kth convergent. Therefore,

$$\left| a - \frac{p_k}{q_k} \right| > \left| \left(\frac{p_k}{q_k} \oplus \frac{p_{k+1}}{q_{k+1}} \right) - \frac{p_k}{q_k} \right| = \left| \frac{p_{k+1} + p_k}{q_{k+1} + q_k} - \frac{p_k}{q_k} \right| = \left| \frac{p_{k+1}\, q_k - p_k\, q_{k+1}}{q_k\, (q_{k+1} + q_k)} \right|$$

$$= \left| \frac{(-1)^k}{q_k\, (q_{k+1} + q_k)} \right| = \frac{1}{q_k\, (q_{k+1} + q_k)},$$

as claimed. $\square$

Now we come to a fundamental result regarding continued fractions for real numbers.

Theorem 2.8 *Any real number $x \in \mathbb{R}$ can be written in exactly one way as a continued fraction $[a_0; a_1, \dots]$ with $a_0 \in \mathbb{Z}$ and positive integer entries $a_j \in \mathbb{N}$, $j \in \mathbb{N}$. This continued fraction is finite if the number is rational, and infinite if it is irrational.*

Remark 2.6 The way it is stated, Theorem 2.8 is not correct, since finite continued fractions cannot be unique without an additional assumption! This can be seen from the simple example

$$[a_0; \,] = a_0 = a_0 - 1 + 1 = a_0 - 1 + \frac{1}{1} = [a_0 - 1; 1].$$

The same argument implies that always

$$[a_0; a_1, \dots, a_n] = [a_0; a_1, \dots, a_n - 1, 1], \tag{2.37}$$

$$[a_0; a_1, \dots, a_n, 1] = [a_0; a_1, \dots, a_n + 1]. \tag{2.38}$$

Hence finite continued fractions that end on "1" have a built-in *ambiguity*. This is resolved by enforcing the convention from the following definition.

Definition 2.8 *(Convention on last digits)* Any *finite* continued fraction

$$[a_0; a_1, \dots, a_n], \qquad a_0 \in \mathbb{Z}, \ a_j \in \mathbb{N}, \quad j \in \mathbb{N},$$

must always satisfy $a_n \neq 1$.

Note that Definition 2.8 is not a restriction on the existence of continued fraction expansions, since any continued fraction that accidentally happens to have the last digit $a_n = 1$ can be rewritten, and even shortened and simplified by (2.37) until the last digit is indeed $\neq 1$. In other words, the convention could also be seen as choosing the continued fraction representation of *minimal length*.

Remark 2.7 Theorem 2.8 shows that the distinction between rational and irrational numbers is simpler in terms of continued fractions than in terms of digit expansions such as binary or decimal digits. Recall that rational numbers are characterized by having either finite or periodic digit expansions, independent of the basis.

Proof (of Theorem 2.8) That rational numbers can be represented by finite continued fractions we already know from Theorem 2.5. So it remains to show the existence of a continued fraction expansion for *irrational* numbers, and in particular the *uniqueness* of their continued fraction expansion.

To that end, we first (re)consider the general method of computing a continued fraction expansion, starting from a number $x \in \mathbb{R} \setminus \mathbb{Q}$. In the first step the only reasonable choice is to set

$$a_0 = \lfloor x \rfloor := \max \{ j \in \mathbb{Z} \: : \: j \leq x \},$$

which either gives $x = a_0$, contradicting the irrationality of x, or there exists some $0 \neq r_1 \in \mathbb{R} \setminus \mathbb{Q}$ such that

$$x = [a_0; r_1] = a_0 + \frac{1}{r_1} \quad \Rightarrow \quad r_1 = \frac{1}{x - a_0} > 1,$$

since $0 < x - a_0 < 1$. Then we continue iteratively, setting

$$a_j = \lfloor r_j \rfloor, \qquad r_{j+1} = \frac{1}{r_j - a_j}, \qquad j = 1, 2, \ldots, \tag{2.39}$$

and noting that the sequences we obtain this way already satisfy $a_0 \in \mathbb{Z}$ and $a_j \in \mathbb{N}$, $j \in \mathbb{N}$, thereby defining a *convergent* continued fraction. The sequence would terminate only if $a_j = r_j$, but then the continued fraction would be finite and $x \in \mathbb{Q}$, i.e., a rational number. Consequently, irrational numbers must have an infinite continued fraction expansion. By construction and using an infinite version of (2.14), we thus obtain

$$x = [a_0; a_1, \ldots, a_{n-1}, r_n] = \frac{r_n \, p_{n-1} + p_{n-2}}{r_n \, q_{n-1} + q_{n-2}}.$$

But then we get, once more from the recurrence relation,

$$x - \frac{p_n}{q_n} = \frac{r_n \, p_{n-1} + p_{n-2}}{r_n \, q_{n-1} + q_{n-2}} - \frac{a_n \, p_{n-1} + p_{n-2}}{a_n \, q_{n-1} + q_{n-2}}$$

$$= \frac{r_n \, p_{n-1} \, q_{n-2} + a_n \, q_{n-1} \, p_{n-2} - r_n \, q_{n-1} \, p_{n-2} - a_n \, p_{n-1} \, q_{n-2}}{(r_n \, q_{n-1} + q_{n-2})(a_n \, q_{n-1} + q_{n-2})}$$

$$= \frac{(p_{n-1} \, q_{n-2} - q_{n-1} \, p_{n-2})(r_n - a_n)}{[(r_n - a_n) \, q_{n-1} + q_n] \, q_n} = \frac{(-1)^n \, (r_n - a_n)}{q_n^2 + (r_n - a_n) \, q_{n-1} q_n},$$

and therefore

$$\left| x - \frac{p_n}{q_n} \right| < \frac{1}{q_n^2}. \tag{2.40}$$

Hence the convergents indeed converge to x, and even with the predicted speed. To summarize, the infinite continued fraction constructed above is a representation for x.

It remains to show uniqueness where we make use of the fact that the representation of $x = [a_0; a_1, \dots] \in \mathbb{R}$ only contains *positive* integers, except maybe for a_0. This already enforces the choice $a_0 = \lfloor x \rfloor$ since

$$x - a_0 = [0; a_1, a_2, \dots] \in (0, 1),$$

except when $x \in \mathbb{Z}$, but that is a trivial case anyway. If now $[a_0; a_1, \dots]$ and $[a_0'; a_1', \dots]$ are two continued fraction expressions of $x \in \mathbb{R}$, then the above reasoning yields that $a_0 = a_0'$. Now suppose that by induction on $k \geq 0$, we have already shown that

$$a_j = a_j' \quad j = 0, \dots, k;$$

hence,

$$p_j = p_j', \, q_j = q_j', \quad j = 0, \dots, k,$$

to conclude that

$$x = \frac{r_{k+1} \, p_k + p_{k-1}}{r_{k+1} \, q_k + q_{k-1}} = \frac{r_{k+1}' \, p_k' + p_{k-1}'}{r_{k+1}' \, q_k' + q_{k-1}'} = \frac{r_{k+1}' \, p_k + p_{k-1}}{r_{k+1}' \, q_k + q_{k-1}}$$

yields

$$[a_{k+1}; a_{k+2}, \dots] = r_{k+1}' = r_{k+1} = [a_{k+1}'; a_{k+2}', \dots],$$

and repeating the above argument we find that $a_{k+1}' = a_{k+1}$, which completes the inductive step. Regardless of whether the continued fractions are finite or infinite, this yields that they must coincide. □

Example 2.1 The construction procedure for continued fractions enables us to easily determine the (irrational) numbers that have a particularly simple infinite continued fraction expansion of the form

$$x = [k; k, \ldots], \qquad k \in \mathbb{N}.$$

They have the property that $r_1 = x$ and therefore

$$x = k + \frac{1}{x} \qquad \Leftrightarrow \qquad x^2 - kx - 1 = 0,$$

that is

$$x = \frac{k + \sqrt{k^2 + 4}}{2}.$$

Since $x > 0$, the negative solution of the quadratic equation can be excluded. In particular, we find that for $k = 1$

$$\frac{1 + \sqrt{5}}{2} = [1; 1, \ldots],$$

which means that the *golden ratio* has the simplest possible continued fraction expansion. In the same fashion, $k = 2$ leads to

$$[2; 2, \ldots] = \frac{2 + \sqrt{8}}{2} = 1 + \frac{2\sqrt{2}}{2} = 1 + \sqrt{2},$$

which yields that $\sqrt{2} = [1; 2, 2, \ldots]$.

Example 2.2 We can extend the same idea and see what we can do with 2-*periodic* continued fractions of the form $x = [k_1; k_2, k_1, k_2, \ldots]$. Now the *fixed point equation* defining x is $r_2 = x$, which leads to

$$x = k_1 + \cfrac{1}{k_2 + \cfrac{1}{x}} = k_1 + \frac{x}{k_2 x + 1} = \frac{(k_1 k_2 + 1) x + k_1}{k_2 x + 1},$$

and we now look for the zeros of

$$k_2 x^2 - k_1 k_2 x - k_1 = k_2 \left(x^2 - k_1 x - \frac{k_1}{k_2} \right),$$

that is,

$$x = \frac{k_1 + \sqrt{k_1 (k_1 + 4/k_2)}}{2}.$$

Again, the numbers are rational plus a plain square root.

2.3.1 Problems

2.8 Compute the continued fraction expansion of $\frac{15}{8}$ and of $\frac{53}{17}$.

2.9 Can the case $k = q$ happen in Corollary 2.5?

2.10 The *Fibonacci numbers* $F_n, n \in \mathbb{N}_0$, are defined as $F_0 = 0$, $F_1 = 1$ and $F_{n+1} = F_n + F_{n-1}$, $n \geq 1$. Show that the unique fraction $\frac{p}{q}$ with *minimal* denominator q whose continued fraction expansion has length n is $\frac{F_{n+2}}{F_{n+1}}$.

2.11 Compute the continued fraction expansion of $\frac{F_{n+2}}{F_{n+1}}, n \geq 1$.

2.12 Implement the routine to compute continued fractions in Matlab or Octave.

2.13 Show that the representation

$$b_j = \frac{j \, p_{k-1} + p_{k-2}}{j \, q_{k-1} + q_{k-2}}, \qquad j = 0, \ldots, a_k,$$

of the intermediate fractions in (2.32) is irreducible.

2.14 Show: any *periodic continued fraction* belongs to $\mathbb{Q} + \sqrt{\mathbb{Q}}$, and hence can be written as $q + r, q, r^2 \in \mathbb{Q}$.

Hint: First show that any $x \in \mathbb{R}$ that can be written as a periodic continued fraction satisfies an equation of the form

$$x = \frac{p(x)}{q(x)}, \qquad p, q \in \mathbb{N}[x], \ \deg p = \deg q = 1.$$

2.15 (*Geometric interpretation of the golden ratio*) The golden ratio is defined by dividing an interval of unit length by a point $x \in [0, 1]$, such that the ratio between the length of the interval $[0, x]$ is the same as the ratio between the subintervals $[0, x]$ and $[x, 1]$, i.e.,

$$\frac{x}{1 - x} = \frac{1}{x} =: \psi.$$

This ratio ψ is called the *golden ratio*. Show that the golden ratio is well defined and compute its continued fraction expansion directly.

2.16 Show that
$$\sqrt{1 + k^2} = [k; 2k, 2k, 2k, \ldots].$$

2.17 A function $f : \mathbb{R} \to \mathbb{R}$ of the form

$$f(x) = \frac{a + bx}{c + dx}, \qquad a, b, c, d \in \mathbb{N}_0, \qquad |bc - ad| = 1,$$

is called a *unimodular function*. Moreover, given a continued fraction $[a_0; a_1, a_2, \dots]$, define

$$f_n(x) := [a_0; a_1, \dots, a_n + x], \qquad x \in \mathbb{R}.$$

Show that

1. f_n is unimodular, $n \in \mathbb{N}$.
2. For $n \in \mathbb{N}$,

$$f(0) = \frac{p_n}{q_n}, \qquad f\left(a_{n+1}^{-1}\right) = \frac{p_{n+1}}{q_{n+1}},$$

and

$$\lim_{x \to \infty} f_n(x) = \frac{p_{n-1}}{q_{n-1}}.$$

2.18 A number $x \in \mathbb{R}$ of the form

$$x = r + s\sqrt{n}, \qquad r, s \in \mathbb{Q}, \quad n \in \mathbb{N} \setminus \mathbb{N}^2$$

is called a *quadratic irrationality*. Show that any unimodular function maps quadratic irrationalities to quadratic irrationalities.

2.4 Convergents as Best Approximants

Knowing that any real number can be represented as an infinite continued fraction and thus can be approximated by a particular sequence of finite continued fractions, namely its convergents, we will further justify their use by showing that continued fractions approximate real numbers *better* than other fractions. Of course with $\mathbb{Q}$ being dense in $\mathbb{R}$, there are lots of fractions that converge to a given $x \in \mathbb{R}$.

Remark 2.8 (*Myths and legends, cf.* [60]) When Christiaan Huygens built his mechanical planetarium, a model of our solar system, he had to find an approximation of the irrational duration of the time it takes planets to complete their orbit using rational numbers, and to make this approximation as good as possible. Rational numbers can be implemented mechanically by *cogwheels* and the transmission is simply the ratio of the number of teeth of the different wheels in the gear, hence a rational number.[8] Therefore good approximations of these irrational orbital time ratios by rational numbers with an as small as possible numerator and denominator were crucial for the mechanical implementation: the larger the numerator and the denominator are, the harder it is to manufacture the cogwheels to achieve a reasonable size. And of course knowing the theory of continued fractions, using convergents was the natural way to attack the problem for Huygens.

---

[8] Even with modern methods such as additive manufacturing and nanotechnology, no one has managed so far to produce a cogwheel with a non-integer number of teeth. The same actually holds true for negative numbers.

A good measure for the *complexity* of a fraction $x \in \mathbb{Q}$ is the size of its denominator: writing x as

$$x = a + \frac{p}{q}, \qquad a \in \mathbb{Z}, \; p, q \in \mathbb{N}, \quad p < q;$$

then the amount of *information* we need to store x is the order of magnitude $\log a + 2 \log q$. This is simply the number of *digits* in the integer part, and in the numerator and denominator of the fractional part. Whether we choose these digits decimally or binary affects it only by a constant and it is not at all relevant. Ignoring the integer part and restricting x to $[0, 1]$, the fundamental quantity for measuring the complexity of a fraction is then the size of its denominator, and the complexity of a rational number is, up to a constant, the size of its denominator in the (minimal) *irreducible* representation.

Remark 2.9 This measure of complexity of a rational number is of relevance and plays an important role in the *symbolic computations* that occur in Computer Algebra systems. This is explained for example in [31].

The goal to approximate a number with fractions of limited complexity is the motivation behind the following definition.

Definition 2.9 A fraction a/b is called *best approximant* or *best approximant of the first kind* to $x \in \mathbb{R}$ if

$$\left| x - \frac{a}{b} \right| \leq \left| x - \frac{c}{d} \right|, \qquad d \leq b.$$

Here we always consider fractions of the form $\mathbb{Z}/\mathbb{N}$ with positive denominators.

What we will show now is that the convergents of the continued fraction expansions are essentially the best approximants.

Theorem 2.9 *Any* best approximant *to a real number* $x \in \mathbb{R}$ *is either a* convergent *of the associated continued fraction expansion or an* intermediate fraction.

Proof Let a/b be a best approximant or an *element of best approximation* in the terminology of Approximation Theory [67] to $x = [a_0; a_1, \dots]$, $a_0 \in \mathbb{Z}$, $a_j \in \mathbb{N}$, $j \in \mathbb{N}$. Then $a/b > a_0$, as otherwise $a/b < a_0 = \lfloor x \rfloor \leq x$ and $a_0/1$ were already a better approximant than a/b which would be a contradiction. Exactly the same type of argument also shows that $\frac{a}{b} < a_0 + 1$, since then $a_0 + 1$ was a better approximant due to $x < a_0 + 1$. Hence

$$a_0 \leq \frac{a}{b} \leq a_0 + 1,$$

and with equality in one of the two cases, the claim is proved: the best integer approximant is then either the convergent a_0 or the intermediate fraction

$$\frac{a_0 + 1}{1} = \frac{p_1 + p_0}{q_1 + q_0},$$

keeping in mind that the initialization $q_0 = 0$, $q_1 = p_0 = 1$, $p_1 = a_0$, holds by definition.

Let us suppose that $a_0 < \frac{a}{b} < a_0 + 1$ and that a/b is neither convergent nor an intermediate fraction. We will show that then there exists an intermediate fraction with a smaller denominator that is even closer to x than a/b. By Proposition 2.3, a/b lies between two intermediate fractions,[9] so that there exist n and k such that either

$$\frac{k\,p_n + p_{n-1}}{k\,q_n + q_{n-1}} < \frac{a}{b} < \frac{(k+1)\,p_n + p_{n-1}}{(k+1)\,q_n + q_{n-1}}$$

or

$$\frac{k\,p_n + p_{n-1}}{k\,q_n + q_{n-1}} > \frac{a}{b} > \frac{(k+1)\,p_n + p_{n-1}}{(k+1)\,q_n + q_{n-1}},$$

depending on the parity of k. Therefore

$$\left| \frac{a}{b} - \frac{k\,p_n + p_{n-1}}{k\,q_n + q_{n-1}} \right| < \left| \frac{(k+1)\,p_n + p_{n-1}}{(k+1)\,q_n + q_{n-1}} - \frac{k\,p_n + p_{n-1}}{k\,q_n + q_{n-1}} \right|$$

$$= \frac{1}{((k+1)q_n + q_{n-1})\,(kq_n + q_{n-1})}.$$

On the other hand, expanding the difference of fractions yields that there exists $c \in \mathbb{N}$ such that[10]

$$0 \neq \left| \frac{a}{b} - \frac{k\,p_n + p_{n-1}}{k\,q_n + q_{n-1}} \right| = \frac{c}{b\,(kq_n + q_{n-1})} \geq \frac{1}{b\,(kq_n + q_{n-1})},$$

which yields

$$\frac{1}{b\,(kq_n + q_{n-1})} < \frac{1}{((k+1)q_n + q_{n-1})\,(kq_n + q_{n-1})}$$

and therefore

$$b > (k+1)\,q_n + q_{n-1}.$$

This shows that the $(k+1)$-intermediate fraction which is closer to x than a/b by construction also has a smaller denominator than a/b. Therefore it is a better approximant to x than a/b which is a contradiction. Therefore a/b must be either a convergent or an intermediate fraction.   □

---

[9] Recall that the intermediate fractions for convergents form a sequence that converges monotonically to x, monotonically increasing if the order of the convergent is even, decreasing if it is odd.

[10] Here is our first encounter with a nice and somewhat subtle argument that we will use several times: the fraction is not zero, hence the numerator is not zero and, since it is an integer, its absolute value must be ≥ 1. Fractions are discrete.

Indeed convergents are even *unique* best approximants if the notion of *best approximation* is formulated in a slightly sharper way. To motivate this concept, we recall what the expression a/b means: it is the (rational) number that when multiplied with b gives the value a. In this respect, x is a good approximation to that number if the difference $|bx - a|$ is as small as possible.

Definition 2.10 A fraction a/b is called *best approximant of the second kind* to $x \in \mathbb{R}$ provided that

$$\frac{c}{d} \neq \frac{a}{b}, \quad 0 < d \leq b \quad \Longrightarrow \quad |bx - a| \leq |dx - c|. \tag{2.41}$$

Best approximants of the second kind are also best approximants of the first kind in the sense of Definition 2.9, as otherwise we would have a fraction c/d, $d \leq b$, such that

$$\left| x - \frac{a}{b} \right| > \left| x - \frac{c}{d} \right|,$$

yielding

$$|bx - a| = b \left| x - \frac{a}{b} \right| > b \left| x - \frac{c}{d} \right| = \frac{b}{d} |dx - c| \geq |dx - c|,$$

which contradicts the assumption that a/b is a best approximant of the second kind. But the converse is not true: not any best approximant of the first kind is also one of the second kind, otherwise the distinction would obviously be pointless. The simplest example is $x = \frac{1}{5}$ and $\frac{a}{b} = \frac{1}{3}$; it is easy to verify that $\frac{1}{3}$ is closer to $\frac{1}{5}$ than its competitors $\left\{ 0, \frac{1}{2}, \frac{2}{3}, 1 \right\}$ of fractions with numerator ≤ 3, but where

$$\left| 3\frac{1}{5} - 1 \right| = \frac{2}{5} > \frac{1}{5} = \left| 1\frac{1}{5} - 0 \right|$$

holds. Best approximants of the second kind play an important role, since they are convergents and only convergents.

Theorem 2.10 *Any best approximant of the second kind to $x \in \mathbb{R}$ is a convergent, and any convergent of the continued fraction expansion of x is a best approximant of the second kind.*

Except for the special case $x = a_0 + \frac{1}{2}$ and convergents of the first order, all best approximants of the second kind are unique.

Proof Let us suppose that $\frac{a}{b}$ is a best approximant of the second kind to $x = [a_0; a_1, \dots]$. If $a/b < a_0 = \lfloor x \rfloor < x$, then $b \geq 1$ yields that

$$|1 \cdot x - a_0| = x - a_0 < x - \frac{a}{b} = \frac{1}{b} |bx - a| < |bx - a|,$$

and $a_0 = a_0/1$ would be a better approximant of the second kind. Hence, the first convergent of order 0 satisfies $\frac{p_0}{q_0} = a_0 \leq a/b$. If a/b is a convergent, the theorem would be trivially true; otherwise it either satisfies $\frac{a}{b} > \frac{p_1}{q_1}$, or it is enclosed between two convergents $\frac{p_{k-1}}{q_{k-1}}$ and $\frac{p_{k+1}}{q_{k+1}}$ due to the monotonic convergence of the convergents, cf. Corollary 2.2. In the first case, we have that $x < \frac{p_1}{q_1} < \frac{a}{b}$ and the monotonicity of the denominators q_k together with $p_1/q_1 \neq a/b$ yield

$$\left| x - \frac{a}{b} \right| > \left| \frac{p_1}{q_1} - \frac{a}{b} \right| = \frac{|b\, p_1 - a\, q_1|}{b\, q_1} \geq \frac{1}{b\, q_1},$$

that is,

$$|bx - a| > \frac{1}{q_1} = \frac{1}{a_1} = \frac{1}{\lfloor x - a_0 \rfloor^{-1}} \geq |1x - a_0|,$$

contradicting the assumption that a/b is a best approximant of the second kind. If, on the other hand, a/b is enclosed between two convergents, we first have

$$\left| \frac{a}{b} - \frac{p_{k-1}}{q_{k-1}} \right| = \frac{|a\, q_{k-1} - b\, p_{k-1}|}{b\, q_{k-1}} \geq \frac{1}{b\, q_{k-1}} \tag{2.42}$$

as well as,[11] by Corollary 2.1,

$$\left| \frac{a}{b} - \frac{p_{k-1}}{q_{k-1}} \right| < \left| \frac{p_k}{q_k} - \frac{p_{k-1}}{q_{k-1}} \right| = \frac{1}{q_k\, q_{k-1}}, \tag{2.43}$$

which allows us to combine (2.42) and (2.43) into $q_k < b$. Moreover, as in (2.42)

$$\left| x - \frac{a}{b} \right| > \left| \frac{p_{k+1}}{q_{k+1}} - \frac{a}{b} \right| \geq \frac{1}{b\, q_{k+1}},$$

which yields, together with (2.36), the estimate

$$|bx - a| > \frac{1}{q_{k+1}} = q_k \frac{1}{q_k\, q_{k+1}} > q_k \left| x - \frac{p_k}{q_k} \right| = |q_k x - p_k|$$

which would make the kth convergent a better approximant, hence another contradiction. As a result all that is left is that a/b is indeed a convergent.

For the converse, we fix the order k of the convergent; we consider the numbers

$$\min_{a \in \mathbb{Z}} |b\, x - a|, \qquad b \in \{1, \ldots, q_k\}, \tag{2.44}$$

---

[11] The appearance of the kth convergent is not a typo; we make use of the fact that the kth convergent lies on the "other" side of x by the alternating property of convergents from Corollary 2.2.

and denote by b^* the value of b for which (2.44) becomes minimal. If the minimum is assumed for several values of b, we take the smallest one so that b^* is unique and well defined. The respective minimized value for a is denoted by

$$a^* = \operatorname{argmin}_{a \in \mathbb{Z}} \left| b^* x - a \right|. \tag{2.45}$$

We first show that a^* in (2.45) is unique. To that end, suppose that we have $a' \neq a^*$ which also satisfies (2.45), and note that

$$\left| x - \frac{a^*}{b^*} \right| = \left| x - \frac{a'}{b^*} \right| \quad \Rightarrow \quad x = \frac{a^* + a'}{2\,b^*}. \tag{2.46}$$

The fraction on the right-hand side of (2.46) has to be irreducible, since otherwise there would be an irreducible representation $x = p/q$ with $q \leq b^*$; therefore $|q\,x - p| = 0$, which yields an unbeatable minimal value of (2.44) and which is also assumed exactly for $a = p$ and $b = q \leq b^* \leq q_k$. Developing the rational number x as a continued fraction, $x = [a_0; a_1, \ldots, a_n]$, where the convention of Definition 2.8 ensures that $a_n \geq 2$, we write it as its final convergent

$$x = \frac{p_n}{q_n};$$

furthermore, irreducibility of the convergents and the fraction in (2.46) yields that

$$\begin{aligned}
p_n &= a^* + a' \\
q_n &= 2b^* = a_n\,q_{n-1} + q_{n-2} \geq 2q_{n-1} + q_{n-2} > 2q_{n-1},
\end{aligned} \tag{2.47}$$

so that $q_{j-1} < b^*$ for any $1 \leq j \leq n$. The situation is special for $n = 1$, since in this case we can obtain $q_1 = b^*$ via $a_1 = 2$, and so $b^* = 1$ due to $q_0 = 1$. This is precisely the special case $x = a_0 + \frac{1}{2}$ for which we have

$$\left| x - (a_0 + 1) \right| = \frac{1}{2} = \left| x - a_0 \right|,$$

so that the best approximant of the second kind is not unique.

If, on the other hand, we are in the situation that $n > 1$, the (2.47) implies that $1 \leq q_{n-1} < b^*$, and then the assumption $a^* \neq a'$, together with (2.46), yield that

$$\begin{aligned}
\left| q_{n-1} x - p_{n-1} \right| &= \left| q_{n-1} \frac{p_n}{q_n} - p_{n-1} \right| = \frac{|q_{n-1}\,p_n - p_{n-1}\,q_n|}{q_n} = \frac{1}{q_n} = \frac{1}{2b^*} \\
&< \frac{1}{2} \leq \frac{|a^* - a'|}{2} = b^* \left| x - \frac{a^*}{b^*} \right| = \left| b^* x - a^* \right|;
\end{aligned}$$

this once more contradicts the assumption that a^*/b^* is the best approximant of the second kind. This eventually proves that a^* is *unique* and therefore a^*/b^* is a unique

best approximant of the second kind to x with minimal denominator. As we have
shown in the first half of the proof, the best approximant of the second kind must be
a convergent: hence $a^*/b^* = p_m/q_m$ for some $m \leq k$, where k is the order that we
fixed in the beginning. If $m = k$, we are done; otherwise two applications of (2.35)
yield that

$$\frac{1}{q_{k-1} + q_k} \leq \frac{1}{q_m + q_{m+1}} < |q_m x - p_m| < |q_k x - p_k| \leq \frac{1}{q_{k+1}}.$$

Then replacing k by $k - 1$ in the above, and using the ubiquitous recursion once
more, we have

$$q_{k-1} + q_{k-2} > q_k = a_k q_{k-1} + q_{k-2},$$

and we end up with $a_k < 1$ which contradicts the assumption that we consider only
continued fractions with positive integer components. This finally shows that p_k/q_k
is a *strict* best approximant of the second kind, which automatically makes it unique,
except in the aforementioned special case. $\square$

2.5 Approximation Order, Quantitative Statements

Having identified convergents or intermediate fractions as best approximants, depending on the kind of approximation we consider, we next address *quantitative* issues,
i.e., the question *how fast* continued fractions converge to a given real number. Of
course, this problem is only nontrivial for irrational numbers.

In the proof of Theorem 2.8, more precisely, in (2.40), we already had an upper
estimate for the *rate of approximation* of the convergents, namely

$$\left| x - \frac{p_n}{q_n} \right| < \frac{1}{q_n^2}.$$

On the other hand, the rational number $a = [0; m, 1, m]$ has the the explicit recursion

$$p_{-1} = 1, \quad p_0 = 0, \quad p_1 = 1, \quad p_2 = 1, \qquad p_3 = m + 1$$
$$q_{-1} = 0, \quad q_0 = 1, \quad q_1 = m, \quad q_2 = m + 1, \quad q_3 = m(m + 2),$$

and thus $a = \frac{m+1}{m(m+2)}$. For this number,

$$\left| a - \frac{p_1}{q_1} \right| = \left| \frac{p_3}{q_3} - \frac{p_1}{q_1} \right| = \frac{1}{m} - \frac{m+1}{m(m+2)} = \frac{1}{m(m+2)} = \frac{1}{q_1^2 (1 + 2/m)},$$

from which we we can already conclude, even if this only a first convergent, that in
general an approximation rate better than q_n^{-2} cannot be expected. More precisely,

for any $\varepsilon > 0$, we have some $m \in \mathbb{N}$ such that $1 < (1 + 2/m)^{-1} < 1 - \varepsilon$, so that the rate of approximation gets as close to $1/q_1^2$ as one wants.

Nevertheless we should not overestimate the relevance of such worst-case estimates since the next result shows that at least half of the convergents improve the rate by a factor of 2.

Proposition 2.5 *If the number $x \in \mathbb{R}$ has a kth convergent, i.e., cannot be written as $x = [x_0; x_1, \ldots, x_m]$ for some $m < k$, then at least one of the following two inequalities holds:*

$$\left| x - \frac{p_{k-1}}{q_{k-1}} \right| < \frac{1}{2\, q_{k-1}^2}, \quad or \quad \left| x - \frac{p_k}{q_k} \right| < \frac{1}{2\, q_k^2}. \tag{2.48}$$

Proof Since x is enclosed by the two convergents, we can once more use (2.9) to conclude that

$$\left| x - \frac{p_{k-1}}{q_{k-1}} \right| + \left| x - \frac{p_k}{q_k} \right| = \left| \frac{p_k}{q_k} - \frac{p_{k-1}}{q_{k-1}} \right| = \frac{1}{q_k\, q_{k-1}},$$

and the inequality between the *arithmetic mean* and the *geometric mean* yields

$$\frac{1}{q_k\, q_{k-1}} = \sqrt{\frac{1}{q_{k-1}^2} \frac{1}{q_k^2}} \leq \frac{1}{2} \left(\frac{1}{q_{k-1}^2} + \frac{1}{q_k^2} \right),$$

and therefore

$$\left| x - \frac{p_{k-1}}{q_{k-1}} \right| + \left| x - \frac{p_k}{q_k} \right| \leq \frac{1}{2\, q_{k-1}^2} + \frac{1}{2\, q_k^2}$$

so that the inequalities in (2.48) cannot be violated simultaneously.   □

Therefore at least one of two successive convergents has an approximation rate not only of $1/q_k^2$, but of $1/(2q_k^2)$, and this statement even has a converse that says that if an approximation is really good, it has to be a convergent.

Theorem 2.11 *Given $x \in \mathbb{R}$. If there exist $a \in \mathbb{Z}$ and $b \in \mathbb{N}$ such that*

$$\left| x - \frac{a}{b} \right| < \frac{1}{2b^2},$$

then a/b is a convergent of the continued fraction expansion of x.

Proof According to Theorem 2.10, it suffices to show that a/b is a best approximant of the second kind. If there would exist $c \in \mathbb{Z}$ and $d \in \mathbb{N}$ such that $|dx - c| < |bx - a| < 1/2b$, then also

$$\left| x - \frac{c}{d} \right| < \frac{1}{2bd},$$

and since by assumption $a/b \neq c/d$, then

$$\frac{1}{bd} \leq \left| \frac{a}{b} - \frac{c}{d} \right| \leq \left| x - \frac{a}{b} \right| + \left| x - \frac{c}{d} \right| < \frac{1}{2b^2} + \frac{1}{2bd} = \frac{b+d}{2b^2d}.$$

This means that

$$2b < b + d \quad \Rightarrow \quad b < d$$

so that a/b is indeed a best approximant of the second kind. $\qquad\square$

Proposition 2.5 can even be improved by considering *three* successive convergents, among which one provides an even better rate of approximation.

Theorem 2.12 *If $x \in \mathbb{R}$ has a convergent of order $k > 1$, then at least one of the following three inequalities is satisfied:*

$$\left| x - \frac{p_{k-2}}{q_{k-2}} \right| < \frac{1}{\sqrt{5}\, q_{k-2}^2}, \quad \left| x - \frac{p_{k-1}}{q_{k-1}} \right| < \frac{1}{\sqrt{5}\, q_{k-1}^2}, \quad \left| x - \frac{p_k}{q_k} \right| < \frac{1}{\sqrt{5}\, q_k^2}.$$
$$(2.49)$$

It is tempting at this point to hope for an extension of this process: maybe among four successive convergents we find an even better rate, then consider five and so on. Unfortunately or fortunately this is not the case, and the counterexample is once more the *golden ratio*

$$x = \frac{1 + \sqrt{5}}{2} = [1; 1, \dots], \quad x = 1 + \frac{1}{x},$$

from Example 2.1. Since

$$x = [1; 1, \dots, 1, r_k], \quad r_k = [1; 1, \dots] = x,$$

we also have that

$$x = \frac{x\, p_k + p_{k-1}}{x\, q_k + q_{k-1}},$$

which implies that

$$\left| x - \frac{p_k}{q_k} \right| = \frac{1}{(x\, q_k + q_{k-1})\, q_k} = \frac{1}{q_k^2\, (x + q_{k-1}/q_k)}.$$

Now formula (2.15) from Proposition 2.1 tells us that

$$\frac{q_k}{q_{k-1}} = [a_k; a_{k-1}, \dots, a_1] = [1; 1, \dots, 1] \to x \quad \text{for } k \to \infty;$$

even if the finite continued fraction above is a "forbidden" form with the last coefficient equal to 1, it still stands for a perfectly well defined number and could also be written as $[1; 1, \ldots, 1, 2]$ with length reduced by 1. Hence,

$$\frac{q_{k-1}}{q_k} = \frac{1}{x} + \varepsilon_k = x - 1 + \varepsilon_k, \qquad \lim_{k \to \infty} \varepsilon_k = 0,$$

and therefore

$$\left| x - \frac{p_k}{q_k} \right| = \frac{1}{q_k^2 (2x - 1 + \varepsilon_k)} = \frac{1}{q_k^2 \left(\sqrt{5} + \varepsilon_k \right)},$$

which means that there cannot be an approximation rate better than $1/\sqrt{5}q_k^2$, regardless of how many successive convergents we consider. In other words: $\sqrt{5}$ appearing as the optimal constant is really a consequence of the golden ratio.

Proof *(of Theorem* 2.12*)* We set

$$\varphi_k := \frac{q_{k-2}}{q_{k-1}}, \qquad \psi_k := r_k + \varphi_k, \qquad k \geq 2,$$

and first prove that

$$k \geq 2, \ \psi_k \leq \sqrt{5}, \ \psi_{k-1} \leq \sqrt{5} \qquad \Rightarrow \qquad \varphi_k > \frac{\sqrt{5} - 1}{2}. \tag{2.50}$$

Since

$$\frac{1}{\varphi_{k+1}} = \frac{q_k}{q_{k-1}} = \frac{a_k \, q_{k-1} + q_{k-2}}{q_{k-1}} = a_k + \frac{q_{k-2}}{q_{k-1}} = a_k + \varphi_k$$

and

$$r_k = [a_k; a_{k+1}, \ldots] = a_k + \frac{1}{[a_{k+1}; a_{k+2}, \ldots]} = a_k + \frac{1}{r_{k+1}},$$

we obtain that

$$\frac{1}{\varphi_{k+1}} - \varphi_k = a_k = r_k - \frac{1}{r_{k+1}} \qquad \Rightarrow \qquad \frac{1}{\varphi_{k+1}} + \frac{1}{r_{k+1}} = r_k + \varphi_k = \psi_k,$$

so that the assumptions in (2.50) yield the inequalities

$$0 \leq r_k + \varphi_k \leq \sqrt{5},$$
$$0 \leq \frac{1}{\varphi_k} + \frac{1}{r_k} \leq \sqrt{5},$$

which in turn imply that

$$5 - \sqrt{5}\left(\varphi_k + \frac{1}{\varphi_k}\right) = \left(\sqrt{5} - \varphi_k\right)\left(\sqrt{5} - \frac{1}{\varphi_k}\right) - 1 \geq \frac{r_k}{r_k} - 1 = 0.$$

Since φ_k is a rational number and $\sqrt{5}$ is irrational, equality cannot be assumed in the above estimate and the inequality is a strict one. Multiplying it by $\varphi_k/\sqrt{5} > 0$ then yields that

$$0 < \sqrt{5}\,\varphi_k - \varphi_k^2 + 1 = -\left(\frac{\sqrt{5}}{2} - \varphi_k\right)^2 + \frac{1}{4},$$

and implies that

$$-\frac{1}{2} < \frac{\sqrt{5}}{2} - \varphi_k < \frac{1}{2},$$

so that

$$\varphi_k > -\frac{1}{2} + \frac{\sqrt{5}}{2} = \frac{\sqrt{5} - 1}{2},$$

as claimed in (2.50).

After these preliminaries, we can turn to the proof itself. To that end, we assume that

$$\left| x - \frac{p_n}{q_n} \right| \geq \frac{1}{\sqrt{5}\, q_n^2}, \qquad n \in \{k-2, k-1, k\},$$

which implies, together with

$$\left| x - \frac{p_n}{q_n} \right| = \left| \frac{r_{n+1}\, p_n + p_{n-1}}{r_{n+1}\, q_n + q_{n-1}} - \frac{p_n}{q_n} \right| = \frac{1}{q_n\,(r_{n+1}\, q_n + q_{n-1})} = \frac{1}{q_n^2\,(r_{n+1} + q_{n-1}/q_n)}$$

$$= \frac{1}{q_n^2\,(r_{n+1} + \varphi_{n+1})} = \frac{1}{q_n^2\,\psi_{n+1}},$$

that

$$\psi_n \leq \sqrt{5}, \quad n = k-1, k, k+1 \qquad \Rightarrow \qquad \varphi_n > \frac{\sqrt{5} - 1}{2}, \quad n = k, k+1,$$

and eventually,

$$a_k = \frac{1}{\varphi_{k+1}} - \varphi_k < \frac{2}{\sqrt{5} - 1} - \frac{\sqrt{5} - 1}{2} = \frac{4 - 5 + 2\sqrt{5} - 1}{2\left(\sqrt{5} - 1\right)} = 1,$$

which is impossible since $a_k \in \mathbb{N}$. Hence we obtained a contradiction and the claim must be true. $\qquad\square$

Let us summarize: for *arbitrary* real numbers, the approximation order of convergents of the continued fraction expansion is bounded essentially by $1/\sqrt{5}q_n^2$. This worst approximation rate occurs for the golden ratio which makes it the most irrational number in the sense that its approximation order by convergents is the worst.

On the other hand, there are irrational numbers that can even be approximated *arbitrarily well* by convergents.

Theorem 2.13 *For any function $\varphi : \mathbb{N} \to \mathbb{R}_+$ there exists $x \in \mathbb{R}$, such that for infinitely many values $q \in \mathbb{N}$, the inequality*

$$\left| x - \frac{p}{q} \right| < \varphi(q)$$

holds.

Proof We construct x by its continued fraction expansion. To that end, we choose $a_0 \in \mathbb{Z}$ arbitrarily, and in addition

$$a_{k+1} > \frac{1}{q_k^2 \varphi(q_k)}, \qquad k \in \mathbb{N}_0, \tag{2.51}$$

which can be done in a lot of ways. Then $x = [a_0; a_1, \dots] \in \mathbb{R}$, and once again using (2.18) from Theorem 2.3,

$$\left| x - \frac{p_k}{q_k} \right| < \frac{1}{q_k q_{k+1}} = \frac{1}{q_k (a_{k+1}q_k + q_{k-1})} < \frac{1}{a_{k+1} q_k^2} < \frac{q_k^2 \varphi(q_k)}{q_k^2} = \varphi(q_k),$$

which even holds for *any* $k \in \mathbb{N}_0$ so that all convergents converge with rate φ. $\square$

The estimate (2.51) that indicated how to choose a_k already tells us what we have to do in order to obtain a number x, such that the convergents approximate them quickly, that is numbers x for which φ decays rapidly: the components a_k in the continued fraction expansion of x have to grow fast. This can be derived from the estimate (2.35) from which we obtain

$$\frac{1}{a_{k+1} q_k^2} > \left| x - \frac{p_k}{q_k} \right| > \frac{1}{q_k (q_{k+1} + q_k)} = \frac{1}{q_k (a_{k+1}q_k + q_{k-1} + q_k)}$$
$$= \frac{1}{q_k^2 (a_{k+1} + 1 + q_{k-1}/q_k)} > \frac{1}{(a_{k+1} + 2) q_k^2} \tag{2.52}$$

which implies an approximation order of $\varphi(q_k) \sim 1/a_{k+1} q_k^2$. This suggests that a good approximation order, i.e., a fast approximation has to do with the rapid growth of the coefficients. And this is indeed the case, since the next result shows that growth is also necessary for a convergence rate better than the worst case in Theorem 2.12.

Theorem 2.14 *Let* $x \in \mathbb{R} \setminus \mathbb{Q}$ *be an irrational number. If the coefficients in the continued fraction expansion of x are bounded, then we have* $c > 0$ *such that*

$$\left| x - \frac{p}{q} \right| < \frac{c}{q^2}, \qquad p \in \mathbb{Z}, \, q \in \mathbb{N}, \tag{2.53}$$

has no solution. Conversely, if the coefficients are unbounded, then we have for any $c > 0$ *infinitely many solutions of* (2.53).

Proof If $\sup\{a_k \, : \, k \in \mathbb{N}_0\} =: M < \infty$, the lower estimate in (2.52) yields that

$$\left| x - \frac{p_k}{q_k} \right| > \frac{1}{(M+2)\,q_k^2}, \qquad k \in \mathbb{N}.$$

For an arbitrary irreducible fraction p/q, we now choose k such that $q_{k-1} < q \leq q_k$, and since all convergents are best approximants of the first and second kind to x, it follows that

$$\left| x - \frac{p}{q} \right| \geq \left| x - \frac{p_k}{q_k} \right| > \frac{1}{(M+2)\,q_k^2} = \frac{1}{(M+2)\,q^2} \left(\frac{q}{q_k} \right)^2 > \frac{1}{(M+2)\,q^2} \left(\frac{q_{k-1}}{q_k} \right)^2$$

$$= \frac{1}{(M+2)\,q^2} \left(\frac{q_{k-1}}{a_k q_{k-1} + q_{k-2}} \right)^2 > \frac{1}{(M+2)\,q^2} \left(\frac{1}{a_k + 1} \right)^2$$

$$> \frac{1}{(M+2)(M+1)^2\,q^2} > \frac{c}{q^2},$$

where the constant c satisfies

$$c < \frac{1}{(M+2)(M+1)^2},$$

an estimate that depends only on the bound M of the components but not on the denominator q.

If, on the other hand, $\sup\{a_k \, : \, k \in \mathbb{N}\} = \infty$, then there exist for any $c > 0$ infinitely many indices k with $a_{k+1} > 1/c$, and we can then apply the upper estimate (2.52) directly for

$$\left| x - \frac{p_k}{q_k} \right| < \frac{1}{a_{k+1}\,q_k^2} < \frac{c}{q_k^2}$$

which yields an infinity of solutions of (2.53). $\square$

2.5.1 Problems

2.19 Construct a number $x \in \mathbb{R}$ such that

$$\left| x - \frac{p_n}{q_n} \right| = O\left(n^{-13}\right), \qquad n \in \mathbb{N},$$

where p_n/q_n is the nth convergent of the continued fraction expansion of x.

2.20 Determine the rate of approximation for the number $[m; m, \ldots]$, $m \geq 1$.

2.6 Algebraic Numbers

An algebraic number is a *zero* or *root* of a *polynomial* with rational or integer coefficients. Since we can always multiply a polynomial that has rational coefficients with the *least common multiple* (*lcm*) of the denominators of these coefficients, the two concepts of the roots of polynomials in $\mathbb{Z}[x]$ or $\mathbb{Q}[x]$ are the same—the zeros are not affected if the polynomial is multiplied by a nonzero constant.

Definition 2.11 A real number $a \in \mathbb{R}$ is called an *algebraic number* of order n, if we have a polynomial f of degree at most n,[12]

$$f \in \mathbb{Z}[x], \qquad f(x) = \sum_{k=0}^{n} f_k \, x^k, \qquad f_k \in \mathbb{Z}, \ k = 0, \ldots, n,$$

such that $f(a) = 0$ and there is no polynomial g of degree $< n$ with $g(a) = 0$. A real number that is not algebraic is called *transcendental*.

Remark 2.10 That an algebraic number a is of order n always means that the polynomial f of *minimal degree* such that $f(a) = 0$ has degree (exactly) n. This way the order of an algebraic number is well defined.

Classical examples for transcendental numbers are e and π, and algebraic numbers are $\sqrt{2}$ and the *golden ratio*. Algebraic numbers are computable in the sense that they allow for symbolic computations by adjoining the polynomial to the base field, cf. [31]. In the end, this may lead to expressions containing RootOf in the symbolic solution of systems of polynomial equations that are hard to interpret, but at least correct. They are known to all users of *computer algebra systems* such as Maple or Mathematica.

What is of interest to us here is the fact that algebraic numbers admit a *slow* approximation by continued fractions; this follows from the following theorem due

---

[12] Sometimes the phrase "polynomial of the degree n" means that the degree is *at most n*; sometimes it means the degree *exactly n*, the latter requiring the additional condition $f_n \neq 0$. This terminology is not unique in the literature, so be careful.

to Liouville that relates the order of an approximation to the order of the algebraic number.

Theorem 2.15 (Liouville) *For any algebraic number $a \in \mathbb{R} \setminus \mathbb{Q}$ of order n, we have a constant $C > 0$, such that*

$$\left| a - \frac{p}{q} \right| > \frac{C}{q^n}, \qquad p \in \mathbb{Z}, \, q \in \mathbb{N}. \tag{2.54}$$

Proof The algebraic number a of order n is a zero of a degree n polynomial $f \in \mathbb{Z}[x]$, and choosing the degree minimally, we can write f as

$$f(x) = (x - a) \, g(x), \qquad g \in \mathbb{R}[x], \, g(a) \neq 0. \tag{2.55}$$

Indeed, if $g(a) = 0$, we can also divide g by $x - a$ to get $f(x) = (x - a)^2 h(x)$; hence,

$$f'(x) = (x - a) \big(2h(x) + (x - a)h'(x) \big) \qquad \Rightarrow \qquad f'(a) = 0,$$

and since $f' \in \mathbb{Z}[x]$, the number a would be of the order (at most) $n - 1$. But $g(a) \neq 0$ implies, by the continuity of polynomials, that there exists some $\delta > 0$ so that

$$g(x) \neq 0, \qquad x \in [a - \delta, a + \delta]. \tag{2.56}$$

Let $p \in \mathbb{Z}$ and $q \in \mathbb{N}$ form a fraction close to a, i.e., they are chosen such that

$$|a - p/q| < \delta. \tag{2.57}$$

Since δ depends only on a, at least if we choose f as the unique *monic* polynomial of minimal degree with $f(a) = 0$, i.e., $f(x) = x^n + \ldots$, all sufficiently good approximants to a must satisfy (2.57). According to (2.56), this implies that $f(p/q) \neq 0$, and substituting $x = p/q$ into $x - a = f(x)/g(x)$, see (2.55), we obtain

$$\frac{p}{q} - a = \frac{f(p/q)}{g(p/q)} = \frac{f_0 + f_1 \dfrac{p}{q} + \cdots + f_n \left(\dfrac{p}{q} \right)^n}{g(p/q)} = \frac{f_0 q^n + f_1 \, p \, q^{n-1} + \cdots + f_n \, p^n}{q^n \, g(p/q)}.$$

The numerator of this fraction is different from zero, since we assumed that a is irrational and thus $a \neq p/q$. Being an integer, the numerator must be ≥ 1 in absolute value, so we can conclude that

$$\left| a - \frac{p}{q} \right| \geq \frac{1}{M \, q^n}, \qquad M = \max_{x \in [a - \delta, a + \delta]} |g(x)|, \tag{2.58}$$

whenever $|a - p/q| \leq \delta$. If, on the other hand, $|a - p/q| > \delta$, then trivially, since $q \geq 1$, we also have $|a - p/q| > \delta/q^n$, and for any constant C with

$$C < \min\left\{\delta, \frac{1}{M}\right\},$$

(2.54) is satisfied.   □

This theorem gives us a simple recipe for constructing transcendental numbers: use rapidly growing continued fraction expansions. For example, we could use

$$a_{k+1} > q_k^{k-1}, \qquad [a_0; a_1, \ldots, a_k] = \frac{p_k}{q_k},$$

since then $a = [a_0; a_1, \ldots]$, according to (2.52), satisfies the inequality

$$\left| a - \frac{p_k}{q_k} \right| < \frac{1}{a_{k+1} q_k^2} < \frac{1}{q_k^{k+1}}.$$

For any fixed numbers C and n, this expression becomes smaller than C/q_k^n as k tends to infinity.

Corollary 2.6 *The* Liouville number

$$\sum_{k=1}^{\infty} 10^{-k!} \tag{2.59}$$

is transcendental.

Proof Let $x_n = 10^{-1} + \cdots + 10^{-n!}$ be the nth partial sum of (2.59) and write

$$x_n = \frac{p}{q}, \qquad p = \sum_{k=1}^{n} \frac{10^{n!}}{10^{k!}} = \sum_{k=1}^{n} 10^{n!-k!}, \qquad q = 10^{n!}.$$

Then,

$$x - \frac{p}{q} = \sum_{k=n+1}^{\infty} 10^{-k!} = 10^{-(n+1)!} \sum_{k=n+1}^{\infty} 10^{(n+1)!-k!} < 1.1 \, 10^{-(n+1)!} = 1.1 \, q^{-n-1},$$

from which the claim follows by applying Theorem 2.15.   □

This, however, is not the full story regarding the approximation order for algebraic numbers. Liouville's theorem Theorem 2.15 says that the order of approximation is *at most* q^{-n} for an algebraic number of order n, and it seems as if the rate of approximation were related to the degree of the order, i.e., the minimal degree of a polynomial that has the number as a root. The higher the degree, the less algebraic the number is, and the better the approximation becomes. At least, this is what the theorem *seems* to suggest.

But (2.54) is just a *lower bound* that decreases faster if the order of the algebraic number is larger, which just says that the estimate becomes weaker and weaker the larger the order of the number becomes, which naturally raises the question whether the decay rate really depends on the order of the algebraic number at all. Can the 500th root $p^{1/500}$ of some $p \in \mathbb{N}$ really be approximated faster than the square root $p^{1/2}$? To that end, we have to look for a *sharp* bound which can be rephrased as follows: is there also an *upper* estimate similar to (2.54), at least with the same rates? This leads to the question whether given an *algebraic number* $x \in \mathbb{R} \setminus \mathbb{Q}$, and some number $\alpha > 0$, the inequality

$$\left| x - \frac{p}{q} \right| < \frac{1}{q^\alpha}, \tag{2.60}$$

can hold for *infinitely* many fractions p/q. In that context, the constant 1 in (2.60) is not a real restriction. Indeed, if (2.60) is satisfied by infinitely many fractions for some constant $C > 0$, then it is satisfied for $C = 1$ for any $\alpha' < \alpha$.

The first results in this direction were given by Thue in 1908 who showed that if (2.60) holds for infinitely many p/q, then $\alpha \le \frac{1}{2}n + 1$, where again n is the order of the algebraic number. In [20] this was even improved to $\alpha \le \sqrt{2n}$, and Siegel conjectured that the α was actually *independent* of n. This was finally verified in [83] as the following famous theorem.

Theorem 2.16 (Thue–Siegel–Roth) *Let $x \in \mathbb{R} \setminus \mathbb{Q}$ be an irrational algebraic number and $\alpha > 0$. If*

$$\left| x - \frac{p}{q} \right| < \frac{1}{q^\alpha}$$

holds for infinitely many fractions p/q, then $\alpha \le 2$.

A proof of the *Thue–Siegel–Roth Theorem* is beyond what we can do here, even if the paper [83] has only 20 pages and does not appear to rely on too heavy a theory. But the proof is extremely tricky and, of course, highly nontrivial. There is, however, a simple consequence of Theorem 2.16 that shows that practically all algebraic numbers have a rational approximation of the "worst possible" type.

Corollary 2.7 *If $x \in \mathbb{R} \setminus \mathbb{Q}$ is an irrational algebraic number and $\varepsilon > 0$, then there exists a constant $C(\varepsilon)$ such that*

$$\left| x - \frac{p}{q} \right| > \frac{C(\varepsilon)}{q^{2+\varepsilon}} \tag{2.61}$$

holds for any fraction p/q.

Proof Theorem 2.16 implies that

$$\left| x - \frac{p}{q} \right| < \frac{1}{q^{2+\varepsilon}}$$

only holds for finitely many fractions $p_1/q_1, \ldots, p_N/q_N$, and we can simply set

$$C(\varepsilon) := \min_{j=1,\ldots,N} q_j^{2+\varepsilon} \left| x - \frac{p_j}{q_j} \right| > 0,$$

to obtain (2.61).   □

To summarize, this shows that *any* algebraic number can only be approximated with a rate of $1/q^2$ using rational numbers, independent of its order. They are all equally bad.

Before we leave the world of numbers, we state a final theorem that shows that any *periodic continued fraction* can be identified with a *square root*, i.e., an algebraic number of order 2. To that end, we consider periodicity in a slightly more generous way, namely as periodicity after a certain index.

Definition 2.12 An infinite continued fraction expansion $[a_0; a_1, a_2, \ldots]$ is called *periodic* if there exists an index $k_0 \in \mathbb{N}_0$ and a *period* $\ell \in \mathbb{N}$ such that $a_{k+\ell} = a_k$ for all $k \geq k_0$.

Theorem 2.17 *Any periodic continued fraction represents an algebraic number of the second order, and any algebraic number of the second order has a periodic continued fraction expansion.*

Proof If x has a periodic expansion, then also

$$r_{k+\ell} = \left[a_{k+\ell}; a_{k+\ell+1}, a_{k+\ell+2}, \ldots \right] = \left[a_k; a_{k+1}, a_{k+2}, \ldots \right] = r_k, \qquad k \geq k_0,$$

holds for some $k_0 \in \mathbb{N}_0$ and some period length $\ell \in \mathbb{N}$. Therefore,

$$x = [a_0; a_1, \ldots] = \frac{r_k \, p_{k-1} + p_{k-2}}{r_k \, q_{k-1} + q_{k-2}} = \frac{r_{k+\ell} \, p_{k+\ell-1} + p_{k+\ell-2}}{r_{k+\ell} \, q_{k+\ell-1} + q_{k+\ell-2}} = \frac{r_k \, p_{k+\ell-1} + p_{k+\ell-2}}{r_k \, q_{k+\ell-1} + q_{k+\ell-2}},$$

and thus

$$(r_k \, p_{k-1} + p_{k-2}) \, (r_k \, q_{k+\ell-1} + q_{k+\ell-2}) - (r_k \, q_{k-1} + q_{k-2}) \, (r_k \, p_{k+\ell-1} + p_{k+\ell-2}) = 0,$$

which is a quadratic equation in r_k with integer coefficients. Therefore r_k and consequently also x is an algebraic number of order 2.

The converse is a bit more work. If $x = [a_0; a_1, \ldots]$ satisfies

$$ax^2 + bx + c = 0,$$

we again write x as

$$x = \frac{r_k \, p_{k-1} + p_{k-2}}{r_k \, q_{k-1} + q_{k-2}},$$

and obtain that

$$0 = a \, (r_k \, p_{k-1} + p_{k-2})^2 + b \, (r_k \, p_{k-1} + p_{k-2})$$
$$(r_k \, q_{k-1} + q_{k-2}) + c \, (r_k \, q_{k-1} + q_{k-2})^2$$
$$= A_k \, r_k^2 + B_k r_k + C_k,$$

where

$$A_k := a \, p_{k-1}^2 + b \, p_{k-1} \, q_{k-1} + c \, q_{k-1}^2, \tag{2.62}$$
$$B_k := 2a \, p_{k-1} \, p_{k-2} + b \, (p_{k-1} \, q_{k-2} + p_{k-2} \, q_{k-1}) + 2c \, q_{k-1} \, q_{k-2}, \tag{2.63}$$
$$C_k := a \, p_{k-2}^2 + b \, p_{k-2} \, q_{k-2} + c \, q_{k-2}^2 = A_{k-1}. \tag{2.64}$$

The *discriminant* $D_k = B_k^2 - 4A_k C_k$ has the value

$$D_k = \left(b^2 - 4ac\right) \underbrace{(p_{k-1} \, q_{k-2} - q_{k-1} \, p_{k-2})^2}_{=1} = b^2 - 4ac =: d,$$

independently of k. Since the discriminant describes the "square root" part of the number, this is already a good sign. Next, we record that

$$\left| x - \frac{p_{k-1}}{q_{k-1}} \right| < \frac{1}{q_{k-1}^2} \quad \Rightarrow \quad p_{k-1} = q_{k-1} x + \frac{\delta_{k-1}}{q_{k-1}}, \quad |\delta_{k-1}| < 1,$$

which we can substitute in (2.62) to obtain

$$A_k = a \left(q_{k-1} x + \frac{\delta_{k-1}}{q_{k-1}} \right)^2 + b \, q_{k-1} \left(q_{k-1} x + \frac{\delta_{k-1}}{q_{k-1}} \right) + c \, q_{k-1}^2$$
$$= \underbrace{(ax^2 + bx + c)}_{=0} q_{k-1}^2 + (2ax + b) \, \delta_{k-1} + a \frac{\delta_{k-1}^2}{q_{k-1}^2},$$
$$|A_k| \le 2|a| \, |x| + |b| + |a| = (2|x| + 1) \, |a| + |b|.$$

According to (2.64) the numbers A_k and $C_k = A_{k-1}$, but also

$$B_k^2 \le D_k + 4 \, |A_k| \, |C_k| \le b^2 + 4|a| \, |c| + [(2|x| + 1) \, |a| + |b|]^2$$

are bounded from above, independently of k. Hence there are only finitely many combinations of (A_k, B_k, C_k) and at least one of them has to be repeated after a while. Thus, there exist k, ℓ satisfying $A_{k+\ell} = A_k$, $B_{k+\ell} = B_k$ and $C_{k+\ell} = C_k$; hence also $r_{k+\ell} = r_k$ and by the construction rule for continued fractions, see the proof of Theorem 2.8, it also follows that $r_{k+n\ell} = r_k$, $k \in \mathbb{N}$. $\qquad \square$

2.6.1 Problems

2.21 Show that every rational number is algebraic. *Hint*: this is very easy.

2.22 Give an explicit continued fraction expansion of a transcendental number. *Hint*: Corollary 2.6.

2.23 Show that if x is an algebraic number of order n, then so is $1/x$.

2.24 Prove that the algebraic numbers are countable.

2.25 ([78], p. 92) *Pell's equation* is the *diophantic equation* asking for *integer* solutions x, y of

$$x^2 - D\,y^2 = 1, \qquad D \in \mathbb{N}, \qquad \sqrt{D} \notin \mathbb{N}. \tag{2.65}$$

Show that

1. any pair of solutions is coprime,
2. for any solution of (2.65), x/y must be a convergent.
3. If k denotes the period length of the continued fraction of $\sqrt{D}$, then any convergent of the form

$$\frac{x}{y} = \frac{p_{nk-1}}{q_{nk-1}}, \qquad n \in \begin{cases} \mathbb{N}, & k \in 2\mathbb{N}, \\ 2\mathbb{N}, & k \in 2\mathbb{N}+1 \end{cases}$$

   is a solution.

2.7 Continued Fractions and Music

The last chapter on number theoretic aspects of continued fractions is concerned with a seemingly unrelated topic: *music* and the concept of *harmony* in the sense of *consonance*. The connections we present here can be found for example in the books [5, 82] and in part even in [46]. Continued fractions will give an answer to the urgent questions:

1. why there are *pentatonic* scales in "simple" music,
2. why the *octave* consists of 12 semitones which is strange since $8 \neq 12 \times \frac{1}{2}$, although only very few people care about this apparent contradiction,
3. what would be the next natural partition of an octave into semitones.

To understand and eventually answer these questions, let us begin with the fundamental atoms of music analysis.

Definition 2.13 A *tone* with *amplitude* $a : \mathbb{R} \to \mathbb{R}$ is a time-dependent periodic event, i.e., there exists some $T > 0$ such that $a(\cdot + T) = a$.

Remark 2.11 According to Definition 2.13, percussionists do not produce a tone, just only *noise*. I feel sorry for them.

This mathematical model for music works only with a *constant* tone of infinite duration and excludes melodies so far. To consider melodies—which we will not do here—would involve concepts from *time-frequency-analysis* like a *Gabor transform/short time Fourier transform*, a *wavelet transform* or the *instantaneous frequency*. These approaches and at least some of their meaning in audio analysis can be found in [71], but we will not dwell on it here. To be perceived within a melody, a tone would have to be at least long enough to perform several periods of oscillation, a fact that can be seen as the musical version of the famous *Heisenberg uncertainty principle*.

Since the tone a is a periodic function with period T by definition, $a\left(\frac{T}{2\pi}\cdot\right)$ is a 2π-periodic function that can be considered on the *torus* $\mathbb{T} = \mathbb{R}/2\pi\mathbb{Z}$, and has a *Fourier series*

$$a\left(\frac{T}{2\pi}\cdot\right) = \frac{a_0}{2} + \sum_{k=1}^{\infty} a_k \cos(k\cdot) + b_k \sin(k\cdot), \qquad (2.66)$$

where

$$\left\{\begin{matrix} a_k \\ b_k \end{matrix}\right\} = \frac{1}{2\pi} \int_{-\pi}^{\pi} a(t) \left\{\begin{matrix} \cos kt \\ \sin kt \end{matrix}\right\} dt. \qquad (2.67)$$

We do not care about the convergence of the Fourier series here; what counts is that its coefficients are a unique *representation* of the tone. Since the sine is only a phase shift of the cosine and thus physiologically more or less irrelevant, cf. [46], one usually assumes that $b_k = 0, k \in \mathbb{N}$, as well as $a_0 = 0$, since a permanent constant air pressure can be compensated by the environment. Defining the *frequency* $\omega = \frac{2\pi}{T}$, a tone can thus be written in the simplest possible way as

$$a(t) = \sum_{k=1}^{N} a_k \cos(k\omega t), \qquad t \in \mathbb{R}, \qquad (2.68)$$

where N corresponds to the number of relevant or audible parts of the tone. The $a_k \cos(k\omega\cdot)$ are called *partial tones* of a and their absolute values define the *timbre* of the tone which depends on and characterizes the instrument, see Fig. 2.2.

Definition 2.14 The *timbre* of a tone, defined as *"...the character or quality of a musical sound or voice, as distinct from its pitch and intensity ..."*, is given as the vector $[a_k : k = 1, \ldots, n]$ of the amplitudes of its audible partial tones.

In principle, the coefficients a_k defining the timbre of a tone are complex numbers, but mostly it is common to consider only their modulus as depicted, for example in Fig. 2.2. The phase of these coefficients is normally neglected.

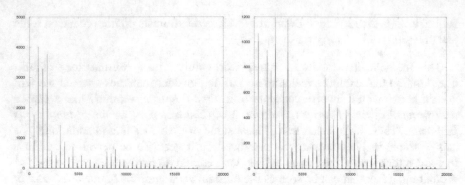

Fig. 2.2 Spectral fingerprint with $|a_k|$ for two bagpipe chanters, a Northern English "Border Pipe" *(left)* and a French "Cornemuse Bechonnet" *(right)*. Reason which one is louder and sounds "sharper"

The second important concept in musical physiology that affects the notion of harmony are the *beats* which are an audible version of the addition theorem

$$\cos \omega x + \cos \omega' x = 2 \cos \frac{\omega + \omega'}{2} x \cos \frac{\omega - \omega'}{2} x, \qquad x \in \mathbb{R}, \qquad (2.69)$$

which says that the sum of two simple tones can be seen and will be perceived as a tone of *average frequency* $\cos\left(\frac{\omega+\omega'}{2}\cdot\right)$, equipped with an *amplitude modulation* $\cos\left(\frac{\omega-\omega'}{2}\cdot\right)$. If the two frequencies are close and the difference is small, then these *beats* can very well be heard as a "wah-wah-wah" effect. This actually was and is the way how musical instruments would be tuned "by ear" without the aid of electronic tuning devices.

This now leads to the concept of consonances and dissonances introduced by Helmholtz [46], which is in fact a property of the partial tones. The maximal consonance is obtained for an *octave* which is the simultaneous sound of a and $a(2\cdot)$ as then, using infinite series for simplicity without caring for their convergence, we get

$$a(t) + a(2t) = \sum_{k=1}^{\infty} a_k \cos(k\omega t) + \sum_{k=1}^{\infty} a_k \cos(2k\omega t) = \sum_{k=1}^{\infty} \tilde{a}_k \cos(k\omega t),$$

where

$$\tilde{a}_{2k+\epsilon} = \begin{cases} a_{2k+\epsilon}, & \epsilon = 1, \\ a_{2k} + a_k, & \epsilon = 0, \end{cases} \qquad k \in \mathbb{N}_0, \ \epsilon \in \{0, 1\},$$

so that we get the same tone, just with a different timbre. The first real consonance is the *fifth*, in which the frequency is the multiple $\frac{3}{2}$ of the base frequency and gives

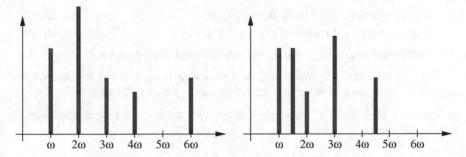

Fig. 2.3 Overlay of the spectra of a tone and its own octave *(left)* and a tone and its fifth *(right)*. Here we assume that both tones have the same distribution of partial tones, i.e., the same coefficients in the "Fourier fingerprint"

$$
a(t) + a\left(\frac{3}{2}t\right)
$$

$$
= \sum_{k=1}^{n} a_k \cos(k\omega t) + \sum_{k=1}^{n} a_{2k} \cos(3k\omega t) + \sum_{k=1}^{n} a_{2k-1} \cos\left(\left(3k - \frac{1}{2}\right)\omega t\right),
$$

where half of the partials merge with those of the base tone and just change the timbre once more, while the other half of the partials create new partial tones in the middle between the original partials. The effect is illustrated in Fig. 2.3.

There are complex physiological explanations already dating back to the 19th century, cf. [46] again, to define the following notion of *dissonance*:

> Two tones are dissonant if some of their partials get close to each other and generate perceptible beats.

Even if they did not have a scientific explanation, the fact itself was already known to the Pythagoreans who gave and used the following definition of harmony.

Definition 2.15 Two tones with frequencies $\omega < \omega'$ are in *harmony* if $\frac{\omega'}{\omega}$ is a fraction with a small numerator and denominator.

Since an octave only considers tones "modulo timbre", it is sufficient to build a musical scale within a single octave, i.e., to find harmonic frequencies $\omega' \in [\omega, 2\omega]$. Let us look for them.

Example 2.3 We already know the two most consonant intervals, namely the octave that corresponds to the fraction $2 = \frac{2}{1}$, the only one with denominator 1 and the fifth which is $\frac{3}{2}$. The next ones are:

1. the two fractions with denominator 3 in $(1, 2)$, namely $\frac{4}{3}$, the *fourth*, but also $\frac{5}{3}$, the often underrated[13] pure sixth.

---

[13] Due to the fact that it does not occur in the standard chord and does not mix too well with the fifth—their ratio is $\frac{10}{9}$ and therefore some sort of minor second which sounds unpleasant. This

2. the normalized fractions with denominator 4, i.e., $\frac{5}{4}$, the *major third*, and the usually not used $\frac{7}{4}$ for which there is no space in the usual 12 semitone octave.

The scale built from small harmonic tones would be called a *pure scale*.

The potential best harmonies within a scale are related to another interesting mathematical concept with again quite an impact in Number Theory.

Definition 2.16 The *Farey fractions* of order n are all reduced fractions between 0 and 1 of denominator at most n:

$$F_n := \left\{ \frac{p}{q} : 0 \le p \le q \le n, \gcd(p, q) = 1 \right\}. \tag{2.70}$$

The *Farey sequence*

$$f_n = \left(\frac{p_j}{q_j} \in F_n : j = 0, \ldots, N \right), \qquad N = N(n) = \#F_n$$

of order n lists all elements of F_n in increasing order.

The "best harmonics" relative to a base tone are therefore found in the ratios from $1 + F_n$ for some small n. This would, in fact, be particularly suitable in the case of a drone instrument which is accompanied by the base tone all the time.

In order to build a musical scale, ideally one suitable for polyphonic music, it is natural to construct it in such a way that, together with a certain tone, it contains tones that are in harmony with this one. Now the fact that the fifth forms the best possible nontrivial harmony is the basis for the original *Pythagorean construction* of a *scale*, i.e., a sequence of tones, according to two construction principles:

1. With every tone, its closest harmonic relative should be included, i.e., the fifth to the tone.
2. Since octaves are only a matter of timbre concerning harmonies, we can always go up and down by an octave without really changing the harmonic effect of the tone.

This construction principle leads to the *Pythagorean spiral* of the tones with frequencies $\omega_n := \left(\frac{3}{2} \right)^n \omega$, $n \in \mathbb{Z}$, where, starting with $\omega_0 := \omega$, we construct the sequence

$$
\begin{array}{ll}
\omega_1 = \frac{3}{2}\omega & \omega_{-1} = \frac{2}{3}\omega \to \frac{4}{3}\omega \\
\omega_2 = \frac{9}{4}\omega \to \frac{9}{8}\omega & \omega_{-2} = \frac{8}{9}\omega \to \frac{16}{9}\omega \\
\omega_3 = \frac{27}{16}\omega \to \frac{27}{32}\omega & \omega_{-3} = \frac{32}{27}\omega \\
\omega_4 = \frac{81}{64}\omega & \omega_{-4} = \frac{64}{81}\omega \to \frac{128}{81}\omega \\
\;\;\vdots & \;\;\vdots
\end{array}
\tag{2.71}
$$

---

makes the sixth somewhat inconvenient for "modern" music that relies a lot on chords, but the interval plays a role in *drone* instruments such as the *bagpipe* and the *hurdy-gurdy*.

where the "$\rightarrow$" indicates that we shifted a tone into the proper base octave by normalizing the fractions into the interval $[1, 2]$, either multiplying by 2 or dividing by 2. The usage of negative steps as well as positive steps has the harmonic advantage that the scale not only considers the fifth but also the fourth.

The name *Pythagorean spiral* reflects the fact that this sequence of tones is infinite and never closes to a circle since $\omega_k = \omega_{k'}$ *modulo octave* would be equivalent to

$$\left(\frac{3}{2}\right)^k \omega = 2^n \left(\frac{3}{2}\right)^{k'} \omega \quad \Leftrightarrow \quad 3^{k-k'} = 2^{n+k-k'}, \tag{2.72}$$

for some $k, k', n \in \mathbb{N}_0$, i.e.,

$$(k - k') \left(\log_2 3 - 1\right) = n,$$

which is impossible except for $k = k'$ and $n = 0$ since 2 and 3 are coprime. But writing $m := k - k'$ for the "width" of the scale spanned between ω_k and $\omega_{k'}$, we can replace the right-hand condition in (2.72) by

$$\min_{m \leq M} \min_n \left| \log_2 \left(\frac{3}{2}\right) - \frac{n}{m} \right| \quad \text{or} \quad \min_{m \leq M} \min_n \left| m \log_2 \left(\frac{3}{2}\right) - n \right| \tag{2.73}$$

to get the *best scale* with at most M tones. The solution of this problem is a best approximant of the first and second kind, respectively, hence a *convergent* of the irrational number $\log_2 \left(\frac{3}{2}\right)$.

Therefore all we have to do is to compute the convergents of this number for which we use a simple octave [21] routine CFconvergent. This gives us

```
>> CFconvergent( log2( 1.5),10 );
n=0        [0]        0 / 1    0.584963
n=1        [1]        1 / 1    0.415037
n=2        [1]        1 / 2    0.339850
n=3        [2]        3 / 5    0.375937
n=4        [2]        7 / 12           0.234600
n=5        [3]        24 / 41          0.678036
n=6        [1]        31 / 53          0.159665
n=7        [5]        179 / 306        0.451282
n=8        [2]        389 / 665        0.041881
n=9        [23]       9126 / 15601     0.409514
n=10       [2]        18641 / 31867    0.334001
```

where the second column shows the components in the continued fraction expansion, the third, the convergent and the fourth, the error $q_n^2 |x - p_n/q_n|$ which should be

Compute convergents

```
%%
%% CFconvergent
%% Compute first n components and convergents, return last
%%
function y=CFconvergent( x,n )
  p1 = q0 = 1; q1 = 0; an = floor(x); p0 = an; xx = x-p0;
  printf( "n=0 \t[%d]\t %d / %d \t%f\n",
            an,p0,q0,abs( x-p0/q0 )*q0^2 );
  y = an;

  for k=1:n
    xx = 1/xx ;
    an = floor( xx );
    xx = xx - an;
    A = [ an 1; 1 0 ] * [ p0,q0 ; p1,q1 ];
    p0 = A(1,1); p1 = A(2,1); q0 = A(1,2); q1 = A(2,2);
    y = [ y,an ];
    printf( "n=%d \t[%d]\t %d / %d \t%f\n",
              k,an,p0,q0,abs( x-p0/q0 )*(q0^2) );
    if ( xx == 0 ) % Continued fraction computed
      break;
    end
  end
end
```

Fig. 2.4 Simple octave program to compute the first n components and convergents for an arbitrary number

less than $\frac{1}{2}$ for a good and less than $\frac{1}{\sqrt{5}} \approx .44\ldots$ for an exceptional convergent, see Proposition 2.5 and Theorem 2.12.

We thus conclude that the convergents $n = 3$ with a scale of 5 tones, the one for $n = 4$ with a scale of 12 tones, and next the one for $n = 6$ with 53 tones are exceptional. They correspond to the well known pentatonic scale, the classical 12 semitone scale, and Bosanquet's enharmonic harmonium whose pictures can be found in [5] and in various sources over the internet. These three scales can be found by quite simple trial and error, and in fact the 53 tones octave was already known in ancient Greece, but we also see that the next good one already comprises 665 tones in the scale. This is at least hard to realize for woodwind instruments as long as they are not supposed to be played by centipedes. Note that the good approximation of transcendental numbers, which $\log_2 3$ is, is a disadvantage here, since it leads to a rapid growth of the denominators of the convergents.

To summarize, continued fractions tell us which scales, built on a Pythagorean spiral, i.e., a sequence of fifths, are almost complete.

This, however, is not the end of mathematical harmony theory, but only the beginning of it. An unprejudiced look at (2.71) shows that the Pythagorean construction has severe drawbacks. Even if each tone has *two* harmonic partners, namely the fifth above and below, its harmonic relationship to most of the other tones is not on such good terms. In fact, it turns out that the Pythagorean thirds are particularly disharmonic which makes chords very unpleasant in this system. Emerging from this observation, there is the whole theory of tuning systems found in print as early as Renaissance times, for example in Stifel [102, 103], one of the oldest mathematics books in Germany, and of course by Mersenne [73]. More on the mathematical relationship with tuning can be found in [5, 23, 82], while [2] gives a nice overview from the musical historian's perspective of this wide field.

We close this section by remarking once more that *harmony* between musical tones is characterized by the fact that their ratio is a rational number with a *small* numerator and denominator; so if the ratio is irrational and hard to approximate by such simple fractions, it is more and more *disharmonic*. Since the worst approximation occurs, as we know for [1; 1, 1, ...], we obtain the following musical "metatheorem":

The most disharmonic interval is the golden ratio.

Despite this number's usual public relations, cf. [72], it has nothing to do with harmony and good proportions in music.

2.8 Problems

2.26 Prove the addition theorem (2.69).

2.27 What could be a reason to tune the sixth at $\frac{8}{5}$ as it happens?

2.28 Show that a fourth is a fifth modulo octave, or at least try to make sense of that statement.

2.29 The deviation δ between tones with frequencies $\omega < \omega'$ is measured in *cent*, for example on modern tuning devices, and is defined as

$$\delta = 1200 \times \log_2 \frac{\omega'}{\omega}.$$

1. Show that the deviation is an additive measure.
2. Compute the deviation of the standard intervals.

2.30 Construct a Pythagorean scale and compute the deviation δ of the tones from the equal temperament. If you find the 12 semitone octave boring, do this for the octave with 53 tones. Feel free to write a program for it.

2.31 The first sets of Farey fractions are

$$F_1 = \{0, 1\}$$
$$F_2 = \left\{0, \frac{1}{2}, 1\right\}$$
$$F_3 = \left\{0, \frac{1}{3}, \frac{1}{2}, \frac{2}{3}, 1\right\}.$$

Compute the Farey sequence f_5.

2.32 Show that three successive elements $\frac{p_{j-1}}{q_{j-1}}$, $\frac{p_j}{q_j}$ and $\frac{p_{j+1}}{q_{j+1}}$ have the property that

$$\frac{p_j}{q_j} = \frac{p_{j-1}}{q_{j-1}} \oplus \frac{p_{j+1}}{q_{j+1}}, \tag{2.74}$$

i.e., each Farey fraction is the mediant of its neighbors.

2.33 Show that two successive Farey fractions are Farey neighbors, that is

$$p_j q_{j+1} - p_{j+1} q_j = -1.$$

2.34 Show that f_{n+1} can be obtained from f_n by computing the mediants

$$\frac{p}{q} = \frac{p_{j-1}}{q_{j-1}} \oplus \frac{p_j}{q_j}$$

and inserting $\frac{p}{q}$ between $\frac{p_{j-1}}{q_{j-1}}$ and $\frac{p_j}{q_j}$ of $\gcd(p, q) = 1$.

2.35 Write a `Matlab` program that lists the Farey sequence of order n.

Chapter 3
Rational Functions as Continued Fractions of Polynomials

Lord, I haven't met so many problems since I gave up algebra.
E. Derr Biggers, Charlie Chan ...

Now it is time to extend the idea of continued fractions with integer entries, and their role in the representation of real numbers to more general situations, in particular to rational functions. Recall that a *rational function* is a function of the form

$$f(x) = \frac{p(x)}{q(x)}, \qquad p, q \in \mathbb{K}[x], \tag{3.1}$$

i.e., quotients of polynomials over a field $\mathbb{K}$. Note that rational functions are closed under addition, multiplication, and division, hence form a *field* just like the rational numbers. It is convenient to consider the slightly more general situation of rational objects and continued fractions over general rings. However, we will soon see that the structure of a *Euclidean Ring* will be necessary to obtain most of the properties that we learned to appreciate in the context of numbers in the preceding chapters; in the long term, we will restrict ourselves to continued fractions and *univariate* polynomials. Multivariate generalizations are, however, available for geometric interpretations of continued fractions, cf. [57].

3.1 A Starting Point with Some New Notation ...

Since all that is needed for defining a continued fraction as a *rational object* is the capability of adding and multiplying objects, it makes sense to consider continued fractions over rings. Since this is the fundamental object we are considering, it is worthwhile to recall the definition.

© The Author(s), under exclusive license to Springer Nature Switzerland AG 2021
T. Sauer, *Continued Fractions and Signal Processing,* Springer Undergraduate Texts in Mathematics and Technology, https://doi.org/10.1007/978-3-030-84360-1_3

Definition 3.1 (*Ring*) A *ring* R is an abelian, i.e., commutative, group with respect to addition, a semigroup with respect to multiplication; the two operations are related by the distributional laws $a(b + c) = ab + ac$ and $(b + c)a = ba + ca$.

Remark 3.1 While addition is always commutative and has an inverse, hence subtraction is well defined, multiplication need neither be commutative nor have an inverse. The ring of matrices is a classical example for that. In our context here, however, we will always assume that R is a *commutative ring* at least, i.e., that also multiplication is a commutative operation.

In this more general context, some new and more flexible notation for continued fractions will be needed.

Definition 3.2 For an arbitrary ring R and $b_1, \ldots, b_n \in R$ as well as $c_1, \ldots, c_n \in R$, we write

$$
\frac{b_1|}{|c_1} + \cdots + \frac{b_n|}{|c_n} := \cfrac{b_1}{c_1 + \cfrac{b_2}{c_2 + \cfrac{b_3}{\ddots + \cfrac{b_{n-1}}{c_{n-1} + \cfrac{b_n}{c_n}}}}} \tag{3.2}
$$

and analogously for infinite continued fractions. If the ring has a neutral element 1 with respect to multiplication, and $b_j = 1$ for all j, the resulting finite or infinite continued fraction $\frac{1|}{|c_1} + \frac{1|}{|c_2} + \cdots$ is called *simple*.

The simplest case of a continued fraction with polynomials in it is an expression of the form

$$
f(x) = [p; m_1, m_2, \ldots, m_n] = p(x) + \cfrac{1}{m_1(x) + \cfrac{1}{m_2(x) + \cfrac{1}{\ddots + \cfrac{1}{m_{n-1}(x) + \cfrac{1}{m_n(x)}}}}},
$$

where, in addition, each component $m_j(x) = a_j x^{k_j}, a_j \in \mathbb{R}, k_j \in \mathbb{N}$, is a *monomial*. Such continued fractions are called *C-continued fractions* in [79, p. 107]. Let us formalize this new notation in an even more general context.

Sticking to polynomials and simple C-continued fractions for a moment, we observe that the "1" appearing in the numerators of the continued fraction above is not a restriction: a more general continued fraction of the form

$$f(x) = p(x) + \cfrac{b_1}{m_1(x) + \cfrac{b_2}{m_2(x) + \cfrac{b_3}{\ddots + \cfrac{b_{n-1}}{m_{n-1}(x) + \cfrac{b_n}{m_n(x)}}}}}$$

$$=: p(x) + \frac{b_1|}{|m_1(x)} + \frac{b_2|}{|m_2(x)} + \cdots + \frac{b_n|}{|m_n(x)}$$

with $b_j \in \mathbb{K} \setminus \{0\}$, $j = 1, \ldots, n$, can also be written in the form

$$f(x) = [p; \widetilde{m}_1, \ldots, \widetilde{m}_n] = p(x) + \frac{1|}{|\widetilde{m}_1(x)} + \cdots + \frac{1|}{|\widetilde{m}_n(x)}, \tag{3.3}$$

where

$$\widetilde{m}_j(x) = m_j(x) \begin{cases} \displaystyle\prod_{\ell=0}^{k} \frac{b_{2\ell}}{b_{2\ell+1}}, & j = 2k+1, \\[2ex] \displaystyle\prod_{\ell=0}^{k} \frac{b_{2\ell+1}}{b_{2\ell+2}}, & j = 2k+2, \end{cases} \qquad b_0 = 1. \tag{3.4}$$

The simplified form (3.4) is easily obtained by normalizing the fractions successively which yields

$$f(x) - p(x) = \frac{b_1|}{|m_1(x)} + \frac{b_2|}{|m_2(x)} + \cdots + \frac{b_n|}{|m_n(x)}$$

$$= \frac{1|}{\left|m_1(x)\,\dfrac{1}{b_1}\right.} + \frac{\dfrac{b_2}{b_1}\Big|}{|m_2(x)} + \frac{b_3|}{|m_3(x)} + \cdots + \frac{b_n|}{|m_n(x)}$$

$$= \frac{1|}{\left|m_1(x)\,\dfrac{1}{b_1}\right.} + \frac{1|}{\left|m_2(x)\,\dfrac{b_1}{b_2}\right.} + \frac{\dfrac{b_1 b_3}{b_2}\Big|}{|m_3(x)} + \cdots + \frac{b_n|}{|m_n(x)}$$

$$= \frac{1|}{\left|m_1(x)\,\dfrac{1}{b_1}\right.} + \frac{1|}{\left|m_2(x)\,\dfrac{b_1}{b_2}\right.} + \frac{1|}{\left|m_3(x)\,\dfrac{b_2}{b_1 b_3}\right.} + \frac{\dfrac{b_2 b_4}{b_1 b_3}\Big|}{|m_4(x)} + \cdots + \frac{b_n|}{|m_n(x)},$$

and so on. Since the fact that the m_j were supposed to be monomials does not play a role in these computations because the normalization also works with continued fractions

$$\frac{b_1|}{|p_1} + \frac{b_2|}{|p_2} + \cdots$$

as long as the "numerators" there are nonzero numbers, which are exactly the elements of a polynomial ring that have a multiplicative inverse. We will get back to this issue later.

Any continued fraction of the form $[p; m_1, \ldots, m_n]$ is a rational function and, at least for univariate polynomials, any rational function can be expanded into a *finite* continued fraction. We will soon see this in a more general context.

3.1.1 Problems

3.1 (*Recurrence relation*) Prove that the numbers p_k, q_k, defined by

$$\frac{c_0|}{|1} + \frac{b_1|}{|c_1} + \cdots + \frac{b_k|}{|c_k} = \frac{p_k}{q_k}, \qquad k = 1, \ldots, n,$$

satisfy the recurrence relation

$$
\begin{aligned}
p_k &= c_k\, p_{k-1} + b_k\, p_{k-2}, \quad p_{-1} = 1,\ p_0 = c_0, \\
q_k &= c_k\, q_{k-1} + b_k\, q_{k-2}, \quad q_{-1} = 0,\ q_0 = 1,
\end{aligned}
\tag{3.5}
$$

that is valid for continued fractions of the form (3.2).

3.2 Show that the convergents in (3.5) satisfy

$$q_k\, p_{k-1} - p_k\, q_{k-1} = (-1)^k \prod_{j=1}^{k} b_j. \tag{3.6}$$

Hint: Try to extend the proof of the case $b_1 = \cdots = b_n = 1$.

3.3 An element $a \in R$ is called a *unit* in R if it has a multiplicative inverse $a^{-1} = \frac{1}{a}$ in R. Let $R^\times$ denote the units of R.

1. Show that $R^\times$ is a multiplicative subgroup of $R \setminus \{0\}$.
2. Show that for $b_1 \in R^\times$ and $c_j \in R$, any continued fraction can be renormalized by

$$\sum_{j=1}^{n} \frac{b_j|}{|c_j} = \sum_{j=1}^{n} \frac{1|}{|c'_j}$$

into a simple continued fraction. Give a formula for c'_j.

3.4 (*General renormalization formula*) Show that

$$\frac{\dfrac{a_1}{c_1}\Big|}{\Big|\dfrac{b_1}{d_1}}+\cdots+\frac{\dfrac{a_n}{c_n}\Big|}{\Big|\dfrac{b_n}{d_n}}$$

$$=\frac{a_1d_1|}{|b_1c_1}+\frac{a_2c_1d_1d_2|}{|b_2c_2}+\cdots+\frac{a_jc_{j-1}d_{j-1}d_j|}{|b_jc_j}+\cdots+\frac{a_nc_{n-1}d_{n-1}d_n|}{|b_nc_n} \quad (3.7)$$

for any nonzero elements $a_j, b_j \in R \setminus \{0\}$, $c_j, d_j \in R^\times$, $j = 1, \ldots, n$, and $n \in \mathbb{N}$.

3.2 Euclidean Rings and Continued Fractions

Let us recall: a *ring* is a structure in which addition, subtraction and multiplication are reasonably defined, and the structure is closed under these operations. Since we need a little bit more to really be able to *compute* continued fraction expansions, we have to introduce some more terminology.

Definition 3.3 (*Euclidean ring*) A ring R is called

1. *integral domain*,[1] if there exist no elements $a, b \in R \setminus \{0\}$ such that $ab = 0$. Any element $a \in R$ for which there exists $b \in R \setminus \{0\}$ such that $ab = 0$ is called a *zero divisor*. An integral domain is a ring where the only zero divisor is the trivial one, the zero element.
2. *Euclidean Ring*, if R is an integral domain and there exists a *Euclidean function* $d : R \to \mathbb{N} \cup \{-\infty\}$ so that for any $p, q \in R$, $q \neq 0$, there exist a *factor* $s \in R$ and a *remainder* $r \in R$ so that we perform a *Euclidean division*, namely

$$p = sq + r, \quad d(r) < d(q). \quad (3.8)$$

We then write $s =: p/q$ and $r =: (p)_q$.

Remark 3.2 (*Properties of the Euclidean functions*)

1. Every Euclidean function satisfies $d(0) < d(a)$ for all $a \in R \setminus \{0\}$. Assuming that there exists some $a \in R \setminus \{0\}$ such that $d(a) \leq d(R)$, then setting $p = q = a$, we get a representation (3.8) for a, i.e.,

$$p = sq + r, \quad s \in R, \quad \Rightarrow \quad r = p - sq = (1 - s)a.$$

And regardless of how we choose s, each of these remainders would satisfy $d((1 - s)r) \geq d(a)$ which contradicts the fact that the ring is Euclidean.
2. Not any Euclidean function has the apparently very natural property

---

[1] In German "nullteilerfrei" or "Integritätsring". The google translation "integrity ring" of the latter may only earn raised eyebrows among mathematicians.

$$d\,(a \cdot b) \geq d(a), \qquad a, b \in R \setminus \{0\}, \tag{3.9}$$

which we know from the classical Euclidean functions: the absolute value for $\mathbb{Z}$ and the degree for $\mathbb{K}[x]$. For any integral domain there exists a special Euclidean function, called the *minimal Euclidean function*, which satisfies (3.9). The minimal Euclidean function d^* is defined as the element-wise minimum of all possible Euclidean functions:

$$d^* : a \mapsto \min\{d(a) : d \text{ Euclidean function}\},$$

cf. [31, Exercise 3.5]. As a result we can and will always assume that we use the minimal Euclidean function and therefore the Euclidean function satisfies (3.9).

3. The value $d(a) = -\infty$ can only occur for $a = 0$, but need not be assumed, i.e., $\{a \in R : d(a) = -\infty\} = \emptyset$ is **not** excluded. Indeed, for $R = \mathbb{Z}$, we have $d(0) = 0$ while for $R = \mathbb{K}[x]$, we have $d(0) = -\infty$.

Example 3.1 (*Euclidean rings*)

1. The integers $\mathbb{Z}$ are a Euclidean ring with $d = |\cdot|$.
2. The univariate polynomials $\mathbb{K}[x]$ form a Euclidean ring with $d(f) = \deg f$, $f \in \mathbb{K}[x]$ where we set $\deg 0 = -\infty$.
3. Any field $\mathbb{K}$ is a Euclidean ring with $d = (1 - \delta_0)$, however not a very interesting one.
4. A somewhat obscure Euclidean function on $\mathbb{Z}$ is $d(3) = 2$ and $d = |\cdot|$ otherwise. This Euclidean function is made Euclidean by choosing the remainder in $\{-1, 0, 1\}$ when dividing by 3. This Euclidean function does **not** satisfy (3.9), since

$$d(-1 \cdot 3) = d(-3) = 3 > 2 = d(3).$$

Nevertheless $d(0)$ is still minimal among all values $d(R)$; as mentioned before it has to be minimal.

Euclidean rings are useful for an obvious reason: the concept allows us to do division with a remainder, and the remainders that we obtain this way are smaller (in the sense of the Euclidean function) or "simpler" than the divisor. Moreover, if we recall that Euclidean division was one of the fundamental operations when computing the continued fraction expansions with integer components, it is clear why we insist on Euclidean rings: they allow us to extend this trick almost literally.

Theorem 3.1 *Let R be a Euclidean ring with unity 1 as the neutral element of multiplication. Then any finite continued fraction $[r_0; r_1, \ldots, r_n]$, $r_j \in R$, is rational over R and any rational element over R can be expanded into a finite continued fraction.*

Definition 3.4 The set of all *rational elements* or *fractions* over the commutative ring R with the usual operations for addition, subtraction, multiplication, and division will be denoted by

$$R^\star := \left\{ \frac{p}{q} \; : \; p \in R, \, q \in R \setminus \{0\} \right\}.$$

In this notation, $\mathbb{Q} = \mathbb{Z}^\star$, and $R^\star$ is a *field* if R is an integral domain with one, see [42].

Proof (of Theorem 3.1) That finite continued fractions are rational over R can be obtained by expanding the definition, or by inductively using the recurrence

$$[r_0; r_1, \ldots, r_n] = r_0 + \frac{1}{[r_1; r_2, \ldots, r_n]},$$

so then this part is quite obvious.

For the converse let $f = p/q, \, p, q \in R, \, q \neq 0$. We set $s_0 = p, s_1 = q$ and run the *Euclidean algorithm*. To that end, we determine r_0 to be such that $s_0 = r_0 s_1 + s_2$, $d(s_2) < d(s_1)$ which is possible since we are working in a Euclidean ring. For $j = 1, 2, \ldots$, we proceed the same way and form

$$s_j = r_j s_{j+1} + s_{j+2}, \qquad d(s_{j+2}) < d(s_{j+1}),$$

to conclude by induction on k that

$$\frac{p}{q} = \left[r_0; r_1, \ldots, r_k, \frac{s_{k+1}}{s_{k+2}} \right], \qquad k \in \mathbb{N}. \tag{3.10}$$

Indeed,

$$\left[r_0; \frac{s_1}{s_2} \right] = r_0 + \frac{s_2}{s_1} = \frac{r_0 s_1 + s_2}{s_1} = \frac{s_0}{s_1} = \frac{p}{q},$$

and since

$$r_k + \frac{s_{k+2}}{s_{k+1}} = \frac{r_k s_{k+1} + s_{k+2}}{s_{k+1}} = \frac{s_k}{s_{k+1}},$$

we also get

$$\left[r_0; r_1, \ldots, r_k, \frac{s_{k+1}}{s_{k+2}} \right] = \left[r_0; r_1, \ldots, r_k + \frac{s_{k+2}}{s_{k+1}} \right] = \left[r_0; r_1, \ldots, r_{k-1}, \frac{s_k}{s_{k+1}} \right] = \frac{p}{q},$$

which proves (3.10). Since $d(s_k)$ is a strictly decreasing sequence in $\mathbb{N}_0 \cup \{-\infty\}$, this procedure has to terminate after finitely many steps, and thus gives us a *finite* continued fraction. $\qquad \square$

This, of course, was not extremely surprising so far, since the name already hints that the *Euclidean* ring and the *Euclidean* algorithm may have something in common and should fit together. But it is getting even better if we assume that R is a *commutative ring* with (multiplicative) *identity* 1. Then the recurrence relation of Theorem 2.1 can simply be copied, which will lead to a lot of interesting formulas for convergents or, *Näherungsbrüche*, as the convergents are called in [78]. The proofs of the preceding chapter can now be transferred literally to the setting of rational elements over arbitrary Euclidean rings and can be summarized as follows.

Theorem 3.2 *The convergents* $\kappa_k := p_k/q_k$, $k \leq n$, *of the finite continued fraction* $[a_0; a_1, \ldots, a_n]$, $a_j \in R$, *fulfill the recurrence relations*

$$\begin{matrix} p_k = a_k p_{k-1} + p_{k-2} \\ q_k = a_k q_{k-1} + q_{k-2} \end{matrix}, \quad \begin{matrix} p_{-1} = 1, & p_0 = a_0, \\ q_{-1} = 0, & q_0 = 1, \end{matrix} \tag{3.11}$$

i.e., $\kappa_0 = a_0/1 = a_0$, *as well as*

$$\frac{p_{k-1}}{q_{k-1}} - \frac{p_k}{q_k} = \frac{(-1)^k}{q_{k-1}\,q_k}, \quad \frac{p_k}{q_k} - \frac{p_{k-2}}{q_{k-2}} = \frac{(-1)^k\,a_k}{q_{k-2}\,q_k}, \tag{3.12}$$

and thus are coprime.

Definition 3.5 Two elements $p, q \in R$ of a commutative ring R with identity are called *coprime*, if $p \in q\,R^\times$ where $R^\times = \left\{ r \in R : r^{-1} \in R \right\}$ denotes the *units* in R.

The units of $\mathbb{Z}$ are $\mathbb{Z}^\times = \{\pm 1\}$; the units among the polynomials $\mathbb{K}[x]$ are $\mathbb{K}[x]^\times = \mathbb{K}^\times = \mathbb{K} \setminus \{0\}$, which means that not all units are identities, not even in the absolute value.

3.2.1 Problems

3.5 Prove that whenever R is an integral domain with unity, then any nonzero element of $R^\star$ has a multiplicative inverse.

3.6 An element $a \in R$ is called a *unit* if it has a multiplicative inverse in the ring, i.e., $a^{-1} \in R$. Show that $d(ab) = d(b)$, $b \in R \setminus \{0\}$ if and only if a is a unit.

3.7 (*Laurent polynomials*) A *Laurent polynomial* is an expression of the form

$$f(x) = \sum_{k \in \mathbb{Z}} f_k x^k, \quad x \in \mathbb{C} \setminus \{0\}, \quad \#\{k : f_k \neq 0\} < \infty, \tag{3.13}$$

where the coefficients can be from any field $\mathbb{K}$, as with polynomials. Show that:

1. any Laurent polynomial can be written as $f(x) = x^k\,p(x)$, $k \in \mathbb{Z}$, $p \in \mathbb{K}[x]$ with $p(0) \neq 0$.

2. the units in the ring Λ of Laurent polynomials are the nonzero multiples of monomials
$$\{c\,x^k : k \in \mathbb{Z}, \, c \in \mathbb{K} \setminus \{0\}\}.$$

3. the function
$$d(f) := \max\{k : f_k \neq 0\} - \min\{k : f_k \neq 0\}$$

is a Euclidean function for Λ.

3.8 Give an explicit algorithm for division with a remainder of Laurent polynomials.

3.9 Show that $\Lambda^\star = \mathbb{K}[x]^\star$.

3.3 Continued Fractions and the Extended Euclidean Algorithm

Continued fractions give us even more, due to their close relationship to the Euclidean algorithm. Whenever the recursion in the proof of Theorem 3.1 stops, which means $s_{n+2} = 0$, then we have computed a[2] greatest common divisor of p and q, cf. [31]. In other words $r_n = \gcd(p, q)$ and the components $p_n = p/r_n$, $q_n = q/r_n$ of the convergent are coprime since everything that can be divided off is found in the gcd. Therefore we have that

$$\frac{p}{q} = [r_0; r_1, \ldots, r_n] = \frac{p_n}{q_n} = \frac{r_n\,p_{n-1} + p_{n-2}}{r_n\,q_{n-1} + q_{n-2}},$$

and using (3.12),

$$q_{n-1}p - p_{n-1}q = r_n\,(q_{n-1}\,p_n - p_{n-1}q_n) = (-1)^{n+1}\,r_n = (-1)^{n+1}\gcd(p, q).$$

In other words the numerator and denominator of the *penultimate convergent* (which is the last true convergent, since the last convergent p_n/q_n is the fraction itself and therefore not really an approximation any more), are the solutions of the *Bézout identity*

$$a\,p + b\,q = \gcd(p, q) \quad \Leftrightarrow \quad a = (-1)^{n+1}q_{n-1}, \quad b = (-1)^n p_{n-1}. \quad (3.14)$$

Thus the Bézout identity can even be seen as a side effect of the computation of the convergents of a continued fraction.

This is no new observation, since the *extended Euclidean algorithm* is well known to compute such a solution, but it still offers one more useful connection. For the extended Euclidean algorithm, we consider a matrix

---

[2] It is "the" for numbers, but for polynomials the greatest common divisor is *not* defined uniquely, and only up to units, i.e., up to a nonzero constant multiple in the case of polynomials.

$$A = \begin{bmatrix} a_1^T \\ a_2^T \end{bmatrix} = \begin{bmatrix} a_{11} \; \ldots \; a_{1n} \\ a_{21} \; \ldots \; a_{2n} \end{bmatrix},$$

compute the *integer division*

$$r := a_{11}/a_{21}, \quad \text{i.e.,} \quad a_{11} = r\, a_{21} + s, \quad d(r) < d(a_{21}), \tag{3.15}$$

and as long as $r = 0$, we can then replace A by

$$A \leftarrow \begin{bmatrix} a_2^T \\ a_1^T - r a_2^T \end{bmatrix} = \begin{bmatrix} 0 & 1 \\ 1 & -r \end{bmatrix} \begin{bmatrix} a_1^T \\ a_2^T \end{bmatrix} = \begin{bmatrix} 0 & 1 \\ 1 & -r \end{bmatrix} A$$

and iterate on.

Definition 3.6 The *extended Euclidean algorithm* generates from a given $A^{(0)} \in \mathbb{R}^{2 \times n}$ a finite matrix sequence $A^{(k)}, k \in \mathbb{N}$, via

$$r = a_{11}^{(k)}/a_{21}^{(k)}, \qquad A^{(k)} = \begin{bmatrix} 0 & 1 \\ 1 & -r \end{bmatrix} A^{(k-1)}, \tag{3.16}$$

using the integer division (3.15).

Since again by (3.15),

$$d\left(a_{21}^{(k)}\right) = d(s) < d\left(a_{21}^{(k-1)}\right),$$

the extended Euclidean algorithm terminates after finitely many iterations.

Indeed the extended Euclidean algorithm can be used to compute convergents.

Theorem 3.3 *If $x = \frac{p}{q} = [a_0; a_1, \ldots, a_n] \in \mathbb{Q}$ and the extended Euclidean algorithm starts with*

$$A^{(0)} = \begin{bmatrix} p & 0 & -1 \\ q & 1 & 0 \end{bmatrix}, \tag{3.17}$$

then we obtain the convergents as

$$[a_0; a_1, \ldots, a_k] = \frac{p_k}{q_k} = \frac{a_{22}^{(k+1)}}{a_{23}^{(k+1)}}, \qquad k = 0, \ldots, n. \tag{3.18}$$

Proof Let $s_k = a_{11}^{(k)}/a_{21}^{(k)}$ denote the quotient in the kth step of the iteration and observe from the proof of Theorem 3.1 that $r_k = a_k, k \in \mathbb{N}_0$. Now the recurrence (3.16) yields

$$a_2^{(k)} = \left(a_1^{(k-1)}\right)^T - a_k \left(a_2^{(k-1)}\right)^T = \left(a_2^{(k-2)}\right)^T - a_k \left(a_2^{(k-1)}\right)^T, \qquad k \geq 2,$$

that is for $j > 1$,

$$a_{2j}^{(k)} = a_{2j}^{(k-2)} - a_k a_{2j}^{(k-1)}, \qquad k \geq 2$$

with

$$a_{22}^{(0)} = 1, \quad a_{22}^{(1)} = -a_1, \quad a_{23}^{(0)} = 0, \quad a_{23}^{(1)} = -1;$$

hence an easy induction shows that

$$a_{22}^{(k)} = (-1)^k p_k, \quad a_{22}^{(k)} = (-1)^k q_k, \qquad k \in \mathbb{N}_0, \tag{3.19}$$

from which (3.18) follows immediately. $\qquad\square$

Example 3.2 The continued fraction expansion for $\frac{37}{26}$ is computed using the extended Euclidean algorithm according to the following scheme:

r_j	37	0	-1	
1	26	1	0	
2	11	-1	-1	$\to 1$
2	4	3	2	$\to \frac{3}{2}$
1	3	-7	-5	$\to \frac{7}{5}$
3	1	10	7	$\to \frac{10}{7}$
	0	-37	-26	$\to \frac{37}{26}$

giving

$$\frac{37}{26} = [1; 2, 2, 1, 3].$$

Example 3.3 The same procedure can also be done with polynomials, for example, computing the continued fraction expansion of the rational function

$$f(x) = \frac{x^5 + 4x^3 + 3x}{x^4 + 3x^2 + 1}.$$

The Euclidean function used here for $\mathbb{Q}[x]$ is the degree. We obtain

r_j	$x^5 + 4x^3 + 3x$	0	-1	
x	$x^4 + 3x^2 + 1$	1	0	
x	$x^3 + 2x$	$-x$	-1	$\to x$
x	$x^2 + 1$	$x^2 + 1$	x	$\to \dfrac{x^2+1}{x}$
x	x	$-x^3 - 2x$	$-x^2 - 1$	$\to \dfrac{x^3+2x}{x^2+1}$
x	1	$x^4 + 4x^2 + 1$	$x^3 + 2x$	$\to \dfrac{x^4+4x^2+1}{x^3+2x}$
	0	$-x^5 - 4x^3 - 3x$	$-x^4 - 3x^2 - 1$	$\to f(x)$

giving

$$\frac{x^5 + 4x^3 + 3x}{x^4 + 3x^2 + 1} = [x; x, x, x, x].$$

It is tempting to assume that continued fraction expansions of rational functions only have affine polynomials as coefficients. Indeed this is the generic case, see Problem 3.11, but the following simple example shows that this assumption nevertheless does not hold true in general.

Example 3.4 For the rational function

$$f(x) = \frac{x^5 + 1}{x^4 + x^3 + x^2 + x + 1},$$

we get that

r_j	$x^5 + 1$	0	-1
$x - 1$	$x^4 + x^3 + x^2 + x + 1$	1	0
$\frac{1}{2}\left(x^4 + x^3 + x^2 + x + 1\right)$	2	$-x - 1$	-1
	0	$x^5 + 1$	$x^4 + x^3 + x^2 + x + 1$

yielding the continued fraction expansion

$$\frac{x^5 + 1}{x^4 + x^3 + x^2 + x + 1} = \left[x - 1; \frac{1}{2}\left(x^4 + x^3 + x^2 + x + 1\right)\right].$$

Remark 3.3 In a symbolic environment like `Maple` or `Maxima` [92], the iteration (3.16) of the extended Euclidean algorithm can be used directly to compute the continued fraction expansion of any rational object over any ring that can be represented symbolically.

3.3.1 Problems

3.10 Prove (3.19).

3.11 (*Euclidean division—generic case*) Let $f, g \in \mathbb{Q}[x]$ be two polynomials with rational coefficients such that $\deg f =: m > n := \deg g$. Show that for any $\varepsilon > 0$, there exists a polynomial $\hat{g}$ of degree m such that

1. $|\hat{g}_j - g_j| < \varepsilon$, $j = 0, \ldots, n$,
2. Euclidean division gives

$$f = p\,\hat{g} + r, \qquad \deg p = m - n, \ \deg r = n - 1.$$

In other words: the generic degree of the remainder is one less than that of the divisor.

Now derive the "standard" properties of the extended Euclidean algorithm with an initialization matrix of the form (3.17) so that it computes the gcd and the solution for the Bézout identity. To conclude we give a classical formal definition.

Definition 3.7 A *greatest common divisor* $\gcd(p, q)$ of $p, q \in R$ is defined by

1. $\gcd(p, q)$ divides p and q,
2. any common divisor of p and q divides $\gcd(p, q)$.

3.12 Assume that the extended Euclidean algorithm starts with $A^{(0)}$ from (3.17). Show that

1. any divisor of p and q divides $a_{21}^{(k)}$, $k \geq 1$,
2. if $a_{21}^{(k)} = 0$, then $a_{11}^{(k)}$ divides p and q,

and use that to conclude that the extended Euclidean algorithm computes $\gcd(p, q)$.

Hint: Use the recurrence $a_{21}^{(k)} + r_k\, a_{21}^{(k-1)} = a_{11}^{(k)}$ and the relationship between $A^{(k)}$ and $A^{(k-1)}$.

3.13 Is $\gcd(p, q)$ as defined in Definition 3.7 unique in any ring? Are there special choices of p and q where it is unique?

3.14 (*Bézout identity*) Assume that the extended Euclidean algorithm starts with $A^{(0)}$ from (3.17). Prove the invariance

$$a_{21}^{(k)} = a_{22}^{(k)} q - a_{23}^{(k)} p, \qquad k \in \mathbb{N}_0,$$

and derive the *Bézout identity*

$$\gcd(p, q) = a_{12}^{(n)} q - a_{13}^{(n)} p \tag{3.20}$$

from it where n is the first index with $a_{21}^{(n)} = 0$.

3.15 (*Inversion in* $\mathbb{F}_p$) Let $p \in \mathbb{N}$ be a prime number and let $a \in \mathbb{F}_p \simeq \{0, \ldots, p - 1\}$ be an element from the finite field $\mathbb{F}_p$. Prove that the extended Euclidean algorithm starting with

$$A = \begin{bmatrix} p & 0 \\ a & 1 \end{bmatrix}$$

stops with $a_1^T = (1, a^{-1})$, where $a^{-1} \in \mathbb{N}$ has the property that $a^{-1} a \in p\mathbb{Z}$.

3.16 (*General inversion*) Show that whenever $p, q \in R$ are coprime, there exists $q' \in R$ that inverts q modulo p and that is $qq' \in pR$.

3.4 One Result of One Bernoulli

It is quite a natural question for continued fractions on arbitrary Euclidean rings like polynomials to find which rational objects can be convergents of some continued fraction; of course we consider here the full sequence of convergents, since for any $p < q$, the first convergent of

$$[0; a, b] = \cfrac{1}{a + \frac{1}{b}} = \frac{b}{a + b}$$

equals $\frac{p}{q}$ as soon as $a = q - p$ and $b = p$. Hence any rational number is a convergent of some continued fraction. So the question is:

> For which *sequences* $c_n \in R^\star$ does there exist a continued fraction which has this sequence as its sequence of convergents?

According to [79] this question had already been answered in 1775 by D. Bernoulli[3] in [7], even for continued fractions of the quite general form

$$r_0 + \frac{s_1|}{|r_1} + \frac{s_2|}{|r_2} + \cdots + \frac{s_n|}{|r_n}, \qquad r_j, s_j \in R^\star \setminus \{0\}. \tag{3.21}$$

Moreover, Bernoulli's theorem was the foundation for Gauss' derivation of his quadrature rules in [32], where there is even an explicit reference to Bernoulli used as the *deus ex machina* for the quadrature method.

Theorem 3.4 (*Bernoulli's theorem*, D. Bernoulli) *A sequence $c_n \in R^\star$ has a continued fraction expansion as*

$$c_n = r_0 + \frac{s_1|}{|r_1} + \frac{s_2|}{|r_2} + \cdots + \frac{s_n|}{|r_n}, \qquad r_j, s_j \in R^\star \setminus \{0\},$$

if and only if $c_{n+1} \neq c_n$, $n \in \mathbb{N}_0$. In this case, the coefficients are given explicitly as

$$r_n = \frac{1}{q_{n-1}} \frac{c_n - c_{n-2}}{c_{n-2} - c_{n-1}}, \qquad s_n = \frac{1}{q_{n-2}} \frac{c_{n-1} - c_n}{c_{n-2} - c_{n-1}}. \tag{3.22}$$

Proof The proof is based on a *recurrence relation* for the convergents

$$\frac{p_k}{q_k} = r_0 + \frac{s_1|}{|r_1} + \frac{s_2|}{|r_2} + \cdots + \frac{s_k|}{|r_k}, \qquad k \in \mathbb{N}_0,$$

of continued fractions of the form (3.21). This recurrence

---

[3] DANIEL BERNOULLI, 1700–1782, son of JOHANN BERNOULLI, brother of NICOLAUS II BERNOULLI. Given the Bernoulli family and their appearance in the mathematics of their time, this extra information is necessary to identify the right person.

$$p_k = r_k\, p_{k-1} + s_k\, p_{k-2}, \qquad p_{-1} = 1, \quad p_0 = r_0$$
$$q_k = r_k\, q_{k-1} + s_k\, q_{k-2}\,, \qquad q_{-1} = 0, \quad q_0 = 1, \tag{3.23}$$

is obtained in the same way as the one in (2.7) in Theorem 2.1, namely by induction on k; the case $k = 0$ is simply the definition of p_0 and q_0, while $k = 1$ is obtained by a straightforward computation:

$$r_0 + \frac{s_1|}{|r_1} = r_0 + \frac{s_1}{r_1} = \frac{r_0\, r_1 + s_1}{r_1} = \frac{r_1\, p_0 + s_1\, p_{-1}}{r_1\, q_0 + s_1\, q_{-1}}.$$

For the inductive step $k \to k + 1$, we again set

$$\frac{p_k'}{q_k'} = r_1 + \frac{s_2|}{|r_2} + \cdots + \frac{s_{k+1}|}{|r_{k+1}},$$

which immediately yields

$$\frac{p_{k+1}}{q_{k+1}} = r_0 + \cfrac{s_1}{r_1 + \cfrac{s_2|}{|r_2} + \cdots + \cfrac{s_{k+1}|}{|r_{k+1}}} = r_0 + \frac{s_1\, q_k'}{p_k'} = \frac{r_0\, p_k' + s_1\, q_k'}{p_k'}$$

and the shifted induction hypothesis then gives

$$p_{k+1} = r_0 \left(r_{k+1} p_{k-1}' + s_{k+1}\, p_{k-2}' \right) + s_1 \left(r_{k+1} q_{k-1}' + s_{k+1} q_{k-2}' \right)$$
$$= r_{k+1} \left(r_0\, p_{k-1}' + s_1\, q_{k-1}' \right) + s_{k+1} \left(r_0\, p_{k-2}' + s_1\, q_{k-2}' \right) = r_{k+1}\, p_k + s_{k+1}\, p_{k-1}$$
$$q_{k+1} = p_k' = r_{k+1}\, p_{k-1}' + s_{k+1}\, p_{k-2}' = r_{k+1}\, q_k + s_{k+1}\, q_{k-1},$$

which proves (3.23). Multiplying the first line by $-q_{k-1}$, the second one by p_{k-1}, and adding everything, we get that

$$p_{k-1}\, q_k - p_k\, q_{k-1} = r_k \left(-p_{k-1}\, q_{k-1} + p_{k-1}\, q_{k-1} \right) - s_k \left(p_{k-2}\, q_{k-1} - p_{k-1}\, q_{k-2} \right)$$
$$= -s_k \left(p_{k-2}\, q_{k-1} - p_{k-1}\, q_{k-2} \right) = s_k\, s_{k-1} \left(p_{k-3}\, q_{k-2} - p_{k-2}\, q_{k-3} \right)$$
$$= \cdots = (-1)^k \prod_{j=1}^{k} s_j \left(p_{-1}\, q_0 - p_0\, q_{-1} \right);$$

hence,

$$p_{k-1}\, q_k - p_k\, q_{k-1} = (-1)^k \prod_{j=1}^{k} s_j. \tag{3.24}$$

This already gives one direction of our theorem: if c_n, $n \in \mathbb{N}$, is a sequence of convergents, then

$$c_n - c_{n-1} = \frac{p_n}{q_n} - \frac{p_{n-1}}{q_{n-1}} = \frac{(-1)^{n+1} s_1 \cdots s_n}{q_n \, q_{n-1}} \neq 0,$$

since $s_j \neq 0$ for all j was assumed.

For the converse, we use the recurrence (3.23) to obtain

$$c_n = \frac{p_n}{q_n} = \frac{r_n \, p_{n-1} + s_n \, p_{n-2}}{r_n \, q_{n-1} + s_n \, q_{n-2}} \qquad \Leftrightarrow \qquad \begin{bmatrix} p_n \\ q_n \end{bmatrix} = \begin{bmatrix} p_{n-1} & p_{n-2} \\ q_{n-1} & q_{n-2} \end{bmatrix} \begin{bmatrix} r_n \\ s_n \end{bmatrix}$$

which can be solved *uniquely* for r_n, s_n since

$$\det \begin{bmatrix} p_{n-1} & p_{n-2} \\ q_{n-1} & q_{n-2} \end{bmatrix} = p_{n-1} \, q_{n-2} - p_{n-2} q_{n-1} = q_{n-1} \, q_{n-2} \left(\frac{p_{n-1}}{q_{n-1}} - \frac{p_{n-2}}{q_{n-2}} \right)$$

$$= q_{n-1} \, q_{n-2} \, (c_{n-1} - c_{n-2}) \neq 0$$

due to our assumption about the c_k and by induction on $q_k, k = n-1, n-2$, respectively. *Cramer's rule* now implies that

$$r_n = \frac{\det \begin{bmatrix} p_n & p_{n-2} \\ q_n & q_{n-2} \end{bmatrix}}{\det \begin{bmatrix} p_{n-1} & p_{n-2} \\ q_{n-1} & q_{n-2} \end{bmatrix}} = \frac{q_n \, q_{n-2} \, (c_n - c_{n-2})}{q_{n-1} \, q_{n-2} \, (c_{n-1} - c_{n-2})} = \frac{q_n}{q_{n-1}} \frac{c_n - c_{n-2}}{c_{n-1} - c_{n-2}}$$

$$s_n = \frac{\det \begin{bmatrix} p_{n-1} & p_n \\ q_{n-1} & q_n \end{bmatrix}}{\det \begin{bmatrix} p_{n-1} & p_{n-2} \\ q_{n-1} & q_{n-2} \end{bmatrix}} = \frac{q_n \, q_{n-1} \, (c_{n-1} - c_n)}{q_{n-1} \, q_{n-2} \, (c_{n-1} - c_{n-2})} = \frac{q_n}{q_{n-2}} \frac{c_{n-1} - c_n}{c_{n-1} - c_{n-2}}.$$

Replacing r_n, s_n by $r'_n = a \, r_n, s'_n = a \, s_n$ for an arbitrary $a \in R \setminus \{0\}$, we still have

$$\frac{p'_n}{q'_n} = \frac{a \, p_n}{a \, q_n} = \frac{p_n}{q_n} = c_n,$$

where we only have to set $a = 1/q_n$ to end up with (3.22).   □

With the last remark in the proof of Theorem 3.4 we return to the simple continued fraction $[r_0; r_1, \ldots, r_n]$, where *simple* means that $s_1 = \cdots = s_n = 1$. Indeed setting $a = 1/s_n$ in the above argument regarding the division, we obtain

$$r'_n = \frac{q_{n-2}}{q_{n-1}} \frac{c_n - c_{n-2}}{c_{n-1} - c_n}, \qquad s'_n = 1.$$

Therefore we have an expansion in the "old" notation as in the form $[a_0; a_1, \ldots]$ for the continued fraction which is slightly more restrictive than (3.21).

Corollary 3.1 (Bernoulli's theorem, normalized) *If the sequence $c_n \in R^\star$, $n \in \mathbb{N}_0$, satisfies $c_n \neq c_{n-1}$, then*

$$c_n = [r_0; r_1, \ldots, r_n], \quad n \in \mathbb{N}_0,$$

where

$$r_n = \frac{q_{n-2}}{q_{n-1}} \frac{c_n - c_{n-2}}{c_{n-1} - c_n}, \quad n \geq 2, \quad r_{-1} = 0, \ r_0 = c_0, \ r_1 = \frac{1}{c_1 - c_0}. \quad (3.25)$$

Proof We can obtain (3.25) directly from (3.12) if we solve for the proper terms, taking into account the assumption $c_n = \frac{p_n}{q_n}$:

$$c_{n-1} - c_n = \frac{(-1)^n}{q_{n-1}\, q_n} \quad \Rightarrow \quad q_n = \frac{(-1)^n}{q_{n-1}\, (c_{n-1} - c_n)}, \quad (3.26)$$

and

$$c_n - c_{n-2} = \frac{(-1)^n\, r_n}{q_{n-2}\, q_n}. \quad (3.27)$$

Solving (3.27) for r_n and substituting (3.26), we finally get

$$r_n = (-1)^n\, q_{n-2} q_n\, (c_n - c_{n-2}) = \frac{q_{n-2}}{q_{n-1}} \frac{c_n - c_{n-2}}{c_{n-1} - c_n},$$

which is (3.25). $\qquad\qquad\qquad\qquad\qquad\qquad\qquad\qquad\qquad\qquad\qquad \square$

Remark 3.4 (*Continued fraction expansions*)

1. The above observation shows that in $R^\star$, the continued fraction expansion (3.21) is *not* unique in general, mainly because R can have too many units. Recall that, for example, in the polynomial ring $\mathbb{K}[x]$ the units consist of $\mathbb{K} \setminus \{0\}$. This leads to the notion of *equivalent continued fractions*: two continued fractions are called *equivalent* if all their convergents coincide.
2. The continued fraction expansion from Corollary 3.1, that is the one with $s_n = 1$, $n \in \mathbb{N}$, plays a particular role in its equivalent family of continued fractions: they are those continued fraction expansions in which the components of the convergent, formed by the *recurrence relation*, are *irreducible*, i.e., those where the convergent is in normalized form. This follows immediately from (3.12); the argument is exactly the same as in Theorem 2.6.
3. In general continued fraction expansions, common divisors of the numerator and the denominator cannot be excluded anymore, see (3.24).

With the help of Bernoulli's theorem, we now can compute the continued fraction expansion in a *power series*, which is the counterpiece to a real number in the world of rational functions. Let us study this with an example.

Example 3.5 The *exponential function* $f(x) = e^x$ has the power series expansion

$$e^x = 1 + x + \frac{x^2}{2} + \frac{x^3}{3!} + \cdots = \sum_{j=0}^{\infty} \frac{x^j}{j!},$$

and we can determine the continued fraction expansion whose nth convergent is the *partial sum*

$$\sum_{j=0}^{n} \frac{x^j}{j!} =: c_n = [r_0; r_1, \ldots, r_n], \qquad r_0, \ldots, r_n \in \mathbb{K}[x], \qquad n \in \mathbb{N}.$$

According to Corollary 3.1, this is possible since $c_n - c_{n-1} = \frac{x^n}{n!} \neq 0$ where for a polynomial $p \in \mathbb{K}[x]$, the expression $p \neq 0$ means that the polynomial is not the neutral element of addition in the ring $\mathbb{K}[x]$ which is of course the *zero polynomial*. The first two values, $r_0 = 1$, $r_1 = 1/x$, and therefore also $q_0 = 1$, $q_1 = 1/x$ yield together with

$$\frac{c_n - c_{n-2}}{c_{n-1} - c_n} = -\left(1 + \frac{n}{x}\right),$$

the values

$$r_2 = -\frac{1}{1/x}\left(1 + \frac{2}{x}\right) = -(x + 2) \quad q_2 = r_2\, q_1 + q_0 = -\frac{x+2}{x} + 1 = 2x^{-1}$$

$$r_3 = \frac{1}{2} + \frac{3}{2}x^{-1} \qquad\qquad q_3 = -3x^{-2}$$

$$r_4 = -\frac{2}{3}x - \frac{8}{3} \qquad\qquad q_4 = 8x^{-2}$$

$$r_5 = \frac{3}{8} + \frac{15}{8}x^{-1} \qquad\qquad q_5 = 15x^{-3}$$

$$r_6 = -\frac{8}{15}x - \frac{48}{15} \qquad\qquad q_6 = -48x^{-3}$$

$$r_7 = \frac{5}{16} + \frac{35}{16}x^{-1} \qquad\qquad q_7 = -105x^{-4}$$

$$r_8 = -\frac{16}{35}x - \frac{128}{35} \qquad\qquad q_9 = 384x^{-4}$$

and so on. The expressions would be a bit nicer for $f(x) = e^{1/x}$ when x is replaced by x^{-1}, see Problem 3.17.

The example already shows that the "natural environment" for continued fractions will most likely be the ring of Laurent polynomials, i.e., all finite sums of the form

$$f(x) = \sum_{k \in \mathbb{Z}} f_k\, x^k, \qquad \#\{k : f_k \neq 0\} < \infty,$$

that also admit negative powers of x, and are therefore not defined any more at $x = 0$. But note that although any Laurent polynomial can be written as $f(x) = x^{-k} p(x)$, $k \in \mathbb{N}_0$, $p \in \mathbb{K}[x]$, the ring has a completely different structure: all nonzero multiples of monomials are now units, since

$$\left(cx^k\right)^{-1} = c^{-1} x^{-k}, \quad c \in \mathbb{K} \setminus \{0\}, \quad k \in \mathbb{Z},$$

and therefore the ring is generated by *units* as a vector space which already implies that the notion of degree is impossible here; see Problem 3.18.

The method of Example 3.5 can be generalized by making a general equivalence between continued fractions and the series over $R^\star$. More precisely we use the following concept which is due to Seidel [96].

Definition 3.8 A series $c_0 + c_1 + \cdots$, $c_j \subset R^\star \setminus \{0\}$ and a continued fraction $r_0 + \dfrac{s_1|}{|r_1} + \cdots$, $r_j, s_j \in R^\star \setminus \{0\}$ are called *equivalent* if

$$\sum_{j=0}^{n} c_j = \frac{p_n}{q_n} = r_0 + \frac{s_1|}{|r_1} + \cdots + \frac{s_n|}{|r_n}, \quad n \in \mathbb{N}. \tag{3.28}$$

Any series has an equivalent continued fraction expansion and vice versa, and the conversion between the two is explicit.

Theorem 3.5 (Euler) *The continued fraction $r_0 + \dfrac{s_1|}{|r_1} + \cdots$ and the series*

$$\sum_{n=0}^{\infty} \frac{(-1)^{n+1}}{q_{n-1} q_n} \prod_{j=1}^{n} s_j, \tag{3.29}$$

and the series $c_0 + c_1 + \cdots$, and the continued fraction

$$\frac{c_0|}{|1} - \frac{\frac{c_1}{c_0}|}{\left|1 + \frac{c_1}{c_0}\right.} - \cdots - \frac{\frac{c_j}{c_{j-1}}|}{\left|1 + \frac{c_j}{c_{j-1}}\right.} - \cdots \tag{3.30}$$

are equivalent.

Proof Equivalence is equivalent to $c_0 = r_0$ and for $n \geq 1$,

$$c_n = \sum_{j=0}^{n} c_j - \sum_{j=0}^{n-1} c_j = \frac{p_n}{q_n} - \frac{p_{n-1}}{q_{n-1}} = \frac{p_n q_{n-1} - p_{n-1} q_n}{q_{n-1} q_n} = \frac{(-1)^{n+1}}{q_{n-1} q_n} \prod_{j=1}^{n} s_j,$$

due to (3.24) from which (3.29) already follows.

For the converse we apply Theorem 3.4 to the sequence

$$a_n = \sum_{j=0}^{n} c_j, \qquad n \in \mathbb{N}_0,$$

which satisfies the conditions of the theorem since $c_j \neq 0$. For $n = 0$, Eq. (3.30) is trivially correct, while for $n = 1$ we have

$$\frac{c_0|}{|1} - \left|\frac{\frac{c_1}{c_0}}{1 + \frac{c_1}{c_0}}\right| = \frac{c_0}{1 - \frac{c_1/c_0}{1 + c_1/c_0}} = \frac{c_0\left(1 + \frac{c_1}{c_0}\right)}{1 + c_1/c_0 - c_1/c_0} = c_0 + c_1$$

which verifies the equivalence of (3.30) for $n = 1$. Then for $n \geq 2$,

$$r_n = \frac{1}{q_{n-1}} \frac{a_n - a_{n-2}}{a_{n-2} - a_{n-1}} = -\frac{1}{q_{n-1}} \frac{c_n + c_{n-1}}{c_{n-1}} = -\frac{1}{q_{n-1}}\left(1 + \frac{c_n}{c_{n-1}}\right) \qquad (3.31)$$

and

$$s_n = \frac{1}{q_{n-2}} \frac{a_{n-1} - a_n}{a_{n-2} - a_{n-1}} = \frac{1}{q_{n-2}} \frac{c_n}{c_{n-1}}; \qquad (3.32)$$

hence, by the recurrence (3.23),

$$q_n = r_n q_{n-1} + s_n q_{n-2} = -\frac{q_{n-1}}{q_{n-1}}\left(1 + \frac{c_n}{c_{n-1}}\right) + \frac{q_{n-2}}{q_{n-2}} \frac{c_n}{c_{n-1}}$$

$$= -\left(1 + \frac{c_n}{c_{n-1}}\right) + \frac{c_n}{c_{n-1}} = -1$$

for any $n \in \mathbb{N}$. Now again, substituting this into (3.31) and (3.32), respectively, gives

$$r_n = 1 + \frac{c_n}{c_{n-1}}, \qquad s_n = -\frac{c_n}{c_{n-1}}, \qquad n \in \mathbb{N}, \qquad (3.33)$$

and verifies the equivalence to (3.30). $\square$

Remark 3.5 Computing the equivalent representation for a power series is nice, but in the next section we will see that at least in some cases, we can do better and thereby determine continued fractions whose convergents cover more coefficients in a given Laurent series.

3.4.1   Problems

3.17 Compute the continued fraction expansion as in Example 3.5 for $e^{1/x}$.

The concept of a graded ring is to some extent the most general structure that is a valid generalization of the degree of a polynomial, namely a decomposition into generalized homogeneous elements in which the degrees add when they are multiplied. Formally the definition is as follows; for more information, see for example [22].

Definition 3.9 (*Graded ring*) A *monoid* is an additive semigroup with neutral element. A *graded ring* is a ring R, together with a monoid Γ, such that

$$R = \bigoplus_{\gamma \in \Gamma} R_\gamma, \quad \text{i.e.,} \quad R_\gamma \cap R_{\gamma'} = \{0\}, \quad \gamma \neq \gamma',$$

and

$$R_\gamma R_{\gamma'} \subseteq R_{\gamma + \gamma'}.$$

3.18 Show that whenever a graded ring has a vector space basis of units, then the grading is trivial: $R = R_0$.

3.19 Compute the (non-normalized) continued fraction expansion for an arbitrary power series and especially for $f(x) = e^x$.

3.5 Power Series and Euler's Continued Fractions

Euler's theorem, Theorem 3.5, can be used to approximate irrational numbers using continued fractions when they are given in power series expansions as they are known for e or π. This approach even goes back to Euler and is based on the equivalence of the series $\sum c_j$, and the continued fraction in (3.30) in the sense that

$$\sum_{j=0}^n c_j = \frac{c_0 |}{|1} - \frac{\frac{c_1}{c_0} |}{\left|1 + \frac{c_1}{c_0}\right.} - \cdots - \frac{\frac{c_n}{c_{n-1}} |}{\left|1 + \frac{c_n}{c_{n-1}}\right.}. \tag{3.34}$$

Normally the formula is written in a slightly different way.

Corollary 3.2 (Euler) *For $c_j \in R \setminus \{0\}$, we have that*

$$1 + \sum_{j=1}^n c_1 \cdots c_j = \frac{1 |}{|1} - \frac{c_1 |}{|1 + c_1} - \cdots - \frac{c_n |}{|1 + c_n}. \tag{3.35}$$

Proof We apply (3.34) to $c_0' = 1$ and $c_j' = c_1 \cdots c_j$, i.e., $c_j'/c_{j-1}' = c_j, j = 1, \ldots, n$, hence

$$1 + \sum_{j=1}^n c_1 \cdots c_j = \sum_{j=0}^n c_j' = \frac{1 |}{|1} + \frac{c_1 |}{|1} - \frac{c_2 |}{|1 + c_2} - \cdots - \frac{c_n |}{|1 + c_n},$$

which is (3.35),   □

Remark 3.6 Yet another popular version of (3.35) is

$$\sum_{j=0}^{n} c_0 \cdots c_j = c_0 \left(1 + \sum_{j=1}^{n} c_1 \cdots c_j \right) = \frac{c_0|}{|1} - \frac{c_1|}{|1 + c_1} - \cdots - \frac{c_n|}{|1 + c_n}. \quad (3.36)$$

Example 3.6 (*Continued fraction expansion of e*) The power series expansion

$$e^x = \sum_{k=0}^{\infty} \frac{x^k}{k!}$$

is of the form (3.35) with $c_j = \frac{x}{j}$, $j \in \mathbb{N}$; hence,

$$e^x = \frac{1|}{|1} - \frac{x|}{|1} - \frac{\frac{x}{2}|}{|1 + \frac{x}{2}} - \cdots - \frac{\frac{x}{n}|}{|1 + \frac{x}{n}} - \cdots .$$

The partial sum can be renormalized as

$$\frac{1|}{|1} - \frac{x|}{|1} - \frac{\frac{x}{2}}{1 + \frac{x}{2}} - \cdots - \frac{\frac{x}{n}}{1 + \frac{x}{n}} = \frac{1|}{|1} - \frac{x|}{|1} - \frac{x|}{|2 + x} - \frac{2x|}{|3 + x} - \cdots - \frac{(n-1)x|}{|n + x};$$

and $x = 1$ gives the infinite continued fraction expansion

$$e = \frac{1|}{|1} - \frac{1|}{|1} - \frac{1|}{|3} - \frac{2|}{|4} - \cdots - \frac{n-1|}{|n+1} - \cdots \quad (3.37)$$

due to Euler.

Remark 3.7 The expansion from (3.37) is not a "nice" continued fraction expansion with positive integer coefficients as was discussed in Sect. 2.3; therefore we also cannot make statements about their quality of approximation. But we have a simple and explicit expression.

3.5.1 Problems

3.20 Compute the continued fraction expansion of π using the series

$$\frac{\pi}{4} = \sum_{n=0}^{\infty} \frac{(-1)^n}{2n + 1}.$$

Derive the series expansion from a Taylor series of $\tan^{-1}$, the inverse function of the tangent function "tan".

3.21 Determine the continued fraction expansion of $f(x) = \log(1 + x)$ and from that the expansion for $\log 2$.

3.6 Padé Approximation

Padé approximation is a process of rational (polynomial) approximation of a given (formal) power series. It is a classical and well-studied subject; see, for example [11, 13]. Its relations to continued fractions are also manifold and again classical; see [79, p. 235ff], but also [68]. For the sake of completeness, we give a very short introduction to the concept, and point out its main connections to continued fractions. Interested readers are invited to look into the vast amount of literature on Padé approximation.

Following [79], we begin with a formal *power series*[4]

$$f(x) = f_0 + f_1 x + f_2 x^2 + \cdots = \sum_{j=0}^{\infty} f_j x^j, \qquad f_0 \neq 0, \qquad (3.38)$$

where "formal" means that we do not care about convergence; the series may have a convergence radius of 0. Still, formal power series can be added to each other, be multiplied by polynomials, and they can be (formally) differentiated which supplies us with a substantial toolbox of operations.

The goal is to approximate (3.38) by a rational function p/q in the best possible way; in the context of formal power series, "best possible" means that the rational function meets as many terms of f as possible.

Remark 3.8 Rational functions

$$r = \frac{p}{q} \qquad p \in \Pi_m, \qquad q \in \Pi_n, \qquad (3.39)$$

are always ambiguous, since one can multiply p and q by the same nonzero number and obtain the same rational function again. The same holds true if $p(0) = q(0) = 0$, and then a power of x could be removed from the numerator and the denominator.

Definition 3.10 (*Rational functions*) We denote by

$$R_{m,n} = \left\{ r = \frac{p}{q} : p, q \in \Pi, \ \deg p = m, \ \deg q = n, \ q \neq 0 \right\} \qquad (3.40)$$

---

[4] Everything could also be formulated in terms of the Laurent series, and this will be the approach of choice in Chap. 4; here we stick to the power series, as the power series are more in the spirit of Padé approximation.

the *space of rational functions* of degrees m, n.

Remark 3.9 Keep in mind that $R_{m,n}$ is not a linear space and is not even *convex*. However it is still a (nonlinear) manifold of dimension $m + n + 1$ and satisfies $R_{m,n} \supseteq R_{m',n'}, m \geq m', n \geq n'$.

Naïvely, one might therefore compare f with a Taylor expansion for p/q at 0 which requires that $q(0) \neq 0$, since otherwise the Taylor series would not be well defined. Let us look at a simple example.

Example 3.7 Let us try to approximate f in the best possible way using a rational function $r \in R_{0,1}$, written in the general form

$$r = \frac{1}{ax + b}, \qquad b \neq 0.$$

From the (formal) Taylor series

$$Tr(x) = \sum_{k=0}^{\infty} \frac{1}{k!}(-1)^k \frac{a^{k-1}}{b^k} x^k,$$

we get that with

$$b = \frac{1}{f_0} \quad \text{and} \quad \frac{a}{b^2} = f_1, \qquad \text{that is} \qquad a = \frac{f_1}{f_0^2},$$

we can capture the first two coefficients of f as long as $f_0 \neq 0$, which was conveniently requested in (3.38). This is the best possible that can be done with two free parameters a and b.

However as soon as things become slightly more complicated we are in trouble.

Example 3.8 Since one of the polynomials can be normalized so that it is *monic*, we take this liberty for the numerator polynomial; we write

$$r = \frac{x + b}{cx + d} \in R_{1,1},$$

and get the Taylor series

$$Tr(x) = \frac{b}{d} - \sum_{k=1}^{\infty} (-1)^k \frac{(d - bc)c^{k-1}}{d^k} x^k;$$

the first three equations are

$$f_0 = \frac{b}{d}, \qquad f_1 = \frac{d - bc}{d^2}, \qquad f_2 = \frac{(d - bc)c}{d^3},$$

and they are well defined since $d \neq 0$; otherwise the Taylor expansion at 0 makes no sense. The first two equations give

$$b = f_0 d, \qquad c = \frac{1 - f_1 d}{f_0},$$

hence $d - bc = f_1 d^2$, and so d has to satisfy

$$f_2 d = \frac{f_1(1 - f_1 d)}{f_0}, \qquad \text{i.e.,} \qquad d = \frac{f_1}{f_2 f_0 + f_1^2},$$

which is undefined if $f_1 \neq 0$ and $f_0 f_2 = -f_1^2$ and yields the forbidden value $d = 0$ if $f_1 = 0$ and $f_0 f_2 \neq -f_1^2$.

Example 3.8 already shows that there is no hope of being able to capture as many terms in a power series, since a rational function has degrees of freedom just by means of a Taylor expansion of a rational function. In addition the equations that determine the coefficients for the "best" polynomial also become nonlinear and thus are highly complex. Therefore another approach, similar to the concept used in the best approximations of the second kind for irrational numbers will be pursued; see Definition 2.10.

Definition 3.11 (*Padé approximation problem*) Given f in the form (3.38), find $p \in \Pi_m$ and $q \in \Pi_n$ such that

$$f(x) q(x) - p(x) = g_{n+m+1} x^{n+m+1} + g_{n+m+2} x^{n+m+2} + \cdots = O\left(x^{m+n+1}\right).$$
$$(3.41)$$

There is always a trivial way to satisfy (3.41) provided there are no degree restrictions on p and q: simply set $p(x) = x^{m+n+1} \tilde{p}(x)$ and $q(x) = x^{m+n+1} \tilde{q}(x)$ for arbitrary p, q; then the left-hand side of (3.41) is trivially a formal power series of the form $O(x^{m+n+1})$. So the degree restrictions are really necessary to obtain a well defined problem.

If we set $g := fq$, then (3.41) means that

$$g_0 = p_0, \ldots, g_m = p_m, \qquad g_{m+1} = \cdots = g_{m+n} = 0,$$

which can be written for $m > n$ as the linear system

$$
\begin{bmatrix}
f_0 & & & & \\
f_1 & f_0 & & & \\
\vdots & \ddots & \ddots & & \\
f_n & \cdots & f_1 & f_0 & \\
\vdots & \ddots & \vdots & \vdots & \\
f_m & \cdots & f_{m-n+1} & f_{m-n} & \\
f_{m+1} & \cdots & f_{m-n+2} & f_{m-n+1} & \\
\vdots & \ddots & \vdots & \vdots & \\
f_{m+n} & \cdots & f_{m+1} & f_m &
\end{bmatrix}
\begin{bmatrix}
q_0 \\
\vdots \\
q_n
\end{bmatrix}
=
\begin{bmatrix}
p_0 \\
\vdots \\
p_m \\
0 \\
\vdots \\
0
\end{bmatrix},
\tag{3.42}
$$

with the convention that $f_j = 0$ for $j < 0$; see also Problem 3.23. In a more convenient block matrix form, (3.42) reads as

$$
\begin{bmatrix} F_1 \\ F_2 \end{bmatrix} q = \begin{bmatrix} p \\ 0 \end{bmatrix}, \qquad F_1 \in \mathbb{R}^{(m+1)\times(n+1)}, \ F_2 \in \mathbb{R}^{n\times(n+1)}.
\tag{3.43}
$$

Remark 3.10 Equation (3.42) shows what would happen if the normalizing condition $f_0 \neq 0$ were not true. Indeed $f_0 = \cdots = f_k = 0$ would simply result in $p_0 = \cdots p_k = 0$, and hence just in a shift of the problem with a numerator polynomial of lesser degree.

According to [79], the following result already goes back to Frobenius [26] and Padé [76].

Lemma 3.1 *The Padé approximation problem* (3.41) *always has a unique solution in* $R_{m,n}$.

Proof The underdetermined homogeneous problem $F_2 q = 0$ has at least one non-trivial solution $q \neq 0$ which directly defines the associated $p = F_1 q$; hence (3.41) has at least one solution. This solution to (3.43) is unique up to normalization if and only if rank $F_1 = n$, or equivalently

$$
\det \begin{bmatrix}
f_{m+1} & \cdots & f_{j+1} & f_{j-1} & \cdots & f_{n-m+1} \\
\vdots & \ddots & \vdots & \vdots & \ddots & \vdots \\
f_{m+n} & \cdots & f_{j+n} & f_{j+n-2} & \cdots & f_m
\end{bmatrix} \neq 0, \qquad j = m-n+1, \ldots, m+1.
$$

If there are two solutions, p, q and $\tilde{p}, \tilde{q}$, of (3.41), then[5]

$$
f(x)\, q(x) - p(x) = O\left(x^{m+n+1}\right) \quad \text{and} \quad f(x)\, \tilde{q}(x) - \tilde{p}(x) = O\left(x^{m+n+1}\right),
$$

hence

---

[5] The two $O(\cdot)$-terms can be different; it is not admissible to claim that $f(x)\, q(x) - p(x) = f(x)\, \tilde{q}(x) - \tilde{p}(x)$.

$$(f(x)\,q(x) - p(x))\,\tilde{q}(x) = O\left(x^{m+n+1}\right), \qquad (f(x)\,\tilde{q}(x) - \tilde{p}(x))\,q(x) = O\left(x^{m+n+1}\right),$$

so that also

$$\tilde{p}(x)\,q(x) - p(x)\,\tilde{q}(x) = (f(x)\,q(x) - p(x))\,\tilde{q}(x) - (f(x)\,\tilde{q}(x) - \tilde{p}(x))\,q(x)$$
$$= O\left(x^{m+n+1}\right),$$

and since the expression on the left-hand side is a polynomial of degree $n + m$, it must be zero from which it follows that $\tilde{p}(x)\,q(x) = p(x)\,\tilde{q}(x)$ or

$$\frac{p}{q} = \frac{\tilde{p}}{\tilde{q}} \tag{3.44}$$

as claimed. $\qquad\qquad\qquad\qquad\qquad\qquad\qquad\qquad\qquad\qquad\qquad\qquad\qquad\qquad\square$

It is worthwhile to keep in mind that the solution is only unique as a *rational function*, and is not necessarily unique with respect to the explicit numerator p and denominator q. We can make it unique by normalizing p and q such that $q(0) = q_0 = 1$, and hence $p(0) = p_0 = f_0$ due to the first equation in (3.42). In fact if $0 = q_0 = \cdots = q_k \neq q_{k-1}$, which can be induced by $F_1 q = 0$, then the lower triangular structure of F_1 from (3.43) implies that $p = F_1 q$ also has the property that $0 = p_0 = \cdots = p_k \neq p_{k+1}$; consequently p and q have a common factor x^k that can be divided off in the normalization. This representation, further normalized to $p(0) = f_0$ and $q(0) = 1$, of the unique solution r for the Padé approximation problem can therefore be of a strictly lower degree $m' < m$ and $n' < n$, and will capture only some part of the coefficients of f. This can, however, be compensated by multiplying both polynomials of the normalized version with an appropriate monomial $(\cdot)^k$. This allows us to summarize the unique solvability of the Padé approximation problem as follows.

Theorem 3.6 (Padé approximation) *Given a formal power series*

$$f(x) = \sum_{j=0}^{\infty} f_j\, x^j, \qquad f_0 = 0,$$

there exist for any $m, n \in \mathbb{N}_0$ *a unique rational function* $r \in R_{m,n}$ *with the normalized representation*

$$r = \frac{p}{q}, \qquad p \in \Pi_m,\ q \in \Pi_n, \qquad p(0) = f_0,\ q(0) = 1, \tag{3.45}$$

and $k \in \mathbb{N}_0$, *such that* $(\cdot)^k\, p \in \Pi_m$, $(\cdot)^k\, q \in \Pi_n$ *and*

$$f(x)\, x^k q(x) - x^k p(x) = O\left(x^{m+n+1}\right).$$

Definition 3.12 (*Padé approximant*) For $m, n \in \mathbb{N}_0$, the above unique solution of the Padé approximation problem is called the *Padé approximant* of order (m, n). Its normalized representation as in (3.45) is written as

$$r_{m,n} = \frac{p_m}{q_n}.$$

Moreover the bi-infinite matrix

$$\left[r_{j,k} : j, k \in \mathbb{N}_0 \right]$$

of rational functions or of pairs of polynomials is called the *Padé table*.

Example 3.9 For the function $f(x) = e^x = 1 + x + \frac{1}{2}x^2 + \cdot$, the first part of the Padé table is

$n \backslash m$	0	1	2
0	$\dfrac{1}{1}$	$\dfrac{1+x}{1}$	$\dfrac{1+x+\frac{1}{2}x^2}{1}$
1	$\dfrac{1}{1-x}$	$\dfrac{1+\frac{1}{2}x}{1-\frac{1}{2}x}$	$\dfrac{1+\frac{3}{2}x+\frac{1}{6}x^2}{1-\frac{1}{3}x}$
2	$\dfrac{1}{1-x+\frac{1}{2}x^2}$	$\dfrac{1+\frac{1}{3}x}{1-\frac{3}{2}x+\frac{1}{6}x^2}$	$\dfrac{1+\frac{1}{2}x+\frac{1}{12}x^2}{1-\frac{1}{2}x+\frac{1}{12}x^2}$

$$(3.46)$$

Continued fractions now come in naturally, and there actually exists a much more general theory on the relationship between the Padé table and continued fractions, cf. [79, p. 256ff]. We still can only scratch the surface here. But to at least illustrate this relationship, we follow [79] and consider the sequence

$$(0, 0), (0, 1), (1, 1), (1, 2), (2, 2), \ldots$$

i.e., $m_{2k} = m_{2k+1} = n_{2k} = k, n_{2k+1} = k + 1$ and assume that the associated rational functions

$$r_k = r_{m_k, n_k} = \frac{p_{m_k, n_k}}{q_{m_k, n_k}}, \qquad k \in \mathbb{N}_0,$$

from the Padé table are all disjoint and *regular*, that is that $\deg p_{m_k, n_k} = m_k$ or $\deg q_{m_k, n_k} = n_k$ or both. This is of course a condition that has to be verified which is a nontrivial task.

But assuming that this is the case, Theorem 3.4 tells us that there is a continued fraction

$$t_0 + \frac{s_1|}{|t_1} + \frac{s_2|}{|t_2} + \cdots$$

whose kth convergent is r_k, i.e.,

$$\frac{p_k}{q_k} = \frac{p_{m_k,n_k}}{q_{m_k,n_k}}, \qquad k \in \mathbb{N}_0.$$

To avoid notational overkill, we introduce $\hat{p}_k := p_{m_k,n_k}$ and $\hat{q}_k := q_{m_k,n_k}$, and write the recurrence of the convergents in matrix form,

$$\begin{bmatrix} \hat{p}_k \\ \hat{q}_k \end{bmatrix} = \begin{bmatrix} \hat{p}_{k-1} & \hat{p}_{k-2} \\ \hat{q}_{k-1} & \hat{q}_{k-2} \end{bmatrix} \begin{bmatrix} s_k \\ t_k \end{bmatrix}; \qquad (3.47)$$

then note that we can solve this if and only if

$$0 \neq \det \begin{bmatrix} \hat{p}_{k-1} & \hat{p}_{k-2} \\ \hat{q}_{k-1} & \hat{q}_{k-2} \end{bmatrix} = \hat{p}_{k-1}\hat{q}_{k-2} - \hat{p}_{k-2}\hat{q}_{k-1} \qquad \Leftrightarrow \qquad \frac{\hat{p}_{k-1}}{\hat{q}_{k-1}} \neq \frac{\hat{p}_{k-2}}{\hat{q}_{k-2}},$$

which was precisely our assumption that $r_{k-1} \neq r_{k-2}$. Solving (3.47) then yields that

$$s_k = \frac{\hat{p}_{k-1}\hat{q}_k - \hat{p}_k\hat{q}_{k-1}}{\hat{p}_{k-1}\hat{q}_{k-2} - \hat{p}_{k-2}\hat{q}_{k-1}}, \qquad t_k = \frac{\hat{p}_k\hat{q}_{k-2} - \hat{p}_{k-2}\hat{q}_k}{\hat{p}_{k-1}\hat{q}_{k-2} - \hat{p}_{k-2}\hat{q}_{k-1}}. \qquad (3.48)$$

Now we must make use of the fact that the fractions come from a Padé table and then compare for $m, n \in \mathbb{N}_0$, $m', n' \geq 0$, $m'n' \neq 0$, products of regular entries from the table as they appear in (3.48)

$$\begin{aligned} p_{m+m',n+n'} q_{m,n} &- p_{m,n} q_{m+m',n+n'} \qquad (3.49) \\ &= \left(f\, q_{m,n} - p_{m,n} \right) q_{m+m',n+n'} - \left(f\, q_{m+m',n+n'} - p_{m+m',n+n'} \right) q_{m,n} \\ &= O\left((\cdot)^{m+n+1} \right). \end{aligned}$$

Since the first polynomial in (3.49) is of the degree at most $m + n + m'$, the second of the degree $m + n + n'$, the polynomial in (3.49) is of the form

$$(\cdot)^{m+n+1} g_{m,n}^{m',n'}, \qquad \deg g_{m,n}^{m',n'} = \max\{m' - 1, n' - 1\}. \qquad (3.50)$$

In (3.48) we are considering pairs with indices (m_k, n_k) together with

$$(m_{k+1}, n_{k+1}) = \begin{cases} (m_k, n_k + 1), & k \in 2\mathbb{N}_0, \\ (m_k + 1, n_k), & k \in 2\mathbb{N}_0, \end{cases}$$

i.e., $m_{k+1} + n_{k+1} = m_k + n_k + 1$, and

$$(m_{k+2}, n_{k+2}) = (m_k + 1, n_k + 1).$$

Now note that

$$(\cdot)^{m+n+1} g_{m,n}^{1,0} q_{m+1,n+1} = p_{m+1,n} q_{m,n} q_{m+1,n+1} - p_{m,n} q_{m+1,n} q_{m+1,n+1}$$
$$(\cdot)^{m+n+1} g_{m,n}^{1,1} q_{m+1,n} = p_{m+1,n+1} q_{m,n} q_{m+1,n} - p_{m,n} q_{m+1,n+1} q_{m+1,n},$$

and subtract the two to obtain

$$(\cdot)^{m+n+1} \left(g_{m,n}^{1,0} q_{m+m'+1,n+n'+1} - g_{m,n}^{1,1} q_{m+m',n+n'} \right)$$
$$= q_{m,n} \left(p_{m+1,n} q_{m+1,n+1} - p_{m+1,n+1} q_{m+1,n} \right) = q_{m,n} (\cdot)^{m+n+2} g_{m+1,n}^{0,1};$$

hence after division by $(\cdot)^{m+n+1}$, then

$$g_{m,n}^{1,0} q_{m+1,n+1} - g_{m,n}^{1,1} q_{m+1,n} + x \, g_{m+1,n}^{0,1} q_{m,n} = 0. \tag{3.51}$$

In the same fashion, one gets

$$g_{m,n}^{0,1} q_{m+1,n+1} - g_{m,n}^{1,1} q_{m,n+1} + x \, g_{m,n+1}^{1,0} q_{m,n} = 0. \tag{3.52}$$

Note that the definition (3.50) implies that all g-expressions in (3.51) and (3.52) are polynomials of degree 0, i.e., numbers, so that a substitution of $x = 0$ in these two identities yields that

$$g_{m,n}^{0,1} = g_{m,n}^{1,0} = g_{m,n}^{1,1}, \tag{3.53}$$

while taking derivatives with respect to x and then setting $x = 0$ results in

$$g_{m,n}^{1,0} \left(q'_{m+1,n+1} - q_{m+1,n} \right)(0) + g_{m+1,n}^{0,1} = 0,$$
$$g_{m,n}^{0,1} \left(q'_{m+1,n+1} - q'_{m,n+1} \right)(0) + g_{m,n+1}^{1,0} = 0. \tag{3.54}$$

With these identities and (3.48), we can now determine the coefficients for the continued fraction expansion, namely $t_0 = f_0$ as well as

$$s_{2k+1} = \frac{p_{k,k} q_{k,k+1} - p_{k,k+1} q_{k,k}}{p_{k,k} q_{k-1,k} - p_{k-1,k} q_{kk}} = -\frac{g_{k,k}^{0,1}}{g_{k-1,k}^{1,0}} (\cdot) = \left(q'_{k,k} - q'_{k,k-1} \right)(0) (\cdot),$$

$$s_{2k+2} = \frac{p_{k,k+1} q_{k+1,k+1} - p_{k+1,k+1} q_{k,k+1}}{p_{k,k+1} q_{k,k} - p_{k,k} q_{k,k+1}} = -\frac{g_{k,k+1}^{1,0}}{g_{k,k}^{0,1}} (\cdot)$$
$$= \left(q'_{k+1,k+1} - q'_{k,k+1} \right)(0) (\cdot),$$

with $s_0 = c_1 x$ and

$$t_{2k+1} = \frac{p_{k,k+1}\, q_{k-1,k} - p_{k-1,k}\, q_{k,k+1}}{p_{k,k}\, q_{k-1,k} - p_{k-1,k}\, q_{kk}} = \frac{g_{k-1,k}^{1,1}}{g_{k-1,k}^{1,0}} = 1,$$

$$t_{2k+2} = \frac{p_{k+1,k+1}\, q_{k,k} - p_{k,k}\, q_{k+1,k+1}}{p_{k,k+1}\, q_{k,k} - p_{k,k}\, q_{k,k+1}} = \frac{g_{k,k}^{1,1}}{g_{k,k}^{1,0}} = 1.$$

This can be summarized as the fundamental connection between continued fractions and the Padé table.

Theorem 3.7 *If the fractions with the indices*

$$m_k = \left\lfloor \frac{k}{2} \right\rfloor, \quad n_k = \left\lfloor \frac{k+1}{2} \right\rfloor, \quad k \in \mathbb{N}_0,$$

in the Padé table for a power series f are all different and regular, then there exists a continued fraction

$$f_0 + \sum_{j=1}^{\infty} \frac{a_j(\cdot)|}{|1}$$

whose convergents are the entries of the Padé table along the sequence:

$$\frac{p_{m_k}}{q_{m_k}} = \frac{p_k}{q_k}, \quad k \in \mathbb{N}_0.$$

In other words: any "reasonable" Padé table for a power series f contains convergents of a continued fraction expansion for f. An obvious question is whether being "reasonable" is a rare or a generic property. In the next chapter we will study the existence of such an associated continued fraction, not for a power series but for a Laurent series. This however is only a change of perspective.

3.6.1 Problems

3.22 Show that $R_{m,n}$ is not *convex*; that is, there exist $r_1, r_2 \in R_{m,n}$ and $\alpha \in [0, 1]$ such that

$$\alpha\, r_1 + (1 - \alpha)\, r_2 \notin R_{m,n}.$$

Is this a rare or a frequent property among functions from $R_{m,n}$?

3.23 Write down the explicit form of (3.42) for $m < n$.

3.24 Determine the first row of the Padé table for arbitrary f. (Easy)

3.25 Verify the table in (3.46) and extend it to $m, n \le 3$.

3.26 Write a `Matlab` program that computes the Padé approximant for a given power series f.

3.27 Prove (3.52).

3.28 ([79, p. 256]) Verify for $m, m', m'' \geq 0$ and $n, n', n'' \geq 0$, the general identity

$$g_{m,n}^{m',n'}\, q_{m+m'+m'',n+n'+n''} - q_{m,n}^{m+m',n+n'}\, q_{m+m',n+n'} + (\cdot)^{m'+n'}\, g_{m+m',n+n'}^{m'',n''}\, q_{m,n} = 0.$$

Chapter 4
Continued Fractions and Gauss

Arithmetic, you see, is useful; without its aid, I should hardly have been able to guess your age.

Charlotte Brontë, Jane Eyre

4.1 Orthogonal Polynomials, Continued Fractions and Moments

In this chapter we will have a look at the close connection between continued fractions and orthogonal polynomials, which is essentially a consequence of the *three-term recurrence* (3.11) that is common to both concepts. This relationship was already used by CARL FRIEDRICH GAUSS in his original development of the so-called *Gauss quadrature*, which is a fundamental topic in Numerical Analysis, and more precisely in *numerical integration*, see [33, 54]. Gauss used two main tools in his approach: polynomial interpolation and the continued fractions expansion of the Laurent series formed from the moments; the existence of this expansion is due to Bernoulli's theorem, hence we follow precisely the theme presented in the preceding chapter.

We now get more specific and explicitly consider the ring $R = \Pi = \mathbb{R}[x]$ of univariate polynomials with *real* coefficients as well as for $n \in \mathbb{N}_0$, the *vector space*

$$\Pi_n = \text{span} \left\{ 1, x, \dots, x^n \right\} = \{ f \in \Pi \ : \ \deg f \le n \}$$

of all polynomials of *degree* at most n. What we also need is an *inner product* that induces the notion of orthogonality.

Definition 4.1 A bilinear form

$$\langle \cdot, \cdot \rangle \ : \ \Pi \times \Pi \to \mathbb{R},$$

on Π is called an *inner product* if it is *symmetric*, $\langle f, g \rangle = \langle g, f \rangle$ and *definite*, i.e.,

© The Author(s), under exclusive license to Springer Nature Switzerland AG 2021
T. Sauer, *Continued Fractions and Signal Processing*, Springer Undergraduate Texts in Mathematics and Technology, https://doi.org/10.1007/978-3-030-84360-1_4

$$\langle f, f \rangle > 0, \qquad f \neq 0.$$

We want the inner product to be induced by a *square positive linear functional*, i.e.,

$$\langle f, g \rangle = L(fg), \tag{4.1}$$

where $L : \Pi \to \mathbb{R}$ is positive for squares:

$$L(f^2) > 0, \quad f \in \Pi. \tag{4.2}$$

Remark 4.1 The most popular and standard case of a square positive linear functional is of course the standard *integral*

$$L(f) = \int_0^1 f(x)\, dx$$

on the unit interval.

Remark 4.2 In one variable, square positive linear functionals and positive linear functionals, i.e., $L(f) > 0$ whenever $f > 0$ are the same since any positive polynomial can be decomposed as a sum of squares. This is no longer the case in several variables which makes the moment problem significantly more difficult, cf. [94].

Definition 4.2 *(Moments)*

1. The *n*th *moment* of the inner product $\langle \cdot, \cdot \rangle$ is defined as

$$\mu_n = L\left((\cdot)^n\right) = \langle 1, (\cdot)^n \rangle, \quad n \in \mathbb{N}; \tag{4.3}$$

    all together, the moments define the *moment sequence* $\mu = (\mu_n : n \in \mathbb{N})$.
2. The *moment matrix* is the bi-infinite matrix

$$M = \left[\langle (\cdot)^j, (\cdot)^k \rangle : j, k \in \mathbb{N}_0\right] = \left[\mu_{j+k} : j, k \in \mathbb{N}_0\right]. \tag{4.4}$$

    which represents an operator acting on real valued sequences.
3. A matrix A of the form $a_{j,k} = a_{j+k}$ is called a *Hankel matrix*, or in the infinite case a *Hankel operator* for the sequence $a = (a_n : n \in \mathbb{N})$.

Of course the simplest way to obtain square positive functionals and avoid the peculiarities of measurability is to choose $a, b \in \mathbb{R}$, $a \leq b$ and $w : [a, b] \to \mathbb{R}$ as a nonzero and positive continuous function, and to set

$$L : C[a, b] \to \mathbb{R}, \qquad L(f) := \int_a^b f(x)\, w(x)\, dx, \tag{4.5}$$

which is a simple Riemann integral known from basic analysis classes; the example in Remark 4.1 is the even simplest choice $w = 1$. However in order to emphasize the algebraic approach here, we will avoid such explicit representations of the square positive linear functional and focus on moment sequences only.

Remark 4.3 Not every sequence $\mu = (\mu_n : n \in \mathbb{N}_0)$ is a moment sequence; for example any sequence with $0 > \mu_0 = L(1) = L(1^2)$ already contradicts square positivity. Therefore it is a straightforward and natural question as to which sequences μ can be moment sequences associated to a square positive linear functional L, and then show how the functional L or maybe even a, b and w can be recovered from a given moment sequence. Questions of this type are associated with the *moment problem* for which there is a substantial amount of literature, cf. [28]. It is getting even more challenging in several variables, see [94].

On Π, inner products induced by square positive functionals and moment matrices are easily seen to be equivalent. Of course any inner product defines a *moment matrix*, and conversely for any two polynomials

$$f(x) = \sum_{j-0}^{n} f_k x^k, \qquad g(x) = \sum_{j-0}^{n} g_k x^k, \qquad n = \max\{\deg f, \deg g\},$$

we simply get that

$$\langle f, g \rangle = f^T M_n g = [f_0, \ldots, f_n] \begin{bmatrix} \mu_0 & \mu_1 & \cdots & \mu_n \\ \mu_1 & \mu_2 & \ddots & \vdots \\ \vdots & \vdots & \ddots & \mu_{2n-1} \\ \mu_n & \mu_{n+1} & \cdots & \mu_{2n} \end{bmatrix} \begin{bmatrix} g_0 \\ \vdots \\ g_n \end{bmatrix}, \qquad (4.6)$$

where the Hankel structure ensures that $\langle f, g \rangle = L(fg)$.

Remark 4.4 In (4.6) we identified a polynomial with its coefficient vector which will become a convenient tool for what will follow. Also in computational practice polynomials are usually stored as their coefficient vectors, and all manipulations will be applied only to the coefficient vectors.

Setting $g = f$ in (4.6) yields for any square positive functional and any $0 \neq f \in \mathbb{R}^{n+1}$ the simple but fundamental observation that

$$0 < L(f^2) = f^T M_n f,$$

which we can summarize as follows.

Theorem 4.1 *If μ is the moment sequence with respect to a square positive linear functional, then all matrices M_n, $n \in \mathbb{N}_0$, are strictly positive definite.*

Remark 4.5 The notion of a matrix A being *positive definite* is always a source of pleasant confusion. While English literature mostly distinguishes between "positive definite" if $x^T A x \geq 0$ and "strictly positive definite" if $x^T A x > 0$ for $x \neq 0$, German literature prefers "positive semidefinite" and "positive definite" for the same properties. To avoid confusion as much as possible, I will even make use of the redundant but unique distinction "positive semidefinite" and "strictly positive definite" in most situations. Also recall that for complex matrices, positive definiteness always includes that the matrix is Hermitian, i.e., $A^H = A$.

Definition 4.3 A sequence $f_n \in \Pi_n \setminus \{0\}, n \in \mathbb{N}$, of nonzero polynomials of increasing degree is called a *sequence of orthogonal polynomials* with respect to the inner product $\langle \cdot, \cdot \rangle$ if

$$\langle f_n, \Pi_{n-1} \rangle = 0, \quad \text{i.e.,} \quad \langle f_n, f \rangle = 0, \quad f \in \Pi_{n-1}. \quad (4.7)$$

The polynomial f_n is called an *orthogonal polynomial* of degree n; it is called an *orthonormal polynomial* provided that $\langle f_n, f_n \rangle = 1$.

The orthogonal polynomials are of the degree *exactly* n, see Problem 4.3, unique up to normalization, and they can be easily determined from the moment matrix. To that end note that for any $g \in \Pi_{n-1}$, we have

$$0 = \langle g, f_n \rangle = \begin{bmatrix} g_0, \ldots, g_{n-1} \end{bmatrix} \begin{bmatrix} \mu_0 & \cdots & \mu_n \\ \vdots & \ddots & \vdots \\ \mu_{n-1} & \cdots & \mu_{2n-1} \end{bmatrix} \begin{bmatrix} f_0 \\ \vdots \\ f_n \end{bmatrix},$$

and since this has to hold for *any* $g \in \Pi_{n-1}$, it follows that

$$0 = \begin{bmatrix} \mu_0 & \cdots & \mu_n \\ \vdots & \ddots & \vdots \\ \mu_{n-1} & \cdots & \mu_{2n-1} \end{bmatrix} \begin{bmatrix} f_0 \\ \vdots \\ f_n \end{bmatrix} = \begin{bmatrix} M_{n-1} & \begin{bmatrix} \mu_n \\ \vdots \\ \mu_{2n-1} \end{bmatrix} \end{bmatrix} \begin{bmatrix} f_0 \\ \vdots \\ f_n \end{bmatrix}.$$

Since M_{n-1} is positive definite, we therefore get a unique nonzero solution of

$$M_{n-1} \begin{bmatrix} f_0 \\ \vdots \\ f_{n-1} \end{bmatrix} = - \begin{bmatrix} \mu_n \\ \vdots \\ \mu_{2n-1} \end{bmatrix} f_n$$

for any $f_n \neq 0$. In particular, $f_n = 1$ leads to the *monic* orthogonal polynomial of the form $x^n + f_{n-1}x^{n-1} + \cdots + f_0$. This approach from Linear Algebra could also be expressed in a somewhat more sophisticated framework in terms of a *Schur complement* of M_{n-1} in M_n; see [52], but also Problem 5.15.

Theorem 4.2 *A sequence $f_n, n \in \mathbb{N}$, is a sequence of orthogonal polynomials with positive leading coefficients for an inner product, if and only if there exist real coefficients $\alpha_n > 0$, $\beta_n \in \mathbb{R}$ and $\gamma_n > 0$, $n \in \mathbb{N}$, such that*

$$f_n = (\alpha_n x + \beta_n) f_{n-1} - \gamma_n f_{n-2}, \qquad n \in \mathbb{N}, \qquad f_0 = 1, \qquad f_{-1} = 0. \qquad (4.8)$$

Remark 4.6 The request $\alpha_n, \gamma_n > 0$ in Theorem 4.2 could be weakened to be $\alpha_n \gamma_n > 0$, since any such modification would only result in changing the sign of the leading terms of f_n in an alternating fashion.

Proof Let f_n, $n \in \mathbb{N}$, be a sequence of orthogonal polynomials. We will show by induction on n that the polynomial

$$g_{n+1}(x) = x\, f_n(x) - \underbrace{\frac{\langle x f_n, f_n \rangle}{\langle f_n, f_n \rangle}}_{=:\beta'_n} f_n - \underbrace{\frac{\sqrt{\langle g_n, g_n \rangle} \langle f_n, f_n \rangle}{\langle f_{n-1}, f_{n-1} \rangle}}_{=:\gamma'_n > 0} f_{n-1}(x), \qquad x \in \mathbb{R},$$

$$(4.9)$$

is different from zero and orthogonal to Π_n, and therefore it must be a positive multiple of f_{n+1}. Indeed for $n = 0$, we obtain that

$$g_1(x) = x\, f_0(x) - \langle x, 1 \rangle f_0 \qquad \Rightarrow \qquad \langle g_1, f_0 \rangle = \langle g_1, 1 \rangle = \langle x, 1 \rangle - \langle x, 1 \rangle = 0,$$

while for the induction step, we first note that for $n \in \mathbb{N}_0$ and any $f \in \Pi_{n-2}$

$$\langle g_{n+1}, f \rangle = \langle f_n, x f \rangle - \beta'_n \langle f_n, f \rangle - \gamma'_n \langle f_{n-1}, f \rangle = 0$$

holds. Using the induction hypothesis, we also get that $g_n = \lambda_n f_n$ with[1]

$$\langle g_n, g_n \rangle = \lambda_n^2 \langle f_n, f_n \rangle \qquad \Rightarrow \qquad \lambda_n = \sqrt{\frac{\langle g_n, g_n \rangle}{\langle f_n, f_n \rangle}}$$

and end up with

$$
\begin{aligned}
\langle g_{n+1}, f_{n-1} \rangle &= \langle x\, f_n, f_{n-1} \rangle - \beta'_n \langle f_n, f_{n-1} \rangle - \gamma'_n \langle f_{n-1}, f_{n-1} \rangle \\
&= \langle f_n, x\, f_{n-1} \rangle - \gamma'_n \langle f_{n-1}, f_{n-1} \rangle \\
&= \langle f_n, g_n + \beta'_{n-1} f_{n-1} + \gamma'_{n-1} f_{n-2} \rangle - \gamma'_n \langle f_{n-1}, f_{n-1} \rangle \\
&= \sqrt{\frac{\langle g_n, g_n \rangle}{\langle f_n, f_n \rangle}} \langle f_n, f_n \rangle - \gamma'_n \langle f_{n-1}, f_{n-1} \rangle = 0,
\end{aligned}
$$

as well as

$$
\begin{aligned}
\langle g_{n+1}, f_n \rangle &= \langle x f_n, f_n \rangle - \beta'_n \langle f_n, f_n \rangle - \gamma'_n \langle f_n, f_{n-1} \rangle \\
&= \langle x f_n, f_n \rangle - \frac{\langle x f_n, f_n \rangle}{\langle f_n, f_n \rangle} \langle f_n, f_n \rangle = 0.
\end{aligned}
$$

---

[1] We choose λ_n to be the positive solution of the quadratic equation, and the negative root $-\lambda_n$ would work equally well, cf. Remark 4.6.

This proves (4.9) and we can even explicitly give the coefficients to be

$$\alpha_n \in \mathbb{R}_+, \qquad \beta_n = -\alpha_n \beta_n', \qquad \gamma_n = \alpha_n \gamma_n',$$

where $\alpha_n > 0$ is a free normalization parameter.

Suppose conversely that f_n is a sequence of polynomials that satisfies (4.8) and let us choose, for simplicity, $\alpha_n = 1$, so that we obtain a sequence of *monic* polynomials $f_n(x) = x^n + \widetilde{f}_n(x)$, $\widetilde{f}_n \in \Pi_{n-1}$. We also assume inductively that we already determined the inner product on $\Pi_{n-1} \times \Pi_{n-1}$, and know the moments $\mu_0, \ldots, \mu_{2n-3}$. Now we consider the polynomials

$$f_n(x) = x \, f_{n-1}(x) + \beta_n \, f_{n-1}(x) - \gamma_n \, f_{n-2}(x),$$

and remark that for $f \in \Pi_{n-3}$, the inner product with f_n is already defined, since

$$\langle f_n, f \rangle := \langle f_{n-1}, x \, f \rangle + \beta_n \, \langle f_{n-1}, f \rangle - \gamma_n \, \langle f_{n-2}, f \rangle$$

only involves polynomials up to degree $2n - 3$. On the other hand, the additional orthogonality conditions and the recurrence relation (4.8) yield

$$
\begin{aligned}
0 = \langle f_n, x^{n-2} \rangle &= \langle x f_{n-1} + \beta_n f_{n-1} - \gamma_n f_{n-2}, x^{n-2} \rangle \\
&= \langle f_{n-1}, x^{n-1} \rangle + \beta_n \underbrace{\langle f_{n-1}, x^{n-2} \rangle}_{=0} - \gamma_n \langle f_{n-2}, x^{n-2} \rangle \\
&= \langle f_{n-1}, x^{n-1} \rangle - \gamma_n \langle f_{n-2}, x^{n-2} \rangle = \langle x^{n-1} + \widetilde{f}_{n-1}, x^{n-1} \rangle - \gamma_n \langle f_{n-2}, x^{n-2} \rangle \\
&= \mu_{2n-2} + \langle \widetilde{f}_{n-1}, x^{n-1} \rangle - \gamma_n \langle f_{n-2}, x^{n-2} \rangle \qquad\qquad (4.10) \\
&= \mu_{2n-2} + \sum_{j=0}^{2n-3} a_{n,j} \, \mu_j \qquad\qquad\qquad\qquad\qquad\qquad (4.11)
\end{aligned}
$$

for some coefficients $a_{n,0}, \ldots, a_{n,2n-3}$, and

$$
\begin{aligned}
0 = \langle f_n, x^{n-1} \rangle &= \langle f_{n-1}, x^n \rangle + \beta_n \langle f_{n-1}, x^{n-1} \rangle - \gamma_n \langle f_{n-2}, x^{n-1} \rangle \\
&= \mu_{2n-1} + \sum_{j=0}^{2n-2} b_{n,j} \, \mu_j, \qquad\qquad\qquad\qquad\qquad\qquad (4.12)
\end{aligned}
$$

for some $b_{n,0}, \ldots, b_{n,2n-2}$. Now (4.11) defines μ_{2n-2} uniquely in terms of its predecessors and then (4.12) does the same for μ_{2n-1}. To summarize, this process defines the moments up to the choice of the normalization $\mu_0 > 0$:

$$\mu_1 = -\beta_1 \mu_0$$
$$\mu_2 = -a_{2,0}\,\mu_0 - a_{2,1}\,\mu_1$$
$$\mu_3 = -\beta_2 \mu_2 - b_{2,0}\,\mu_0 - b_{2,1}\,\mu_1$$

up to

$$\mu_{2n-2} = -\sum_{j=0}^{2n-3} a_{n,j}\,\mu_j$$

$$\mu_{2n-1} = -\sum_{J=0}^{2n-2} b_{n,j}\,\mu_j.$$

It remains to show that the inner product is definite, that is that $\langle f_n, f_n \rangle > 0$ for $n \in \mathbb{N}_0$ which we will prove once again by induction on n, where the easy case $n = 0$ is the assumption $\mu_0 > 0$. Next we consider

$$\langle f_n, f_n \rangle = \langle f_n, x f_{n-1} \rangle = \langle f_n, x^n \rangle = \mu_{2n} + \langle \widetilde{f}_n, x^n \rangle \tag{4.13}$$

and replacing n in (4.10) by $n + 1$, we can use

$$\mu_{2n} + \langle \widetilde{f}_n, x^n \rangle = \gamma_{n+1} \langle f_{n-2}, x^{n-2} \rangle,$$

together with the induction hypothesis to obtain

$$\langle f_n, f_n \rangle = \langle f_n, x^n \rangle = \mu_{2n} + \langle \widetilde{f}_n, x^n \rangle = \gamma_n \langle f_{n-1}, x^{n-1} \rangle = \gamma_n \langle f_{n-1}, f_{n-1} \rangle > 0; \tag{4.14}$$

hence the symmetric bilinear form is positive and therefore an inner product. $\square$

Remark 4.7 A closer inspection of (4.14) even yields an explicit formula for $\langle f_n, f_n \rangle$, namely,

$$\langle f_n, f_n \rangle = \gamma_n \langle f_{n-1}, f_{n-1} \rangle = \gamma_n \gamma_{n-1} \langle f_{n-2}, f_{n-2} \rangle = \cdots = \left(\prod_{j=1}^{n} \gamma_j \right) \langle f_0, f_0 \rangle$$

$$= \mu_0 \prod_{j=1}^{n} \gamma_j.$$

Therefore if we divide (4.8) by γ_n and set $f_0 = \mu_0^{-1}$, we get a recurrence that produces an *orthonormal polynomial* on each level.

The recurrence (4.8) for orthogonal polynomials is not unique. As long as we multiply every f_n by a nonzero constant, $\widetilde{f}_n := c_n f_n$, we still obtain a sequence of orthogonal polynomials with

$$\left\langle \tilde{f}_j, \tilde{f}_k \right\rangle = c_j c_k \langle f_j, f_k \rangle = c_j c_k \delta_{jk} \langle f_j, f_j \rangle, \qquad j, k \in \mathbb{N}_0.$$

The modified recurrence then takes the form

$$\tilde{f}_n = c_n f_n = c_n (\alpha_n x + \beta_n) f_{n-1} - \gamma_n f_{n-2}$$
$$= \frac{c_n}{c_{n-1}} (\alpha_n x + \beta_n) \tilde{f}_{n-1} - \frac{1}{c_{n-2}} \gamma_n \tilde{f}_{n-2},$$

yielding the modified coefficients

$$\tilde{\alpha}_n = \frac{c_n}{c_{n-1}} \alpha_n, \qquad \tilde{\beta}_n = \frac{c_n}{c_{n-1}} \beta_n, \qquad \tilde{\gamma}_n = \frac{1}{c_{n-2}} \gamma_n. \qquad (4.15)$$

Remark 4.8 There are two special types of orthogonal polynomials that can be recognized from the coefficients in the three-term recurrence:

1. *monic orthogonal polynomials*: $\alpha_n = 1$ and $f_0 = 1$,
2. *orthonormal polynomials*: $\gamma_n = 1$ and $f_0 = \mu_0^{-1/2}$, see Remark 4.7.

This way we can always get orthogonal polynomials as convergents of continued fractions. And there is not even something to prove any more; we just have to compare the respective three-term recurrences. The following corollary even parameterizes all possible orthogonal polynomials in terms of certain infinite continued fractions; the nth orthogonal polynomial is then the denominator of the nth convergent.

Corollary 4.1 (Orthogonal polynomials in terms of continued fractions) *The orthogonal polynomials with parameters $\alpha_n, \beta_n, \gamma_n$ in the recurrence (4.8) are obtained as a denominator of the convergents of the continued fractions*

$$\frac{-\gamma_1 |}{|(\alpha_1 x + \beta_1)} - \frac{\gamma_2 |}{|(\alpha_2 x + \beta_2)} - \frac{\gamma_3 |}{|(\alpha_3 x + \beta_3)} + \cdots$$

or

$$\left[0; -\frac{\alpha_1 x + \beta_1}{\gamma_1}, -\frac{\alpha_2 x + \beta_2}{\gamma_2}, \cdots \right],$$

respectively. Conversely, the denominators of all continued fractions of the form

$$[0; -\alpha_1 x + \beta_1, -\alpha_2 x + \beta_2, \dots], \qquad \alpha_j > 0, \ \beta \in \mathbb{R},$$

are a system of orthogonal polynomials for an appropriate inner product $\langle \cdot, \cdot \rangle$.

Remark 4.9 Orthogonal polynomials can also be defined in several variables, but the geometric and algebraic issues are significantly more intricate [19]. Recurrence relations can be defined, but are based on matrices of an increasing block size [113]; by far not all properties that we will list here can be recovered. In addition, the study of multivariate moment problems is also quite recent [94], and involves intricate positivity issues of multivariate polynomials. Since polynomials in several variables are *not* a Euclidean ring, we cannot construct multivariate continued fractions to speak of with the approaches here, and therefore we will not touch the issue any further in this book.

We found out that any sequence of orthogonal polynomials for a strictly square positive linear functional can be written as denominators of convergents of an infinite continued fraction. But what does this continued fraction mean or represent? In other words, what is the analogy with the real number represented by an infinite continued fraction with positive integer coefficients? To answer these questions, we will introduce the concept of the *Laurent series* which is usually more popular in Complex Analysis [50, 99, 100].

Definition 4.4 *(Laurent series and convergence)*

1. The *Laurent series* $\lambda(x)$ associated to a sequence $(\lambda_j : j \in \mathbb{N}_0)$ is defined as

$$\lambda(x) = \sum_{j=0}^{\infty} \lambda_j x^{-j}. \tag{4.16}$$

2. A sequence $\lambda_n(x)$, $n \in \mathbb{N}$, whose elements are Laurent series is said to be *convergent* to a Laurent series $\lambda^*(x)$, if for any $k \in \mathbb{N}_0$, there exists $n_0 \in \mathbb{N}$ such that for all $n \geq n_0$ there exists λ_n such that

$$\lambda_n(x) - \lambda^*(x) = x^{-k} \tilde{\lambda}_n(x), \qquad \text{i.e.,} \qquad \lambda_{n,j} = \lambda_j^*, \quad j = 0, \ldots, k-1. \tag{4.17}$$

Remark 4.10 Note that Definition 4.4 deals with a *formal* Laurent series or, equivalently the associated infinite sequences. We are not even interested in the radius of convergence in (4.16), since (4.17) is a purely formal comparison of coefficients in the sequence of Laurent series which could as well be formulated entirely in the context of sequences. The advantage of using Laurent series compared to relying only on sequences will become evident soon when we will start to multiply them.

To make it clear: *convergence* in Definition 4.4 means that after a certain index, the first k terms of any Laurent series in the sequence coincide with the first k terms of the limit, and that occurs for any $k \in \mathbb{N}_0$. But we still do not care about whether λ^* or some λ_n are analytic functions.

A first, very simple but surprisingly fundamental observation is that any reciprocal of a polynomial can be expanded into a power series with a lot of zero initial coefficients.

Lemma 4.1 *For $p \in \Pi_n$ with $p_n \neq 0$, one has*

$$\frac{1}{p(x)} = \sum_{j=n}^{\infty} \lambda_j \, x^{-j} =: \lambda(x).$$

Proof We write $p(x) = p_0 + p_1 x + \cdots + p_n x^n$, $n = \deg p$, and set $1/p(x) = \lambda(x)$, which yields

$$1 = p(x)\,\lambda(x) = \left(\sum_{j=0}^{n} p_j x^j\right)\left(\sum_{k=0}^{\infty} \lambda_j x^{-k}\right) = \sum_{j=0}^{n}\sum_{k=0}^{\infty} p_j \lambda_k x^{j-k}$$

$$= \sum_{j=-\infty}^{n} x^j \sum_{k-\ell=j} p_k \lambda_\ell = \sum_{j=-\infty}^{n} x^j \sum_{\ell=-j}^{n-j} p_{j+\ell}\lambda_\ell,$$

where $\lambda_{-n} = \cdots = \lambda_{-1} = 0$. A comparison of the coefficients gives

$$\sum_{k=-j}^{n-j} p_{j+k}\lambda_k = \delta_{j,0} = \begin{cases} 0, & j \neq 0, \\ 1, & j = 0, \end{cases}$$

in particular

$$0 = p_n \lambda_0$$
$$0 = p_{n-1} \lambda_0 + p_n \lambda_1$$
$$\vdots$$
$$0 = p_1 \lambda_0 + \cdots + p_n \lambda_{n-1},$$

which we can write in matrix form and make use of $p_n \neq 0$ to see that

$$0 = \begin{bmatrix} p_n & & \\ \vdots & \ddots & \\ p_1 & \cdots & p_n \end{bmatrix}\begin{bmatrix} \lambda_0 \\ \vdots \\ \lambda_{n-1} \end{bmatrix} \quad\Rightarrow\quad \lambda_0 = \cdots = \lambda_{n-1} = 0.$$

The other coefficients are obtained by successively solving the systems

$$\begin{bmatrix} 1 \\ 0 \\ \vdots \end{bmatrix} = \begin{bmatrix} p_n & & & \\ \vdots & \ddots & & \\ p_0 & \cdots & p_n & \\ & \ddots & & \ddots \end{bmatrix}\begin{bmatrix} \lambda_n \\ \lambda_{n+1} \\ \vdots \end{bmatrix},$$

that determine $\lambda_n, \lambda_{n+1}, \ldots$ *uniquely.* $\qquad\square$

There is a shorthand version of Lemma 4.1 that can be phrased as

$$\frac{1}{p(x)} = O\left(x^{-\deg p}\right), \qquad p \in \Pi. \tag{4.18}$$

Now we get our polynomial-rational analogue of real numbers.

Definition 4.5 An infinite continued fraction $[0; a_1, a_2, \dots]$, $a_j \in \Pi \setminus \Pi_0$ is called *convergent*, if there exists a *Laurent series* $\lambda(x)$ such that

$$\lim_{n \to \infty} [0; a_1, \dots, a_n] = \lim_{n \to \infty} \frac{p_n(x)}{q_n(x)} = \lambda(x)$$

in the sense of Definition 4.4.

Remark 4.11 *(Convergence of continued fractions)*

1. Definition 4.5 still lives entirely in the context of *formal* Laurent series.
2. Definition 4.5 makes sense. Since $p_0 = 0$ and $p_1 = 1$, it follows that $\deg q_n > \deg p_n$, and thus by Lemma 4.1,

$$\frac{p_n(x)}{q_n(x)} = p_n(x) \sum_{j=\deg q_n}^{\infty} \lambda_j x^{-j} = \sum_{j=\deg q_n - \deg p_n}^{\infty} \tilde{\lambda}_j x^{-j},$$

   hence any convergent of the continued fraction can be represented as a Laurent series.
3. One could also expand the rational functions with respect to positive powers of x which would give the *Taylor series*. However one would then need a slightly different notion of continued fractions, see [79].
4. We can illustrate the idea behind the convergence of continued fractions of polynomials by recalling how the objects are generated: we expand a finite segment into a rational function, then transfer that into a Laurent series and consider the limit of this sequence of Laurent series in the sense of Definition 4.4:

$$[0; a_1, \dots] \leftarrow [0; a_1, \dots, a_n] = \frac{p_n}{q_n} = \lambda_n \to \lambda, \qquad n \to \infty.$$

Indeed there are plenty of convergent continued fractions in the sense of Definition 4.5, in particular those that we already know from three-term recurrences with at least *linear* components.

Theorem 4.3 *Any continued fraction of the form* $[0; r_1, \dots]$, $r_j \in \Pi$, $\deg r_j \geq 1$, $j \in \mathbb{N}$, *converges to a Laurent series* $\lambda(x)$ *in such a way that*

$$\lambda(x) - \frac{p_n(x)}{q_n(x)} = O\left(x^{-d_{n+1} - d_n}\right), \tag{4.19}$$

that is

$$\frac{p_n(x)}{q_n(x)} = \lambda_0 + \cdots + \lambda_{d_{n+1}+d_n-1} \, x^{-d_{n+1}-d_n+1} + \cdots , \qquad (4.20)$$

where $d_n := \deg q_n$, $n \in \mathbb{N}_0$.

Proof In the formal Laurent series

$$\lambda(x) - \frac{p_n(x)}{q_n(x)} = \sum_{j=n}^{\infty} \left(\frac{p_{j+1}(x)}{q_{j+1}(x)} - \frac{p_j(x)}{q_j(x)} \right) = \sum_{j=n}^{\infty} \frac{(-1)^j}{q_{j+1}(x)\, q_j(x)} = \sum_{j=d_{n+1}+d_n}^{\infty} \gamma_j \, x^{-j}$$

$$=: \gamma(x)$$

all coefficients γ_j are well defined since any γ_j depends only on finitely many values q_k. Then convergence follows since

$$\frac{p_{n+k}(x)}{q_{n+k}(x)} - \frac{p_n(x)}{q_n(x)} = O\left(x^{-d_n-d_{n+1}}\right), \qquad k \in \mathbb{N},$$

and thus we have an analogy to a Cauchy sequence. This carries over to the limit series $\lambda(x)$ and gives (4.19). $\qquad \square$

Returning to orthogonal polynomials, this particularly implies that continued fractions with *affine coefficients* of degree ≤ 1 always converge, and that the order of convergence can be given in a simple way.

Corollary 4.2 *Any continued fraction of the form* $[0; r_1, \ldots], r_j \in \Pi_1 \setminus \Pi_0, j \in \mathbb{N},$ *converges to a Laurent series* $\lambda(x)$ *in such a way that*

$$\lambda(x) - \frac{p_n(x)}{q_n(x)} = O\left(x^{-2n-1}\right). \qquad (4.21)$$

These continued fractions converge rapidly in the sense that the number of coefficients captured is twice the degree of the denominator and thus the continued fractions fit particularly well with the Laurent series λ, and as a result we should have a closer look at them. The theory could even be developed in a more general framework of continued fractions with $r_j \in \Pi \setminus \Pi_0$, but we will restrict ourselves to continued fractions with factors of degree 1, i.e., $r_j(x) = \alpha_j x + \beta_j$, $\alpha_j \neq 0$, for which we have $\deg q_n = \deg p_n + 1 = n$. Moreover, the good representations of that type for a given Laurent series get a special name.

Definition 4.6 The infinite (simple) continued fraction

$$[0; r_1, \ldots] = \frac{1|}{|r_1} + \frac{1|}{|r_2} + \cdots = \sum_{k=1}^{\infty} \frac{1|}{|r_k}, \qquad r_j \in \Pi_1 \setminus \Pi_0$$

is called *associated* to the *Laurent series* $\lambda(x)$ if

$$\lambda(x) - \frac{p_n(x)}{q_n(x)} = O\left(x^{-2n-1}\right), \qquad n \in \mathbb{N},$$

that is

$$\frac{p_n(x)}{q_n(x)} = \sum_{j=0}^{2n} \lambda_j \, x^{-j} + \sum_{j=2n+1}^{\infty} \gamma_{n,j} \, x^{-j}, \qquad n \in \mathbb{N}. \tag{4.22}$$

It would be too optimistic to assume that *all* Laurent series have associated continued fractions,[2] but it will actually turn out that a description of the Laurent series for which there exists an associated continued fraction is even more interesting and will involve the concept of a *Hankel matrix* which we already know from Eq. (4.4) in Definition 4.2. The following result is the fundamental theorem on associated continued fractions for a Laurent series. It is an improvement over Bernoulli's theorem in the sense that it characterizes the Laurent series that have a continued fraction expansion whose coefficients are of degree *exactly* one.

Theorem 4.4 *A Laurent series* $\lambda(x)$ *has an* associated *continued fraction* $[0; r_1, \dots]$, $r_j(x) := \alpha_j x + \beta_j \in \Pi_1 \setminus \Pi_0$, *if and only if* $\lambda_0 = 0$ *and*

$$\det \Lambda_n \neq 0, \qquad \Lambda_n = \begin{bmatrix} \lambda_1 & \lambda_2 & \cdots & & \lambda_n \\ \lambda_2 & \lambda_3 & \ddots & & \vdots \\ \vdots & \ddots & \ddots & & \lambda_{2n-2} \\ \lambda_n & \cdots & \lambda_{2n-2} & \lambda_{2n-1} \end{bmatrix}, \qquad n \in \mathbb{N}. \tag{4.23}$$

Proof The continued fraction is associated to $\lambda(x)$ if and only if for any $n \in \mathbb{N}$, we have

$$\begin{aligned} \frac{p_n(x)}{q_n(x)} &= \lambda_0 + \cdots + \lambda_{2n} x^{-2n} + \gamma_{n,2n+1} x^{-2n-1} + \gamma_{n,2n+2} x^{-2n-2} + \cdots, \\ \frac{p_{n+1}(x)}{q_{n+1}(x)} &= \lambda_0 + \cdots + \lambda_{2n} x^{-2n} + \lambda_{2n+1} x^{-2n-1} + \lambda_{2n+2} x^{-2n-2} + \cdots. \end{aligned} \tag{4.24}$$

Subtracting the second equation in (4.24) from the first one and keeping in mind that this equals $(-1)^{n+1} / (q_{n+1} q_n)$, we find that

$$\begin{aligned} \frac{(-1)^{n+1}}{q_{n+1}(x) \, q_n(x)} &= \frac{p_n(x)}{q_n(x)} - \frac{p_{n+1}(x)}{q_{n+1}(x)} \\ &= \left(\gamma_{n,2n+1} - \lambda_{2n+1}\right) x^{-2n-1} + \left(\gamma_{n,2n+2} - \lambda_{2n+2}\right) x^{-2n-2} + \cdots. \end{aligned}$$

Next we note that the recursion formula (3.11) for the denominators of the convergents of the simple continued fraction takes the form

---

[2] So finally here is a difference with real numbers and their continued fraction expansions.

$$q_n(x) = r_n(x)\, q_{n-1}(x) + q_{n-2}(x) = (\alpha_n x + \beta_n)\, q_{n-1}(x) + q_{n-2}(x), \qquad n \in \mathbb{N},$$
$$(4.25)$$

from which we can conclude by induction that

$$q_n(x) = \left(\prod_{j=1}^{n} \alpha_j\right)\left(x^n + x^{n-1}\sum_{j=1}^{n}\frac{\beta_j}{\alpha_j}\right) + r(x), \qquad r \in \Pi_{n-2}. \qquad (4.26)$$

Indeed we have $q_0 = 1$, $q_1 = \alpha_1 x + \beta_1$, and, in general the induction hypothesis, tracking only the two terms of highest degree in q_{n+1} yields that we have the following:

$$q_n(x) = (\alpha_n x + \beta_n)\left(\prod_{j=1}^{n-1}\alpha_j\right)\left(x^{n-1} + x^{n-2}\sum_{j=1}^{n-1}\frac{\beta_j}{\alpha_j} + \cdots\right) + q_{n-2}(x)$$

$$= \left(\prod_{j=1}^{n}\alpha_j\right)\left(x^n + x^{n-1}\sum_{j=1}^{n-1}\frac{\beta_j}{\alpha_j}\right) + \beta_n\left(\prod_{j=1}^{n-1}\alpha_j\right)x^{n-1} + r(x)$$

$$= \left(\prod_{j=1}^{n}\alpha_j\right)\left(x^n + x^{n-1}\sum_{j=1}^{n-1}\frac{\beta_j}{\alpha_j} + \frac{\beta_n}{\alpha_n}x^{n-1}\right) + r(x)$$

$$= \left(\prod_{j=1}^{n}\alpha_j\right)\left(x^n + x^{n-1}\sum_{j=1}^{n}\frac{\beta_j}{\alpha_j}\right) + r(x),$$

for some $r \in \Pi_{n-2}$ which proves (4.26). Therefore

$$q_{n+1}(x)\, q_n(x)$$
$$= \alpha_{n+1}\left(\prod_{j=1}^{n}\alpha_j\right)^2 x^{2n+1} + \left(\prod_{j=1}^{n}\alpha_j\right)^2\left(\beta_{n+1} + 2\alpha_{n+1}\sum_{j=1}^{n}\frac{\beta_j}{\alpha_j}\right)x^{2n} + \cdots. \quad (4.27)$$

By Lemma 4.1, we have that

$$\frac{1}{q_{n+1}(x)\, q_n(x)} = \alpha_{n+1}^{-1}\left(\prod_{j=1}^{n}\alpha_j\right)^{-2} x^{-2n-1} + \cdots;$$

hence comparing coefficients implies

$$(-1)^{n+1}\left(\gamma_{n,2n+1} - \lambda_{2n+1}\right) = \alpha_{n+1}^{-1}\left(\prod_{j=1}^{n}\alpha_j\right)^{-2},$$

from which we conclude the following two equivalent identities

$$\alpha_{n+1} = \frac{(-1)^{n+1}}{\left(\gamma_{n,2n+1} - \lambda_{2n+1}\right)\left(\alpha_1 \cdots \alpha_n\right)^2}, \qquad \gamma_{n,2n+1} - \lambda_{2n+1} = \frac{(-1)^{n+1}}{\alpha_{n+1}\left(\alpha_1 \cdots \alpha_n\right)^2}.$$
(4.28)

Let us summarize what we obtained so far: The existence of an associated continued fraction with $r_j \in \Pi_1 \setminus \Pi_0$ is equivalent to the validity of (4.28) with all $\alpha_j \neq 0$, which in turn is equivalent to $\gamma_{n,2n+1} \neq \lambda_{2n+1}$.

To see what this means, we multiply the first line of (4.24) by $q_n(x)$ which[3] leads to

$$
\begin{aligned}
p_n(x) &= \left(\sum_{j=0}^{2n} \lambda_j \, x^{-j} + \sum_{j=2n+1}^{\infty} \gamma_{n,j} \, x^{-j}\right)\left(\sum_{k=0}^{n} q_{n,k} \, x^k\right) \\
&= \sum_{j=0}^{2n}\sum_{k=0}^{n} \lambda_j \, q_{n,k} \, x^{k-j} + \sum_{j=2n+1}^{\infty}\sum_{k=0}^{n} \gamma_{n,j} \, q_{n,k} \, x^{k-j} \\
&= \sum_{k=0}^{n}\sum_{j=k-2n}^{k} \lambda_{k-j} \, q_{n,k} x^j + \gamma_{n,2n+1} \, q_{n,n} x^{-n-1} + O\left(x^{-n-2}\right) \\
&= \sum_{j=-n}^{n} x^j \sum_{k=j}^{j+2n} \lambda_{k-j} \, q_{n,k} + \gamma_{n,2n+1} \, q_{n,n} x^{-n-1} + O\left(x^{-n-2}\right);
\end{aligned}
$$

hence

$$p_n(x) = \sum_{j=0}^{n} \eta_{-j} x^j + \sum_{j=1}^{n} \eta_j x^{-j} + \eta_{n+1} x^{-n-1} + O\left(x^{-n-2}\right), \qquad (4.29)$$

where

$$
\begin{bmatrix} p_{n,n} \\ \vdots \\ p_{n,0} \end{bmatrix} = \begin{bmatrix} \eta_{-n} \\ \vdots \\ \eta_0 \end{bmatrix} = \begin{bmatrix} \lambda_0 & & \\ \vdots & \ddots & \\ \lambda_n & \cdots & \lambda_0 \end{bmatrix} \begin{bmatrix} q_{n,n} \\ \vdots \\ q_{n,0} \end{bmatrix}
\qquad (4.30)
$$

and

$$
\begin{bmatrix} \eta_1 \\ \vdots \\ \eta_n \\ \eta_{n+1} \end{bmatrix} = \begin{bmatrix} \lambda_{n+1} & \lambda_n & \cdots & \lambda_1 \\ \vdots & \vdots & \ddots & \vdots \\ \lambda_{2n} & \lambda_{2n-1} & \cdots & \lambda_n \\ \gamma_{n,2n+1} & \lambda_{2n} & \cdots & \lambda_{n+1} \end{bmatrix} \begin{bmatrix} q_{n,n} \\ \vdots \\ q_{n,0} \end{bmatrix}.
\qquad (4.31)
$$

---

[3] Using the *convention* that $0 = \lambda_j = p_k$, $j, k < 0$ or $k > n$, respectively.

Since the left-hand side of (4.29) is a polynomial, a comparison of coefficients yields that $\eta_1 = \cdots = \eta_{n+1} = 0$; then since $q \neq 0$, the determinant of the matrix in (4.31) is 0. From (4.28) we now determine that

$$\gamma_{n,2n+1} = \lambda_{2n+1} + \frac{(-1)^{n+1}}{\alpha_{n+1} (\alpha_1 \cdots \alpha_n)^2}$$

and substitute this in the matrix from (4.31) with its columns rearranged in reverse order which gives

$$0 = \det \begin{bmatrix} \lambda_1 & \cdots & \lambda_n & \lambda_{n+1} \\ \vdots & \ddots & \vdots & \vdots \\ \lambda_n & \cdots & \lambda_{2n-1} & \lambda_{2n} \\ \lambda_{n+1} & \cdots & \lambda_{2n} & \gamma_{n,2n+1} \end{bmatrix}$$

$$= \det \begin{bmatrix} \lambda_1 & \cdots & \lambda_n & \lambda_{n+1} \\ \vdots & \ddots & \vdots & \vdots \\ \lambda_n & \cdots & \lambda_{2n-1} & \lambda_{2n} \\ \lambda_{n+1} & \cdots & \lambda_{2n} & \lambda_{2n+1} \end{bmatrix} + \left(\alpha_{n+1} \prod_{j=1}^{n} \alpha_j^2 \right)^{-1} \det \begin{bmatrix} \lambda_1 & \cdots & \lambda_n & 0 \\ \vdots & \ddots & \vdots & \vdots \\ \lambda_n & \cdots & \lambda_{2n-1} & 0 \\ \lambda_{n+1} & \cdots & \lambda_{2n} & (-1)^{n+1} \end{bmatrix}$$

$$= \det \Lambda_{n+1} + (-1)^{n+1} \left(\alpha_{n+1} \prod_{j=1}^{n} \alpha_j^2 \right)^{-1} \det \Lambda_n,$$

that is,

$$\det \Lambda_{n+1} = (-1)^n \frac{\det \Lambda_n}{\alpha_{n+1} (\alpha_1 \cdots \alpha_n)^2}, \tag{4.32}$$

and

$$\alpha_{n+1} = (-1)^n \left(\prod_{j=1}^{n} \alpha_j^2 \right)^{-1} \frac{\det \Lambda_n}{\det \Lambda_{n+1}}. \tag{4.33}$$

Now we can draw the final conclusions. If the continued fraction has the coefficients $r_j(x) = \alpha_j(x) + \beta_j$ in $\Pi_1 \setminus \Pi_0$, and is associated to a Laurent series $\lambda(x)$, then (4.32) yields inductively on n that $\det \Lambda_n \neq 0$. This follows immediately from (4.32) and the assumption that the r_j are of the degree exactly 1 and therefore $\alpha_j \neq 0$.

For the converse we first verify the requirement for λ_0: observing in (4.30) that $\deg p_n = n - 1$, then $0 = \eta_{-n} = \lambda_0 q_{n,n}$; furthermore $\deg q_n = n$ already implies that $q_{n,n} \neq 0$ and therefore $\lambda_0 = 0$. To prove the existence of the recursion, we use the fact that (4.33) gives an explicit recursive formula for the coefficients α_n, and the right-hand side shows that $\alpha_n \neq 0$ and hence all components $r_j = \alpha_j x + \beta_j$ are non-constant polynomials as long as the determinant condition is valid. □

Remark 4.12 For $\gamma_n = -1$, the recursion for the convergents of the associated continued fraction is that for the *orthonormal* polynomial sequence, at least up to a multiple of μ_0, see Remark 4.8.

In the proof of Theorem 4.4 there is no need to explicitly consider the values β_n, since the only relevant point for the existence of an associated continued fraction was that $\alpha_n \neq 0$ and this was verified. For the sake of completeness and since we need them later on, we will also derive those coefficients using the same methods as we used for computing the α_n. Indeed the coefficients β_j are determined by looking at the second nonzero term θ_{2n+2} in the Laurent expansion of $1/(q_{n+1}q_n)$, which is according to (4.27) and Lemma 4.1 obtained by solving the system

$$\begin{bmatrix} 1 \\ 0 \end{bmatrix} = \begin{bmatrix} p_{2n+1} & \\ p_{2n} & p_{2n+1} \end{bmatrix} \begin{bmatrix} \theta_{2n+1} \\ \theta_{2n+2} \end{bmatrix},$$

$$p_{2n+1} = \alpha_{n+1} (\alpha_1 \cdots \alpha_n)^2,$$

$$p_{2n} = (\alpha_1 \cdots \alpha_n)^2 \left(\beta_{n+1} + 2\alpha_{n+1} \sum_{j=1}^{n} \frac{\beta_j}{\alpha_j} \right) ;$$

hence $\theta_{2n+1} = 1/p_{2n+1}$ and

$$\theta_{2n+2} = -\frac{p_{2n}}{p_{2n+1}} \theta_{2n+1} = -\frac{p_{2n}}{p_{2n+1}^2} = - \left(\prod_{j=1}^{n+1} \alpha_j \right)^{-2} \left(\beta_{n+1} + 2\alpha_{n+1} \sum_{j=1}^{n} \frac{\beta_j}{\alpha_j} \right).$$

As before a comparison of coefficients yields

$$(-1)^{n+1} \left(\gamma_{n,2n+2} - \lambda_{2n+2} \right) = - \left(\prod_{j=1}^{n+1} \alpha_j \right)^{-2} \left(\beta_{n+1} + 2\alpha_{n+1} \sum_{j=1}^{n} \frac{\beta_j}{\alpha_j} \right) \qquad (4.34)$$

or

$$\gamma_{n,2n+2} = \lambda_{2n+2} + (-1)^n \left(\prod_{j=1}^{n+1} \alpha_j \right)^{-2} \left(\beta_{n+1} + 2\alpha_{n+1} \sum_{j=1}^{n} \frac{\beta_j}{\alpha_j} \right).$$

Proposition 4.1 *The coefficients β_n in the recurrence formula of the associated continued fraction for the Laurent series $\lambda(x)$ of Theorem 4.4 are uniquely defined.*

Proof The proof of Theorem 4.4 has already shown that the α_n are well defined and nonzero. Then (4.34) gives a way to recursively determine the coefficients: suppose $\beta_1, \ldots, \beta_n$ have already been computed, then p_n and q_n are known, hence also $\gamma_{n,2n+2}$, and (4.34) yields that

$$\beta_{n+1} = (-1)^n \left(\prod_{j=1}^{n+1} \alpha_j \right)^2 \left(\gamma_{n,2n+2} - \lambda_{2n+2} \right) - 2\alpha_{n+1} \sum_{j=1}^{n} \frac{\beta_j}{\alpha_j},$$

which uniquely defines β_{n+1}.   □

For an explicit computation of the β_j, we extend the above idea and make use of the fact that also

$$
0 = \begin{bmatrix} \eta_1 \\ \vdots \\ \eta_n \\ \eta_{n+2} \end{bmatrix} = \begin{bmatrix} \lambda_1 & \lambda_2 & \cdots & & \lambda_{n+1} \\ \lambda_2 & \lambda_3 & \ddots & & \vdots \\ \vdots & \ddots & \ddots & & \lambda_{2n-1} \\ \lambda_n & \cdots & \lambda_{2n-1} & & \lambda_{2n} \\ \lambda_{n+2} & \cdots & \gamma_{n,2n+1} & & \gamma_{n,2n+2} \end{bmatrix} \begin{bmatrix} q_{n,0} \\ \vdots \\ q_{n,n} \end{bmatrix},
$$

which yields with the same arguments as above that

$$
0 = \det \begin{bmatrix} \lambda_1 & \lambda_2 & \cdots & & \lambda_{n+1} \\ \lambda_2 & \lambda_3 & \ddots & & \vdots \\ \vdots & \ddots & \ddots & & \lambda_{2n-1} \\ \lambda_n & \cdots & \lambda_{2n-1} & & \lambda_{2n} \\ \lambda_{n+2} & \cdots & \gamma_{2n+1} & & \gamma_{n,2n+2} \end{bmatrix}.
$$

Expanding the determinant gives

$$
0 = \left(\prod_{j=1}^{n+1} \alpha_j \right)^{-2} \left(\beta_{n+1} + 2\alpha_{n+1} \sum_{j=1}^{n} \frac{\beta_j}{\alpha_j} \right) \det \begin{bmatrix} \lambda_1 & \cdots & \lambda_n & \lambda_{n+1} \\ \vdots & \ddots & \vdots & \vdots \\ \lambda_n & \cdots & \lambda_{2n-1} & \lambda_{2n} \\ 0 & \cdots & 0 & (-1)^n \end{bmatrix}
$$

$$
+ \det \begin{bmatrix} \lambda_1 & \cdots & \lambda_n & \lambda_{n+1} \\ \vdots & \ddots & \vdots & \vdots \\ \lambda_n & \cdots & \lambda_{2n-1} & \lambda_{2n} \\ \lambda_{n+2} & \cdots & \gamma_{n,2n+1} & \lambda_{2n+2} \end{bmatrix}
$$

as a first step. Observing that

$$
\det \begin{bmatrix} \lambda_1 & \cdots & \lambda_n & \lambda_{n+1} \\ \vdots & \ddots & \vdots & \vdots \\ \lambda_n & \cdots & \lambda_{2n-1} & \lambda_{2n} \\ \lambda_{n+2} & \cdots & \gamma_{n,2n+1} & \lambda_{2n+2} \end{bmatrix}
$$

$$
= \frac{1}{\alpha_{n+1}} \left(\prod_{j=1}^{n} \alpha_j \right)^{-2} \det \begin{bmatrix} \lambda_1 & \cdots & \lambda_n & \lambda_{n+1} \\ \vdots & \ddots & \vdots & \vdots \\ \lambda_n & \cdots & \lambda_{2n-1} & \lambda_{2n} \\ 0 & \cdots & (-1)^n & 0 \end{bmatrix} - \det \begin{bmatrix} \lambda_1 & \cdots & \lambda_n & \lambda_{n+1} & 0 \\ \vdots & \ddots & \vdots & \vdots & \vdots \\ \lambda_n & \cdots & \lambda_{2n-1} & \lambda_{2n} & 0 \\ \lambda_{n+1} & \cdots & \lambda_{2n} & \lambda_{2n+1} & 1 \\ \lambda_{n+2} & \cdots & \lambda_{2n+1} & \lambda_{2n+2} & 0 \end{bmatrix}
$$

and defining the matrices

$$\Lambda'_n := \begin{bmatrix} \lambda_1 & \cdots & \lambda_{n-1} & 0 \\ \vdots & \ddots & \vdots & \vdots \\ \lambda_{n-2} & \cdots & \lambda_{2n-4} & 0 \\ \lambda_{n-1} & \cdots & \lambda_{2n-3} & 1 \\ \lambda_n & \cdots & \lambda_{2n-2} & 0 \end{bmatrix} \in \mathbb{R}^{n \times n}, \qquad n \geq 2,$$

we can conclude that

$$(-1)^n \left(\prod_{j=1}^{n+1} \alpha_j \right)^2 \det \Lambda'_{n+2} = \left(\beta_{n+1} + 2\alpha_{n+1} \sum_{j=1}^{n} \frac{\beta_j}{\alpha_j} \right) \det \Lambda_n - \alpha_{n+1} \det \Lambda'_{n+1}. \tag{4.35}$$

Using (4.33), we can write

$$(-1)^n \left(\prod_{j=1}^{n+1} \alpha_j \right)^2 = \alpha_{n+1}^2 \frac{1}{\alpha_{n+1}} \frac{\det \Lambda_n}{\det \Lambda_{n+1}} = \alpha_{n+1} \frac{\det \Lambda_n}{\det \Lambda_{n+1}},$$

and substitute this in the left-hand side of (4.35) to obtain

$$\alpha_{n+1} \frac{\det \Lambda_n \det \Lambda'_{n+2}}{\det \Lambda_{n+1}} = \left(\beta_{n+1} + 2\alpha_{n+1} \sum_{j=1}^{n} \frac{\beta_j}{\alpha_j} \right) \det \Lambda_n - \alpha_{n+1} \det \Lambda'_{n+1}. \tag{4.36}$$

Now we summarize these findings by showing that the continued fraction expansion can be expressed entirely in terms of the determinants of the matrices Λ_n or submatrices thereof.

Corollary 4.3 *The recurrence coefficients for the convergents and therefore the coefficients $r_n(x) = \alpha_n x + \beta_n$, $n = 1, 2, \ldots$ of the continued fraction expansion are given recursively for $n \geq 1$ as*

$$\alpha_{n+1} = (-1)^n \left(\prod_{j=1}^{n} \alpha_j \right)^{-2} \frac{\det \Lambda_n}{\det \Lambda_{n+1}} = -\frac{1}{\alpha_n} \frac{(\det \Lambda_n)^2}{\det \Lambda_{n+1} \det \Lambda_{n-1}}, \tag{4.37}$$

$$\beta_{n+1} = \alpha_{n+1} \left(\frac{\det \Lambda'_{n+2}}{\det \Lambda_{n+1}} + \frac{\det \Lambda'_{n+1}}{\det \Lambda_n} - 2 \sum_{j=1}^{n} \frac{\beta_j}{\alpha_j} \right), \tag{4.38}$$

$$\gamma_{n+1} = -1, \tag{4.39}$$

initialized with

$$\alpha_1 = \frac{1}{\lambda_1}, \qquad \beta_1 = -\frac{\lambda_2}{\lambda_1^2}. \tag{4.40}$$

Proof For the computations of α_1 and β_1, we again use the idea of Lemma 4.1 and the resulting requirement

$$\begin{bmatrix} 1 \\ 0 \end{bmatrix} = \begin{bmatrix} \alpha_1 \\ \beta_1 \ \alpha_1 \end{bmatrix} \begin{bmatrix} \lambda_1 \\ \lambda_2 \end{bmatrix} \quad \Rightarrow \quad \alpha_1 = \frac{1}{\lambda_1}, \quad \beta_1 = -\alpha_1 \frac{\lambda_2}{\lambda_1} = -\frac{\lambda_2}{\lambda_1^2},$$

which gives (4.40). To verify the second identity in (4.37), we consider

$$\frac{\alpha_{n+1}}{\alpha_n} = (-1)^n \left(\prod_{j=1}^{n} \alpha_j \right)^{-2} \frac{\det \Lambda_n}{\det \Lambda_{n+1}} (-1)^{n-1} \left(\prod_{j=1}^{n-1} \alpha_j \right)^{2} \frac{\det \Lambda_n}{\det \Lambda_{n-1}}$$

$$= -\frac{1}{\alpha_n^2} \frac{(\det \Lambda_n)^2}{\det \Lambda_{n+1} \ \det \Lambda_{n-1}},$$

from which (4.37) follows by multiplication with α_n. Then (4.38) is a straightforward rearrangement of (4.36). $\qquad\square$

The application of Theorem 4.4 to the quadrature problem and the construction of Gauss' formula is established in the next result that now really connects orthogonal polynomials and continued fractions—and also provides Gauss' implicit definition of orthogonal polynomials.

Theorem 4.5 *Let μ be the* moment *sequence for a square positive linear functional. Then the orthogonal polynomials for this functional are the denominators q_n, $n \in \mathbb{N}$, of the convergents of the continued fraction associated to the* Laurent *series*

$$\mu(x) = \sum_{j=1}^{\infty} \mu_{j-1} x^{-j}.$$

Proof The matrices $\Lambda_n = M_{n-1}$, $n \in \mathbb{N}$, are *strictly* positive definite and therefore all have positive determinants. We now have an associated continued fraction. Due to (4.31) and the comparison of coefficients in (4.29), we moreover have that

$$0 = \begin{bmatrix} \lambda_1 & \cdots & \lambda_n & \lambda_{n+1} \\ \vdots & \ddots & \vdots & \vdots \\ \lambda_n & \cdots & \lambda_{2n-1} & \lambda_{2n} \end{bmatrix} \begin{bmatrix} q_{n,0} \\ \vdots \\ q_{n,n} \end{bmatrix} = \begin{bmatrix} \mu_0 & \cdots & \mu_{n-1} & \mu_n \\ \vdots & \ddots & \vdots & \vdots \\ \mu_{n-1} & \cdots & \mu_{2n-2} & \mu_{2n-1} \end{bmatrix} \begin{bmatrix} q_{n,0} \\ \vdots \\ q_{n,n} \end{bmatrix}$$

$$= \begin{bmatrix} \langle 1, q \rangle \\ \vdots \\ \langle (\cdot)^{n-1}, q \rangle \end{bmatrix},$$

which implies orthogonality of the polynomials. $\qquad\square$

Remark 4.13 The result of Theorem 4.5 seems to contradict (4.37), since in Theorem 4.2 we characterized orthogonal polynomials by a recurrence relation with *positive* recurrence coefficients, while the formula in (4.37) obviously leads to coefficients with an *alternating* sign. But the orthogonal polynomials with $\alpha_j \gamma_j > 0$ are only *one* possible choice; for example the sequence $\tilde{q}_n := (-1)^n q_n$ satisfies

$$
\begin{aligned}
\tilde{q}_{n+1}(x) &= (-1)^{n+1} q_n(x) = (-1)^{n+1} \left((\alpha_n x + \beta_n) q_n(x) - \gamma_n q_{n-1}(x) \right) \\
&= (-1)^{n+1} \left((\alpha_n x + \beta_n) (-1)^n \tilde{q}_n(x) - \gamma_n (-1)^{n-1} \tilde{q}_{n-1}(x) \right) \\
&= (-\alpha_n x - \beta_n) \tilde{q}_n(x) - \gamma_n (-1)^{n-1} \tilde{q}_{n-1}(x),
\end{aligned}
$$

so that $\tilde{q}_n$ has the recurrence coefficients $(-\alpha_n, -\beta_n, \gamma_n)$. The sequence generated by the rules (4.37) and (4.38) is, by the way, $(-1)^{\lfloor n/2 \rfloor} q_n$.

Remark 4.14 There is yet another way to obtain the parameters in the recurrence relation which is by determining α_n via (4.32), and then β_n using the vector[4] $q_n = [q_{n,j} : j = 0, \ldots, n]$ of coefficients for q_n together with the identity

$$
\begin{aligned}
0 = \langle q_{n+1}, q_n \rangle &= \alpha_{n+1} \langle (\cdot) q_n, q_n \rangle + \beta_{n+1} \langle q_n, q_n \rangle + \langle q_{n-1}, q_n \rangle \\
&= \alpha_{n+1} q_n^T \begin{bmatrix} \mu_1 & \cdots & \mu_{n+1} \\ \vdots & \ddots & \vdots \\ \mu_{n+1} & \cdots & \mu_{2n+1} \end{bmatrix} q_n + \beta_{n+1} q_n^T M_n q_n,
\end{aligned}
$$

which gives

$$
\beta_{n+1} = -\alpha_{n+1} \frac{q_n^T \widetilde{M}_n q_n}{q_n^T M_n q_n}, \qquad \widetilde{M}_n = \begin{bmatrix} \mu_1 & \cdots & \mu_{n+1} \\ \vdots & \ddots & \vdots \\ \mu_{n+1} & \cdots & \mu_{2n+1} \end{bmatrix}. \tag{4.41}
$$

The "disadvantage" of this method in contrast to the direct recursions of Corollary 4.3 lies in the fact that it also needs to compute the coefficients of the orthogonal polynomials q_n and to evaluate the bilinear forms $x \mapsto x^T \widetilde{M}_n x$ and $x \mapsto x^T M_n x$ at $x = q_n$.

4.1.1 Problems

4.1 Show that any square positive functional defines an inner product.

4.2 Show that the functional L from (4.5) is square positive.

4.3 Show that any orthogonal polynomial $f_n \in \Pi_n$ in (4.7) must satisfy $\deg f_n = n$.

---

[4] Using the same symbol for the polynomial and the vector of its coefficients is quite reasonable. After all it reflects the way how polynomials are usually represented on a computer: as the vector of their coefficients.

4.4 *(Nonnegative polynomials)* Let $p \in \mathbb{R}[x]$ be a polynomial. Show that $p \geq 0$, i.e., $p(x) \geq 0$, $x \in \mathbb{R}$, if and only if there exist $q_1, q_2 \in \mathbb{R}[x]$ such that

$$p = q_1^2 + q_2^2.$$

Hint: Use the (complex) factorization and the fact that the product of a sum of squares is a sum of squares again which, however, needs a proof of its own.

4.5 Use Problem 4.4 to show that a function is square positive if and only if it is positive.

Remark 4.15 The equivalence of Problem 4.5 only holds in one variable; in several variables this is no longer true which complicates things significantly, cf. [94].

4.6 *(Singular Hankel matrix)* Show that

$$\det \begin{pmatrix} 1 & \cdots & x^n \\ \vdots & \ddots & \vdots \\ x^n & \cdots & x^{2n} \end{pmatrix} = 0, \qquad x \in \mathbb{R}, \quad n \in \mathbb{N}.$$

4.7 Do the neighborhoods

$$T_n(\lambda) = \left\{ \lambda' : \lambda - \lambda' = O\left(x^{-n-1}\right) \right\}, \qquad n \in \mathbb{N},$$

define a *topology* on

$$\Lambda_- := \left\{ \lambda(x) = \sum_{j=0}^{\infty} \lambda_j x^{-j} : \lambda_j \in \mathbb{R} \right\}?$$

Are they open or closed?

4.8 *(Orthogonal polynomials)* Let

$$M_n = QR$$

be the *QR factorization* of M_n where $Q \in \mathbb{R}^{(n+1) \times (n+1)}$ is an *orthogonal matrix*, i.e., $Q^T Q = I$ and

$$R = \begin{bmatrix} * & \cdots & * \\ & \ddots & \vdots \\ & & * \end{bmatrix}$$

is an *upper triangular matrix*. Show that the last column of Q contains the coefficients of the orthonormal polynomial of degree n.

4.9 Let f_n be a sequence of orthonormal polynomials of the form

$$f_n(x) = a_n x^n + b_n x^{n-1} + \cdots.$$

Show that the coefficients in the three-term recurrence (4.8) are obtained as

$$\alpha_n = \frac{a_n}{a_{n-1}},$$

$$\beta_n = \alpha_n \left(\frac{b_n}{a_n} - \frac{b_{n-1}}{a_{n-1}} \right),$$

$$\gamma_n = \frac{a_n \, a_{n-2}}{\lambda_{n-1}^2}.$$

4.2 Gauss Quadrature

What does all that have to do with Gauss? It is really getting time to answer this question. Of course the connection is that continued fractions were the key in the original method to determine the so-called *Gaussian quadrature* formula in [32].

Definition 4.7 *(Quadrature formula)* A *quadrature formula* of order n with *weights* ω_j and *knots* x_j, $j = 0, \ldots, n$ is an expression of the form

$$Q_n(f) = \sum_{j=0}^{n} \omega_j \, f(x_j). \tag{4.42}$$

It is called *real* if all parameters ω_j, x_j, $j = 0, \ldots, n$, are real numbers.

Quadrature is a classical topic and already dates back to Newton and his predecessors like Wallis, so it is even older than the formal notion of an integral or analysis. The intuition of a quadrature formula is to approximate a square positive linear functional L, especially an integral of the form

$$L(f) = \int_a^b f(x) \, w(x) \, dx, \qquad w \geq 0, \tag{4.43}$$

as accurately as possible. The topic is covered in all textbooks on Numerical Analysis, see in particular [33].

Usually the quality of a quadrature formula is measured in terms of *exactness*, which is defined as an exact integration of a subspace of all polynomials up to a certain degree.

Definition 4.8 A quadrature formula Q_n is said to have *exactness k* with respect to a functional L if

$$0 = L(f) - Q_n(f) = L(f) - \sum_{j=0}^{n} \omega_j f(x_j), \qquad f \in \Pi_k. \tag{4.44}$$

In the sequel we will again assume that L is a square positive linear functional.

Remark 4.16 Intuitively it is not necessarily clear why the exact integration of a high degree polynomial is a desirable quantity, since it gives no direct information about the error of this approximate computation. Recovering this well known connection is the content of Problem 4.12.

The maximal possible exactness of a quadrature formula Q_n of order n as defined in (4.42) is $2n + 1$. *Maximal* means that there is no quadrature formula of order n with $n + 1$ weights and knots that is exact on Π_{2n+2}, at least if the functional is square positive (on the polynomials), i.e., $L(f^2) > 0$, $f \in \Pi \setminus \{0\}$. This is easily seen by considering the polynomial

$$f(x) = (x - x_0)^2 \cdots (x - x_n)^2 \in \Pi_{2n+2},$$

which is a square, $f = g^2$, and therefore satisfies

$$L(f) = L(g^2) > 0 = \sum_{j=0}^{n} \omega_j f(x_j) = Q_n(f),$$

so that (4.44) fails for this f.

Now given points $x_0, \ldots, x_n$ or a polynomial $w(x) = (x - x_0) \cdots (x - x_n)$, the weights ω_j, $j = 0, \ldots$, are determined in Gauss' approach as

$$\omega_j = L(\ell_j), \qquad \ell_j = \prod_{k \neq j} \frac{\cdot - x_k}{x_j - x_k} = \frac{w}{w'(x_j)(\cdot - x_j)}, \qquad j = 0, \ldots, n. \tag{4.45}$$

This is what is called an *interpolatory quadrature formula*. Indeed given any f, defined at least at the locations $x_0, \ldots, x_n$, the polynomial

$$p = \sum_{j=0}^{n} f(x_j) \ell_j$$

interpolates f at $x_0, \ldots, x_n$, i.e., $p(x_j) = f(x_j)$, $j = 0, \ldots, n$, and the linearity of the functional yields that

$$L(p) = L\left(\sum_{j=0}^{n} f(x_j) \ell_j \right) = \sum_{j=0}^{n} f(x_j) L(\ell_j) = \sum_{j=0}^{n} f(x_j) \omega_j = Q_n(f);$$

therefore the quadrature formula is obtained by integrating the interpolant once the knots are known. Interpolatory quadrature will be discussed in the problem section; right now it suffices to say that this simply was the natural and obvious way for Gauss to construct a quadrature formula.

Writing $w(x) = w_0 + w_1 x + \cdots + w_n x^n + x^{n+1}$, we get that[5]

$$
\begin{aligned}
w'(x_j) \, \ell_j(x) &= \frac{w(x)}{x - x_j} = \frac{w(x) - w(x_j)}{x - x_j} \\
&= \frac{w_1 (x - x_j) + \cdots + w_n \left(x^n - x_j^n\right) + \left(x^{n+1} - x_j^{n+1}\right)}{x - x_j} \\
&= \sum_{k=1}^{n+1} w_k \frac{x^k - x_j^k}{x - x_j} = \sum_{k=1}^{n+1} w_k \sum_{m=0}^{k-1} x^m \, x_j^{k-1-m},
\end{aligned}
$$

or in Gauss' original style of presentation,

$$
\begin{aligned}
w'(x_j) \, \ell_j(x) = \; & x^n + x_j \, x^{n-1} + \; x_j^2 \, x^{n-2} + \cdots + \quad x_j^n \\
& + w_n \, x^{n-1} + w_n \, x_j \, x^{n-2} + \cdots + \; w_n \, x_j^{n-1} \\
& + w_{n-1} \, x^{n-2} + \cdots + w_{n-1} \, x_j^{n-2} \\
& \qquad\qquad\qquad \ddots \qquad\qquad \vdots \\
& \qquad\qquad\qquad\qquad\qquad + \quad w_1 \\
= \; & \sum_{k=0}^{n} x^k \frac{w(x_j)}{x_j^{k+1}} + O\left(x_j^{-1}\right);
\end{aligned}
$$

hence

$$
\begin{aligned}
w_j'(x_j) \, L(\ell_j) &= \mu_n + \mu_{n-1}(x_j + w_n) + \cdots + \mu_0 \left(x_j^n + w_n \, x_j^{n-1} + \cdots + w_1\right) \\
&= \sum_{k=0}^{n} \mu_k \left(x_j^{n-k} + \sum_{m=k+1}^{n} w_k \, x_j^{n-k}\right) =: \widetilde{w}(x_j), \qquad \widetilde{w} \in \Pi_n,
\end{aligned}
$$

which yields

$$
\omega_j = \frac{\widetilde{w}(x_j)}{w'(x_j)}, \qquad j = 0, \ldots, n. \tag{4.46}
$$

This formula allows us to determine the *weights* of the *quadrature formula directly* from the moments once we fix the knots, and we now have the polynomial w and its coefficients.

---

[5] What follows now is following the original paper by Gauss as closely as possible; only the notation is slightly modernized.

Now let $\lambda_k = Q_n\left((\cdot)^k\right)$ denote the moments of the quadrature formula; write $\theta_k = \mu_k - \lambda_k$ for the moments of the error $E = L - Q_n$ and let $\lambda(x)$ and $\theta(x)$ be the associated *Laurent series*. Using the (formal) identity

$$\frac{1}{x - \xi} = \sum_{j=1}^{\infty} \frac{\xi^{j-1}}{x^j},\tag{4.47}$$

we note that

$$\lambda(x) = \sum_{k=1}^{\infty} \frac{Q_n\left((\cdot)^{k-1}\right)}{x^k} = \sum_{k=1}^{\infty} x^{-k} \sum_{j=0}^{n} \omega_j x_j^{k-1} = \sum_{j=0}^{n} \omega_j \sum_{k=1}^{\infty} x_j^{k-1} x^{-k} = \sum_{j=0}^{n} \frac{\omega_j}{x - x_j},$$

from which we conclude that

$$\theta(x) = \mu(x) - \lambda(x) = \mu(x) - \sum_{j=0}^{n} \frac{\omega_j}{x - x_j}\tag{4.48}$$

has to hold. By construction, the quadrature formula is *interpolatory*, yielding $\theta_0 = \cdots = \theta_n = 0$ and therefore

$$O\left(x^{-1}\right) = w(x)\,\theta(x) = w(x)\,\mu(x) - \sum_{j=0}^{n} \omega_j \frac{w(x)}{x - x_j}$$

$$= w(x)\,\mu(x) - \underbrace{\sum_{j=0}^{n} \omega_j\, w'\left(x_j\right)\,\ell_j(x)}_{\in \Pi_n}.$$

Now we are at the point where Gauss uses the magic of continued fractions in [32]: if we choose specifically $w(x) = q_{n+1}(x)$ as the denominator of the $(n+1)$-convergent of $\mu(x)$ which exists by Theorem 4.5, then

$$\mu(x) = \frac{p_{n+1}(x)}{q_{n+1}(x)} + O\left(x^{-2n-3}\right) \quad \Rightarrow \quad q_{n+1}(x)\,\mu(x) = p_{n+1}(x) + O\left(x^{-n-2}\right),$$

and therefore

$$O\left(x^{-1}\right) = w(x)\,\theta(x) = \underbrace{p_{n+1}(x) - \sum_{j=0}^{n} \omega_j\, w'\left(x_j\right)\,\ell_j(x)}_{=:p(x)} + O\left(x^{-n-2}\right);$$

hence the polynomial p must satisfy $p = 0$ which yields

$$w(x)\,\theta(x) = O\left(x^{-n-2}\right) \quad \Rightarrow \quad \theta(x) = \frac{O\left(x^{-n-2}\right)}{w(x)} = O\left(x^{-2n-3}\right),$$

and consequently,

$$0 = \theta_0 = \cdots = \theta_{2n+1}. \tag{4.49}$$

In other words the quadrature formula whose knots are the zeros of q_{n+1} provides the desired *exactness*.

Remark 4.17 This way of determining the quadrature knots makes **no** use of any sort of integral; it is purely algebraic, and only applies formal manipulations to the formal Laurent series associated to the moment sequence.

In our enthusiasm about this really beautiful construction,[6] we forgot one important point: q_{n+1} has to have *real* and *simple* zeros; otherwise the whole approach makes no sense. We actually made the implicit assumption of real and simple zeros several times. But fortunately the zeros have the desired properties which is ensured by the following proposition that again relates directly to continued fractions, and more precisely to their associated recurrence.

Proposition 4.2 *If a sequence* f_n, $n \in \mathbb{N}$, *of polynomials satisfies a recurrence as in* (4.8), *then each* f_n *has simple and real zeros.*

In a standard numerical analysis lecture, at least in the way it is usually presented nowadays, one would first refer to Theorem 4.2 and then rely on the well-known fact that orthogonal polynomials have only real and simple zeros, cf. [33]. This proof, however, usually relies on an integral representation of the functional which is not available when we start only with a moment sequence. This can be somewhat compensated by using the fact that any *positive polynomial*, i.e., any polynomial $p \neq 0$ with $p(x) \geq 0$, $x \in \mathbb{R}$, can be decomposed into a sum of squares, thereby relating positive and square positive functions.

On the other hand the proof of Proposition 4.2 to be given here will follow a more direct, algebraic and still elementary approach based on Sturm chains which will be introduced next in Sect. 4.3.

4.2.1 Problems

4.10 Prove formula (4.45), which is to show that for $j = 0, \ldots, n$, the polynomial

$$\ell_j = \prod_{k \neq j} \frac{\cdot - x_k}{x_j - x_k}, \qquad j = 0, \ldots, n, \tag{4.50}$$

can be written as

---

[6] After all it is due to Gauss, so what else should we expect?

$$\ell_j = \frac{w}{w'(x_j)\,(\cdot - x_j)}, \qquad w = \prod_{j=0}^{n} (\cdot - x_j).$$

4.11 Prove that the polynomials ℓ_j from (4.50) are *fundamental polynomials* for the Lagrange interpolation of degree n, i.e., the polynomial

$$p = \sum_{j=0}^{n} f(x_j)\,\ell_j$$

1. belongs to Π_n
2. and interpolates f at the *sites* x_j,

$$f(x_j) = p(x_j), \qquad j = 0, \dots, n.$$

4.12 *(Quadrature error)* Prove the following: if Q_n is a quadrature formula of order n and exactness $k \geq 0$ with *nonnegative* weights for the square positive functional L of the form (4.43), then

$$|L(f) - Q_n(f)| \leq 2\mu_0 \frac{(b-a)^{k+1}}{(k+1)!} \max_{x \in [a,b]} \left| f^{(m+1)}(x) \right|$$

holds for any $f \in C^{m+1}[a, b]$.

Hint: Apply the well-known error formula for Lagrange interpolation that ensures for $f \in C^{n+1}[a, b]$ and any $x \in [a, b]$ that there exists $\xi \in [a, b]$ such that

$$f(x) - p(x) = \frac{f^{(n+1)}(\xi)}{(n+1)!} \prod_{j=0}^{n} (\xi - x_j),$$

where p stands for the interpolation polynomial at $x_0, \dots, x_n$.

4.13 Prove (4.47).

4.14 *(Golub and Welsch [37])* Show that if $v_0, \dots, v_n$ are the eigenvectors of M_n, then the weights of the Gauss quadrature rule can be computed as

$$w_j = \mu_0 \frac{v_{j1}^2}{v_j^T v_j}, \qquad v_j = \begin{bmatrix} v_{j1} \\ \vdots \\ v_{jn} \end{bmatrix}, \qquad j = 0, \dots, n.$$

4.2.1.1 Interpolatory Quadrature

The Gauss quadrature formula from (4.45) has a special property that deserves a name of its own.

Definition 4.9 *(Interpolatory quadrature formula)* A quadrature formula Q_n for a functional L is called *interpolatory* if

$$\omega_j = L\left((\cdot)^j\right), \qquad j = 0, \ldots, n. \tag{4.51}$$

4.15 Show that any *interpolatory* quadrature formula of order n has exactness n.

4.16 Show that

1. any quadrature formula of order n and (almost maximal) exactness $2n$ satisfies $\omega_j > 0, j = 0, \ldots, n$.
2. any quadrature formula of order n and exactness $\geq n$ has to be interpolatory.

4.2.1.2 Interpolation

Finally we have to look at an *algebraic* approach to interpolation; this entails considering it only as a way to interpolate polynomials by polynomials using the already known concept of *Euclidean division*.

4.17 *(Algebraic interpolation)* Given $x_0 < \cdots < x_n \in \mathbb{R}$, define the polynomial

$$\omega = \prod_{j=0}^{n} (\cdot - x_j).$$

Show that the remainder r of the division

$$f = q\,\omega + r, \qquad \deg r < \deg \omega = n + 1, \tag{4.52}$$

is the unique degree reducing interpolant[7] to f at the points $x_0, \ldots, x_n$.

4.18 *(Algebraic interpolation II)*
  To continue Problem 4.17, now let the sites x_j be arbitrary and denote by $m_j = \#\{1 \leq k \leq n : x_k = x_j\}$ the *multiplicity* of x_j. Show that the remainder r in (4.52) solves the *Hermite interpolation problem*

$$r^{(k)}(x_j) = f^{(k)}(x_j), \qquad k = 0, \ldots, m_j - 1, \qquad j = 0, \ldots, n. \tag{4.53}$$

4.19 Given $x_j \in \mathbb{C}$ and $\mu_j \in \mathbb{N}$, $j = 0, \ldots, m$, set $n = \mu_0 + \cdots + \mu_m - 1$. Show that the matrix

---

[7] The interpolant p to a polynomial f is called *degree reducing* if $\deg p \leq \deg f$. Degree reducing interpolation operators are a fundamental issue in several variables, cf. [8, 86].

$$V := \begin{bmatrix} 1 & 0 & \cdots & 0 & \cdots & 1 & 0 & \cdots & 0 \\ x_0 & 1 & \cdots & 0 & \cdots & x_m & 1 & \cdots & 0 \\ x_0^2 & 2x_0 & \cdots & 0 & \cdots & x_m^2 & 2x_m & \cdots & 0 \\ \vdots & \vdots & \ddots & \vdots & & \vdots & \vdots & \ddots & \vdots \\ x_0^{\mu_0} & \mu_0 x_0^{\mu_0-1} & \cdots & \mu_0! & \cdots & x_m^{\mu_m} & \mu_m x_0^{\mu_m-1} & \cdots & \mu_m! \\ \vdots & & \ddots & \vdots & & \vdots & & \ddots & \vdots \\ x_0^n & n x_0^{n-1} & \cdots & \frac{n!}{(n-\mu_0)!} x_0^{n-\mu_0} & \cdots & x_m^n & n x_m^{n-1} & \cdots & \frac{n!}{(n-\mu_m)!} x_m^{n-\mu_m} \end{bmatrix}$$

$$(4.54)$$

is nonsingular.

Definition 4.10 *(Vandermonde matrix for the Hermite interpolation problem)* The matrix

$$V = \left[\frac{d^\ell(\cdot)^j}{dx^\ell}(x_j) : \begin{array}{l} j = 0, \ldots, n \\ \ell = 0, \ldots, \mu_k - 1, \ k = 0, \ldots, m, \end{array} \right] \qquad (4.55)$$

in (4.54) is called the *Vandermonde matrix* for the *Hermite interpolation problem* at the x_j with *multiplicity* μ_j.

4.3 Sturm Chains

Sturm chains give a method to *count* the zeros or sign changes in a polynomial within a given interval without having to *determine* them. This is done by counting the sign changes of a certain sequence of numbers which makes them a useful and fairly popular tool in the numerics for univariate polynomials; as a result they can be found in various places in the literature. Here we follow the terminology and notation from [28].

Definition 4.11 A finite sequence $f_0, \ldots, f_n \in \Pi$ of polynomials is called a *Sturm chain* or *Sturm sequence* for an open, closed, bounded or unbounded interval I, if

1. at each *zero* of f_k, the polynomials f_{k+1} and f_{k-1} have an opposite *sign* on I:

$$f_k(x) = 0 \quad \Rightarrow \quad f_{k-1}(x) f_{k+1}(x) < 0, \qquad k = 1, \ldots, n-1, \ x \in I. \quad (4.56)$$

2. the polynomial f_0 has no zero in I, i.e., $0 \notin f(I)$.

Remark 4.18 The second condition in Definition 4.11 means that the continuous function f_0 has to be either strictly positive or strictly negative on I. Since in addition $f_0, \ldots, f_n$ is a Sturm sequence if and only if $-f_0, \ldots, -f_n$ is a Sturm sequence, we could replace this requirement by $f_0(I) > 0$ without an essential loss of generality.

What this concept has to do with zeros becomes clear, if for some $x \in \mathbb{R}$ we count in $V(x)$ how often a true or *proper sign change* occurs in the vector

$(f_0(x), \ldots, f_n(x))$; a *proper sign change* means that zero values in the vector are ignored or erased from the vector so that we only count strict sign changes from $+$ to $-$ or from $-$ to $+$.

Example 4.1 *(Sign changes)* The sequence $+, -, +$ has two sign changes, any sequence $+, 0, \ldots, 0, -$ has one, and the sequence $+, 0, +$ has no sign changes.

Then we let x vary and consider $V(x)$ as a function in x. As long as $f_j(x) \neq 0$, $j = 0, \ldots, n$, the value of V is constant on $[x - \varepsilon, x + \varepsilon]$ for a sufficiently small $\varepsilon > 0$, again due to the continuity of polynomials. If, however, f_k, $1 < k < n$, has a zero at x, i.e., $f_k(x) = 0$, then because of (4.56), either f_{k+1} or f_{k-1} has the same sign as f_k restricted to the half-open interval $[x - \varepsilon, x)$ left of x, and the same also holds true for the half-open interval $(x, x + \varepsilon]$ right of x. But this means that $V(x)$ remains unchanged:

$$V(x - \varepsilon) = V(x + \varepsilon) = V(x) = V(y), \qquad y \in [x - \varepsilon, x + \varepsilon].$$

In other words: $V(x)$ changes only if f_n changes its sign relative to f_{n-1}. If f_{n-1} and f_n have a joint sign change at x, then V is again constant on $[x - \varepsilon, x + \varepsilon]$; otherwise the number of sign changes increases or decreases depending on whether f_{n-1} and f_n had the same or opposite sign at $x - \varepsilon$, respectively. This is depicted in the following table:

	$x - \varepsilon$	x	$x + \varepsilon$			$x - \varepsilon$	x	$x + \varepsilon$
f_n	$\pm$	0	$\mp$	or	f_n	$\pm$	0	$\mp$
f_{n-1}	$\pm$	$\pm$	$\pm$		f_{n-1}	$\mp$	$\mp$	$\mp$
V	k	k	$k+1$		V	k	$k-1$	$k-1$

Remark 4.19 Keep in mind that this process only counts locations where f has a *proper* sign change; zeros without a sign change, for example double zeros are not counted and do not play a role.

In the end we track this behavior along the full interval I, and taking into account that changes become active on the right of the zero, we get the following result.

Theorem 4.6 (Zero counting) *For a Sturm sequence* $f_0, \ldots, f_n$, *define*[8]

$$\sigma_+(f, I) := \#Z_+(f, I) := \#\{x \in I \;:\; f(x - \varepsilon) > f(x) = 0 > f(x + \varepsilon)\},$$

and

$$\sigma_-(f, I) := \#Z_-(f, I) := \#\{x \in I \;:\; f(x - \varepsilon) < f(x) = 0 < f(x + \varepsilon)\}.$$

Then we get for $I = [a, b)$ *that*

---

[8] The notation is slightly sloppy, but here $x - \varepsilon$ always includes "for all sufficiently small $\varepsilon > 0$".

$$\sigma_+\left(\frac{f_n}{f_{n-1}}, I\right) - \sigma_-\left(\frac{f_n}{f_{n-1}}, I\right) = V(b) - V(a). \tag{4.57}$$

Proof Set $f := f_n/f_{n-1}$. If $f(a) = 0$, then $V(a + \varepsilon) = V(a) \pm 1$ depending on whether a belongs to Z_+ or to Z_-. Then $V(x)$ is piecewise constant and increases by 1 on Z_+ and decreases by 1 on Z_-. Eventually

$$V(b) = V(a) + \sigma_+\left(\frac{f_n}{f_{n-1}}, I\right) - \sigma_-\left(\frac{f_n}{f_{n-1}}, I\right),$$

from which (4.57) follows immediately.   □

This is already all we need in order to show that polynomials which obey a three-term recurrence always have simple real zeros.

Proposition 4.3 *For any polynomial sequence f_n, $n \in \mathbb{N}_0$, defined by a three-term recurrence relation of the form*

$$f_0 = 1, \qquad f_{n+1}(x) = (x + \beta_n)\, f_n(x) - \gamma_n\, f_{n-1}(x), \qquad \gamma_n > 0, \qquad n \in \mathbb{N}_0, \tag{4.58}$$

the following holds:

1. *Each finite sequence $f_0, \dots, f_n$ is a* Sturm chain *for any interval $I \subseteq \mathbb{R}$.*
2. *The polynomial f_n has exactly n simple real zeros, that is*

$$\#Z_{\mathbb{R}}\,(f_n) = n, \qquad Z_I(f) = \{x \in I \;:\; f(x) = 0\}.$$

Hence

$$f_n(x) = \prod_{j=1}^{n} \left(x - \xi_j\right), \qquad \xi_1 < \cdots < \xi_n. \tag{4.59}$$

Remark 4.20 According to Theorem 4.2, the recurrence relations from (4.58) are *exactly* the recurrence for *monic* orthogonal polynomials with respect to a square positive linear functional, and it shows that those have simple and real zeros. The proof, however, is purely algebraic and does not use any underlying functionals or measures. But on the other hand, the standard proof with orthogonal polynomials for an integral also shows as a side effect that the zeros of the orthogonal polynomials are always located *inside* the interval of integration.

Remark 4.21 Usually elements of a three-term recurrence of the form (4.58) are not the standard examples for a Sturm chain, but one uses the sequence generated by running the Euclidean algorithm to compute the gcd of f and f' for a polynomial f. We will encounter this fact in Lemma 7.4.

Proof *(of Proposition 4.3)* That $f_0 = 1$ has no zeros is obvious. If for some $n \in \mathbb{N}$, the point x is such that $f_n(x) = 0$, then the recurrence (4.58) yields

$$f_{n+1}(x) = -\gamma_n \, f_{n-1}(x);$$

then f_{n+1} and f_{n-1} either have the opposite sign at x, i.e., $f_{n+1}(x) \, f_{n-1}(x) < 0$, or it could happen that $f_{n+1}(x) = f_n(x) = f_{n-1}(x) = 0$. In the latter case we would also have that

$$f_{n-2}(x) = \frac{f_n(x) - (x + \beta_{n-1}) \, f_{n-1}(x)}{\gamma_{n-1}} = 0;$$

then repeating the argument, we eventually conclude that $0 = f_{n-3}(x) = \cdots = f_0(x)$, contradicting $f_0 = 1$. Therefore $f_{n+1}(x) \, f_{n-1}(x) < 0$, and since n was arbitrary, any finite sequence $f_0, \ldots, f_n$ generated by (4.58) is a Sturm chain.

This allows us to apply Theorem 4.6. Since σ_+ and σ_- only capture the real zeros of f and count them only once, we have for $I = [a, b)$, $a < b \in \mathbb{R}$, that

$$\left| \sigma_+ \left(\frac{f_n}{f_{n-1}}, I \right) - \sigma_- \left(\frac{f_n}{f_{n-1}}, I \right) \right| \le \sigma_+ \left(\frac{f_n}{f_{n-1}}, I \right) + \sigma_- \left(\frac{f_n}{f_{n-1}}, I \right) \le \#Z_{\mathbb{R}}(f_n) \le n. \tag{4.60}$$

Since all the polynomials are *monic*, i.e., $f_k(x) = x^k + \cdots$, it follows that

$$\lim_{x \to -\infty} f_k(x) = (-1)^k \infty, \qquad \lim_{x \to +\infty} f_k(x) = \infty;$$

hence,

$$\lim_{x \to -\infty} \operatorname{sgn} \begin{bmatrix} f_n(x) \\ f_{n-1}(x) \\ \vdots \\ f_0(x) \end{bmatrix} = (-1)^n \begin{bmatrix} 1 \\ -1 \\ \vdots \\ (-1)^n \end{bmatrix}, \qquad \lim_{x \to +\infty} \operatorname{sgn} \begin{bmatrix} f_n(x) \\ f_{n-1}(x) \\ \vdots \\ f_0(x) \end{bmatrix} = \begin{bmatrix} 1 \\ 1 \\ \vdots \\ 1 \end{bmatrix},$$

and we can conclude that

$$\lim_{a \to -\infty} V(a) = n, \qquad \lim_{b \to +\infty} V(b) = 0.$$

As a result, for sufficiently small a and sufficiently large b,

$$n = |V(b) - V(a)| = \left| \sigma_+ \left(\frac{f_n}{f_{n-1}}, I \right) - \sigma_- \left(\frac{f_n}{f_{n-1}}, I \right) \right|.$$

Substituting this into (4.60), we get that

$$n \le \#Z_{\mathbb{R}}(f_n) \le n \quad \Rightarrow \quad \#Z_{\mathbb{R}}(f_n) = n,$$

as claimed. $\square$

Actually the proof tells us even more. Since σ_+ and σ_- are nonnegative numbers, the identity

$$-n = V(b) - V(a) = \sigma_+ \left(\frac{f_n}{f_{n-1}}, \mathbb{R} \right) - \sigma_- \left(\frac{f_n}{f_{n-1}}, \mathbb{R} \right)$$

can only be obtained if

$$\sigma_+ \left(\frac{f_n}{f_{n-1}}, \mathbb{R} \right) = 0 \quad \text{and} \quad \sigma_- \left(\frac{f_n}{f_{n-1}}, \mathbb{R} \right) = n.$$

Hence *all* sign changes of f_n/f_{n-1} are sign changes from $-$ to $+$. But this can only be obtained if f_{n-1} changes its sign between two sign changes of f_n. With this insight we can summarize the findings of this section in the following way.

Theorem 4.7 *If a polynomial sequence f_n, $n \in \mathbb{N}_0$, is defined by the recurrence (4.58), then f_n has n simple real zeros, $n \in \mathbb{N}$, and the zeros of f_n and f_{n-1} are nested.*

This is a well-known property of orthogonal polynomials, cf. [18, 33], but we now know that actually it is a property of polynomials that satisfy a three-term recurrence relation with certain signs of the coefficients; hence we now also have a property of the convergents of certain continued fractions, namely those with affine polynomials in them that have positive leading coefficients. That these continued fractions happen to produce orthogonal polynomials, probably not by accident, is again the result stated in Theorem 4.2.

4.3.1 Problems

4.20 Let $f \in \Pi$ be a polynomial. Show that $g := f/\gcd(f, f')$ has the same zeros as f but all zeros of g are simple.

4.21 Given $f \in \Pi$, set $r_0 = f, r_1 = f'$, and compute the sequence

$$r_{n-1} = s\, r_n + r_{n+1}, \qquad n \geq 1, \tag{4.61}$$

of the Euclidean algorithm, ending with $r_m = \gcd(f, f')$. Show that the functions

$$f_j := \frac{r_{m+j}}{r_m}, \qquad j = 0, \dots, n - m,$$

are polynomials and form a Sturm chain.

4.22 For $f(x) = f_0 + f_1 x + \cdots + f_n x^n$, give a recurrence relation for the coefficients of the polynomials in (4.61).

4.23 Let $f_n \in \Pi_n$ be an orthogonal polynomial with respect to $\int_a^b w\,dx$ with $w > 0$. Let $\xi_1, \ldots, \xi_m$ be the zeros of f_n in $[a, b]$. Consider the integral

$$\int_a^b f(x)\,(x - \xi_1) \cdots (x - \xi_m)\,w(x)\,dx,$$

and use that to conclude that $m = n$. In other words: *all zeros of orthogonal polynomials lie inside the area of integration.*

4.4 Computing the Zeros of Polynomials

A fundamental point left open by Gauss in [32] as well as by Prony in his 1795 paper [81] is *how* the zeros of the polynomials should really be *computed*. We will briefly discuss that issue in this chapter.

The "classical" method for finding a zero of a polynomial is probably *Newton's method* which entailed starting with an initial guess x_0, and then goes to the iteration

$$x_n = x_{n-1} - \frac{f(x_{n-1})}{f'(x_{n-1})}, \qquad n \geq 1. \tag{4.62}$$

As is well known, Newton's method shares the problem of most iterative fixpoint methods: *local convergence*. This means that the initial value x_0 must be chosen sufficiently close to the zero in order for the iteration to converge to the solution. Information about that and the speed of convergence can be found in any textbook on Numerical Analysis, see for example [33, 54].

With the proper starting value however, Newton's method can be easily shown to have convergence guaranteed, where the limit will be the "extremal" zeros of the polynomial. The following classical result from Numerical Analysis is not as well known as it should be which is the reason why we include it here.

Proposition 4.4 *Let*

$$f = \prod_{j=1}^n (\cdot - \zeta_j), \qquad \zeta_1 \leq \cdots \leq \zeta_n,$$

be a polynomial with real zeros. Then the Newton iteration (4.62) *converges to ζ_0 for any $x_0 \leq \zeta_0$ and to ζ_n for any $x_0 \geq \zeta_n$.*

Proof By Rolle's theorem, all zeros of the derivative f' lie between ζ_1 and ζ_n, hence also all of f'', and so on. Since f has no further zeros, it follows that either $f(x) > 0$ for $x > \zeta_n$ or $f(x) < 0$ for $x > \zeta_n$. In the first case $f(x) > 0$, choose some $x > \zeta_n$. By Taylor's formula, we then have that

$$0 < f(x) = f(\zeta_n) + f'(\xi)(x - \zeta_n) = f'(\xi)(x - \zeta_n)$$

for some $\xi \in [\zeta_n, x]$, hence $f'(\xi) > 0$, and since f' has no zero $> \zeta_n$, it follows that $f'(x) > 0, x > \zeta_n$. Iterating this argument and repeating it analogously for $f(x) < 0$, $x > \zeta_n$, we obtain that

$$f^{(j)}(x) f(x) > 0, \qquad x > \zeta_n, \qquad j = 0, \ldots, n, \qquad (4.63)$$

i.e., all derivatives of f have the same sign as f itself. Substituting this into the iteration (4.62) yields for $x_{n-1} > \zeta_n$ that

$$x_n = x_{n-1} - \frac{f(x_{n-1})}{f'(x_{n-1})} < x_{n-1}, \qquad n \geq 1;$$

hence the sequence is monotonically decreasing. On the other hand assume that $x_{n-1} \geq \zeta_n$; then there exists $\xi \in [\zeta_n, x_{n-1}]$ such that

$$0 = f(\zeta_n) = f(x_{n-1}) + (\zeta_n - x_{n-1}) f'(x_{n-1}) + \frac{(\zeta_n - x_{n-1})^2}{2} f''(\xi)$$

$$= (x_{n-1} - x_n) f'(x_{n-1}) + (\zeta_n - x_{n-1}) f'(x_{n-1}) + \frac{(\zeta_n - x_{n-1})^2}{2} f''(\xi)$$

$$= (\zeta_n - x_n) f'(x_{n-1}) + \frac{(\zeta_n - x_{n-1})^2}{2} f''(\xi)$$

and since $f'(x_{n-1})$ and $f''(\xi)$ have the same sign, this requires that $\zeta_n - x_n \leq 0$, hence $x_n \geq \zeta_n$. To summarize, the sequence x_n is monotonically decreasing and bounded from below by ζ_n, hence converges to a limit x^*, and since

$$\left| \frac{f(x_n)}{f'(x_n)} \right| = |x_{n+1} - x_n| \to 0,$$

it follows that

$$f(x^*) = \lim_{n \to \infty} f(x_n) = 0,$$

hence $x^* = \zeta_n$. The proof for $x_0 \leq \zeta_1$ works analogously. $\qquad \square$

This suggests a tempting method for finding the zeros of polynomials: get some upper and lower estimates for the largest and smallest zero, respectively, for example using *Gershgorin circles*, cf. [36, 52]. Then determine these extremal zeros with Newton's method, and then Proposition 4.4 ensures the convergence. Once these two zeros are found, divide by the associated linear factors and apply the same process to

$$\tilde{f} := \frac{f}{(\cdot - \zeta_1)(\cdot - \zeta_n)}.$$

This procedure has only one minor disadvantage. It does not work in practice due to the following reasons:

1. An iterative scheme like Newton's method does not really find the *exact* zeros, only good approximations to it.
2. Besides inaccuracies due to roundoff errors in the computation, the division process is affected by the inaccuracies of determining ζ_1 and ζ_n which leads to a polynomial $\tilde{f}$ whose zeros are not $\zeta_2, \ldots, \zeta_{n-1}$ any more.
3. This process of *error accumulation* very quickly leads to computed zeros that do not have much to do with the original ones.

The unpleasant numerical effects of division can be made milder by switching to *Maehly's method* of implicit division which is numerically more stable, but worst of all is the way zeros of polynomials depend on their coefficients which is highly unstable. In fact, and as nicely described in [111], the nasty numerical behavior of what is now famous as the "*Wilkinson monster*"

$$f(x) = (x - 1) \cdots (x - 20)$$

was even one of the main motivations for the introduction of *Roundoff Error Analysis* as it is known nowadays [49]. Therefore other methods are more suitable to determine the zeros of a polynomial, especially those based on (linear) algebra.

To that end, let

$$f(x) = x^{n+1} + \sum_{j=0}^{n} f_j x^j = (x - \zeta_0) \cdots (x - \zeta_n)$$

be a monic polynomial of degree $n + 1$. The *principal ideal*

$$\langle f \rangle := f \cdot \Pi = \{ f\, p : p \in \Pi \}$$

defines a set of equivalence classes $\Pi / \langle f \rangle$ where $p \equiv q$ if and only if $p - q \in \langle f \rangle$ is a multiple of f. The unique representer for the equivalence class of $p \in \Pi$ is the remainder of division $r \in \Pi_n$ from

$$p = qf + r, \qquad \deg r < \deg f, \tag{4.64}$$

which once more brings us back to the realm of the Euclidean Ring.

Definition 4.12 *(Modulo operation)* The remainder r of division of p by f will be denoted by $(p)_{\langle f \rangle}$ and called p modulo $\langle f \rangle$ or p modulo f for short.

Although $\Pi / \langle f \rangle \simeq \Pi_n$ is the same set as the polynomials of degree at most n and inherits the vector space properties, it has even more structure, namely that of an *algebra* once we define a multiplication on $\Pi / \langle f \rangle$.

Definition 4.13 *(Multiplication)* The *multiplication* of $p, q \in \Pi / \langle f \rangle$ is defined as

$$p \cdot q := (pq)_{\langle f \rangle} \in \Pi / \langle f \rangle. \tag{4.65}$$

For fixed $q \in \Pi / \langle f \rangle$, the operation

$$\Pi / \langle f \rangle \ni p \mapsto pq = (pq)_{\langle f \rangle}$$

is a *linear* operation in p and can be represented with respect to any basis $B = \{b_0, \ldots, b_n\}$ of $\Pi / \langle f \rangle \simeq \Pi_n$ by a *matrix* $M_{q,B} \in \mathbb{C}^{(n+1) \times (n+1)}$ defined by

$$q \cdot b_j = \sum_{k=0}^{n} (M_{q,B})_{kj} b_k, \qquad j = 0, \ldots, n.$$

This matrix will be called the *companion matrix* or *multiplication table* for $\Pi / \langle f \rangle$ with respect to the basis B.

Example 4.2 *(Frobenius companion matrix)* For the basis $B = \{1, \ldots, x^n\}$ and $q(x) = x$, we get that $x^k q(x) = x^{k+1}$, $k = 0, \ldots, n - 1$, and

$$x^n q(x) = x^{n+1} = f(x) - \sum_{j=0}^{n} f_j x^j \equiv - \sum_{j=0}^{n} f_j x^j;$$

hence

$$M_{(\cdot),B} = \begin{bmatrix} 0 & & & -f_0 \\ 1 & & & -f_1 \\ & \ddots & & \vdots \\ & & 1 & -f_n \end{bmatrix}.$$

This is the well-known *Frobenius companion matrix* of the monic polynomial f.

Varying the basis also varies the multiplication table and reveals valuable information about f.

Example 4.3 Another basis of Π_n in the case of simple zeros $\zeta_j \neq \zeta_k$ is the set of fundamental polynomials for interpolation,

$$\ell_j = \prod_{k \neq j} (\cdot - \zeta_j), \qquad j = 0, \ldots, n;$$

see Problem 4.28. Since for any x,

$$x \, \ell_j(x) = x \prod_{k \neq j} (x - \zeta_j) = (x - \zeta_j) \prod_{k \neq j} (x - \zeta_j) + \zeta_j \prod_{k \neq j} (x - \zeta_j)$$
$$= f(x) + \zeta_j \, \ell_j(x) \equiv \zeta_j \ell_j(x),$$

we can immediately conclude that

$$M_{(\cdot),B} = \begin{bmatrix} \zeta_0 & & \\ & \ddots & \\ & & \zeta_n \end{bmatrix}$$

in this case.

Based on these examples, we can easily identify a way to find zeros of polynomials using Linear Algebra.

Theorem 4.8 *Let $f \in \Pi_{n+1}$ be a monic polynomial with simple zeros $\zeta_0, \ldots, \zeta_n$. Then all multiplication tables for $\Pi/\langle f \rangle$ have the eigenvalues $\zeta_0, \ldots, \zeta_n$ and the associated eigenvectors are $\ell_0, \ldots, \ell_n$.*

Proof If B, B' are any two bases of $\Pi/\langle f \rangle$, then there exists a nonsingular matrix T providing the *basis transform*; that is $B' = TB$ and

$$M_{q,B'} = T^{-1}M_{q,B}T, \qquad q \in \Pi/\langle f \rangle;$$

hence all multiplication tables are similar and therefore they have the same eigenvalues. From Example 4.3, we know the diagonal member of this class and its diagonal values are the eigenvalues and the ℓ_j are the respective eigenvectors. $\qquad\square$

Now we can return to the issue of this chapter: how to compute the knots of a Gaussian quadrature formula for a square positive linear functional? In particular, we return to the case where the Hankel matrix M_n is positive definite one.

We already know how to find the knots for this quadrature formula: they are the zeros of the polynomial q_{n+1} appearing in the convergent $\frac{p_{n+1}}{q_{n+1}}$ of the associated continued fraction. Therefore q_{n+1} is tied to the denominators $q_0, \ldots, q_n$ of the previous convergents by the three-term recurrence relation from (4.8). Normalizing them so that they are monic leads to the modified recurrence relation

$$q_k = (\cdot + \beta_k)\, q_{k-1} - \gamma_k\, q_{k-2}, \qquad k = 1, \ldots, n, \qquad q_0 = 1, \qquad q_{-1} = 0.$$

Since $q_k \in \Pi_k$, $q_k(x) = x^k + \cdots$, these polynomials are linearly independent and form a basis of $\Pi_n \simeq \Pi/\langle q_{n+1} \rangle$. For this basis, we have that

$$\begin{aligned}(\cdot)q_k &= ((\cdot + \beta_{k+1})q_k - \gamma_{k+1}\, q_{k-1}) - \beta_{k+1}\, q_k + \gamma_{k+1}\, q_{k-1} \\ &= q_{k+1} - \beta_{k+1}\, q_k + \gamma_{k+1}\, q_{k-1}, \qquad k = 0, \ldots, n,\end{aligned}$$

which yields that

$$
M_{(\cdot),B} = \begin{bmatrix} -\beta_1 & \gamma_2 & & & & \\ 1 & -\beta_2 & \gamma_3 & & & \\ & 1 & -\beta_3 & \gamma_4 & & \\ & & \ddots & \ddots & \ddots & \\ & & & 1 & -\beta_n & \gamma_{n+1} \\ & & & & 1 & -\beta_{n+1} \end{bmatrix} \in \mathbb{R}^{n+1 \times n+1} \tag{4.66}
$$

for $B = \{q_0, \ldots, q_n\}$. This is a *tridiagonal matrix* for which special numerical methods are available, cf. [36], in order to find its eigenvalues.

The coefficients in (4.66) can be obtained from the ones given in Corollary 4.3, just taking into account that those are the coefficients for an *orthonormal* convergent while here we need monic ones. This procedure will be described in more detail in Sect. 6.6 and can also be applied for computing Gaussian quadrature nodes.

4.4.1 Problems

4.24 *(Maehly's method)* Show that for $f = (\cdot - \zeta_1) \cdots (\cdot - \zeta_n)$ and $k \le n$, one has

$$
\left(\frac{f}{(\cdot - \zeta_1) \cdots (\cdot - \zeta_k)} \right)' = \frac{f'}{(\cdot - \zeta_1) \cdots (\cdot - \zeta_k)} - \frac{f}{(\cdot - \zeta_1) \cdots (\cdot - \zeta_k)} \sum_{j=1}^{k} \frac{1}{\cdot - \zeta_j},
$$

and use this result to devise a variant of Newton's method with implicit factorization of the zeros.

4.25 Write a `Matlab` program that implements Newton's method and factorizations to compute the zeros of polynomials. What results do you obtain for the "Wilkinson monster". Compare them to Maehly's method.

4.26 Suppose that $f \in \Pi$ has simple zeros and show that polynomial interpolation of degree $\deg f - 1$ at these zeros defines a projection operator from Π to $\Pi/\langle f \rangle$.

4.27 Show that the multiplication operation $(p, q) \mapsto pq$ on $\Pi/\langle f \rangle$ is a well defined bilinear form; in particular it satisfies a distributional law.

4.28 Given $\zeta_0, \ldots, \zeta_n$ with $\zeta_j \ne \zeta_k$, $j \ne k$, show that the polynomials

$$
\ell_j = \prod_{k \ne j} (\cdot - \zeta_j), \qquad j = 0, \ldots, n+1,
$$

are linearly independent and hence a basis of Π_n.

4.29 The *divided differences* of a function g are defined recursively as

$$[x_0] g = g(x_0)$$

$$\left[x_0, \ldots, x_{n+1}\right] g = \frac{\left[x_0, \ldots, x_n\right] g - \left[x_1, \ldots, x_{n+1}\right] g}{x_0 - x_{n+1}}.$$

Show

1. the *Newton interpolation formula*: the polynomial

$$\sum_{j=0}^{n} \left[x_0, \ldots, x_j\right] g \ (\cdot - x_0) \cdots (\cdot - x_j - j) \tag{4.67}$$

   is the interpolant of degree n for g at the points $x_0, \ldots, x_n$.
2. that $\left[x_0, \ldots, x_{n+1}\right] g$ is symmetric in the arguments $x_0, \ldots, x_n$.
3. that

$$[x_0, \ldots, x_k] p = 0, \qquad p \in \Pi_{k-1}.$$

4.30 Given any numbers $\xi_0, \ldots, \xi_{n+1} \in \mathbb{R}$, $\xi_j \neq \xi_k$, $j \neq k$, consider the basis

$$B = \left\{ \prod_{j=0}^{k-1} (\cdot - \xi_j) : k = 0, \ldots, n \right\}$$

of $\Pi / \langle f \rangle$, $f \in \Pi$, $\deg f = n$. Show that

$$[\xi_0, \ldots, \xi_{n+1}] f \neq 0$$

and that

$$M_{(\cdot), B} = \begin{bmatrix} \xi_0 & & & -\frac{[\xi_0] f}{[\xi_0, \ldots, \xi_{n+1}] f} \\ 1 & \xi_1 & & -\frac{[\xi_0, \xi_1] f}{[\xi_0, \ldots, \xi_{n+1}] f} \\ & \ddots & \ddots & -\frac{[\xi_0, \xi_1] f}{[\xi_0, \ldots, \xi_{n+1}] f} \\ & & 1 & \xi_n & -\frac{[\xi_0, \ldots, \xi_n] f}{[\xi_0, \ldots, \xi_{n+1}] f} \end{bmatrix}.$$

What happens if $\xi_0, \ldots, \xi_n$ are the zeros of f?

Chapter 5
Continued Fractions and Prony

The equations narrowed [...] until they became just a few
expressions that appeared to move and sparkle with a life of
their own. This was maths without numbers, pure as lightning.
T. Pratchett, Men at arms

5.1 Prony's Problem

In this chapter we relate continued fractions to yet another, seemingly unrelated
problem which was considered and solved by GASPARD CLAIR FRANÇOIS MARIE
RICHE DE PRONY[1] [81] as early as 1795. In modern language, the task is to recover
a function of a particular structure, namely an *exponential sum,*

$$f(x) = \sum_{j=1}^{s} f_j\, e^{\omega_j x}, \qquad \omega_j \in \mathbb{R} + i\mathbb{T}, \qquad f_j \neq 0, \tag{5.1}$$

with distinct frequencies ω_j from samples of the function. In other words we want
to determine f from finitely many function values, which we assume to be equally
distributed, and hence as $f(0), \ldots, f(N)$, $N \in \mathbb{N}$. Of course, N will depend on s,
at least if we want to obtain a reconstruction of f—and why should we be satisfied
with less? Originally Prony did not formulate the problem in terms of the frequencies
ω_j, but he used the formulation

$$f(x) = \sum_{j=1}^{s} f_j\, \rho_j^x, \qquad \rho_j \in \mathbb{C} \setminus \{0\}, \qquad f_j \neq 0,$$

---

[1] In his publications he modestly gave his name just as "G. Prony".

© The Author(s), under exclusive license to Springer Nature Switzerland AG 2021
T. Sauer, *Continued Fractions and Signal Processing*, Springer Undergraduate Texts
in Mathematics and Technology, https://doi.org/10.1007/978-3-030-84360-1_5

which we already saw in (1.11). Of course the two are equivalent and related by $\omega_j = \log \rho_j$, but most "modern" formulations use the form (5.1), often choosing ω_j as purely imaginary numbers, i.e., $\omega_j = iy_j$, $y_j \in \mathbb{R}$. There is also a straightforward way to express Prony's problem as a *sparse polynomial interpolation problem*, cf. [4, 91], even in several variables, but although this is interesting, it is unfortunately not within the scope of this chapter.

Remark 5.1 (*Normalizations*)

1. We normalize the jth *frequency* ω_j, $j = 1, \ldots, s$, to

$$\mathbb{R} + i\mathbb{T} = \mathbb{R} + i\,(\mathbb{R}/2\pi\mathbb{Z}) \simeq \mathbb{R} + i[-\pi, \pi),$$

   to avoid ambiguities in the representation (5.1) that may make the problem unsolvable, for example generating functions like $\sin(\pi\cdot) = \frac{1}{2i}\left(e^{i\pi\cdot} - e^{-i\pi\cdot}\right)$ that cannot be recovered from any subset of $\mathbb{Z}$.
2. The request that the *coefficients* f_j are nonzero makes the representation *sparse*, that is it contains no "phantom" frequencies with zero weight.
3. Sampling at the integers $0, \ldots, N$ is not a restriction, since the simple computation

$$f(ax + b) = \sum_{j=1}^{s} f_j\, e^{\omega_j\,(ax+b)} = \sum_{j=1}^{s} \left(e^{\omega_j b} f_j\right) e^{(a\omega_j)x} =: \sum_{j=1}^{s} \tilde{f}_j\, e^{\tilde{\omega}_j x}$$

   shows that any *affine transformation* of the sampling locations only changes the coefficients and the frequencies, namely the *parameters*; it does not affect the *structure* or the possibility of solving the problem. In other words any *equidistant sampling* based on nodes $x_0 + kh$, $k = 0, \ldots, N$, $x_0 \in \mathbb{R}$, $h > 0$, can be easily reduced to sampling at the integers $0, \ldots, N$ without changing the nature of the problem.

The interesting part of Prony's problem consists of recovering the frequencies. Once these are known all that remains is to solve the *linear system*

$$f(k) = \sum_{j=1}^{s} f_j\, e^{\omega_j k}, \qquad k = 0, \ldots, N,$$

that can be written in matrix form as

$$\begin{bmatrix} f(0) \\ \vdots \\ f(N) \end{bmatrix} = \begin{bmatrix} 1 & \cdots & 1 \\ e^{\omega_1} & \cdots & e^{\omega_s} \\ \vdots & \ddots & \vdots \\ e^{N\omega_1} & \cdots & e^{N\omega_s} \end{bmatrix} \begin{bmatrix} f_1 \\ \vdots \\ f_s \end{bmatrix}, \tag{5.2}$$

or more compactly as

$$[f(j) : j = 0, \ldots, N] = V^T [f_j : j = 1, \ldots, s], \qquad (5.3)$$

respectively. The matrix V from (5.3) is very well known in Numerical Analysis.

Definition 5.1 Given a finite set $X = \{x_0, \ldots, x_m\} \subset \mathbb{R}$ and $n \in \mathbb{N}_0$, the associated *Vandermonde matrix* is defined as

$$V = V_n(X) = \left[x_j^k : \begin{matrix} j = 0, \ldots, m \\ k = 0, \ldots, n \end{matrix} \right] \in \mathbb{R}^{(m+1) \times (n+1)}. \qquad (5.4)$$

Remark 5.2 The terminology of Definition 5.1 is not used in a standardized way in the literature. Sometimes the name "Vandermonde matrix" is even used for an arbitrary *collocation matrix* of the form

$$\left[\phi_k(x_j) : \begin{matrix} j = 0, \ldots, m \\ k = 0, \ldots, n \end{matrix} \right]$$

where $\phi_0, \ldots, \phi_n$ are arbitrary functions, while sometimes the name is reserved only for the special case where $\phi_j = (\cdot)^j$ is a monomial. And to make the confusion complete, it sometimes stands for the transpose of the matrix in (5.4), i.e., for the matrix appearing in the linear equation (5.3).

Vandermonde matrices appear naturally in the *polynomial interpolation problem* that consists of finding a polynomial

$$p = \sum_{j=0}^{n} p_j(\cdot)^j \in \Pi_n$$

of degree at most n, such that $p(x_j) = y_j$, $j = 0, \ldots, m$, for some given values $y_j \in \mathbb{R}$. This problem can again be transformed into the linear system

$$\begin{bmatrix} y_0 \\ \vdots \\ y_m \end{bmatrix} = \left[x_j^k : \begin{matrix} j = 0, \ldots, m \\ k = 0, \ldots, n \end{matrix} \right] \begin{bmatrix} p_0 \\ \vdots \\ p_n \end{bmatrix} = V_n(X) \begin{bmatrix} p_0 \\ \vdots \\ p_n \end{bmatrix}. \qquad (5.5)$$

If $m = n$ and all the x_j are distinct, then the unique *interpolation polynomial* p has the explicit form

$$p = \sum_{j=0}^{n} y_j \prod_{k \neq j} \frac{\cdot - x_k}{x_j - x_k};$$

hence the system (5.5) has a unique solution and therefore the matrix $V_n(X)$ is nonsingular.

In general the *Vandermonde matrix* $V = V_N(e^\Omega)$ from (5.3) has rank s whenever the ω_j are all distinct and $N \geq s - 1$, so that the coefficients are already uniquely determined from s samples as soon as the frequencies are known. Finding the frequencies however is a *nonlinear* problem and the more interesting part of the story.

Prony's ingenious trick to compute the frequencies is based on the following simple procedure: let $p(x) = p_0 + p_1 x + \cdots + p_m x^m$ be a polynomial of degree $m \geq n$, and for fixed $0 \leq j \leq N - m$, consider

$$\sum_{k=0}^{m} f(j+k)\, p_k = \sum_{k=0}^{m} \sum_{\ell=1}^{s} f_\ell\, e^{\omega_\ell\,(j+k)}\, p_k = \sum_{\ell=1}^{s} f_\ell e^{\omega_\ell\, j} \sum_{k=0}^{m} p_k\, (e^{\omega_\ell})^k$$

$$= \sum_{\ell=1}^{s} f_\ell e^{\omega_\ell\, j}\, p\,(e^{\omega_\ell}).$$

In matrix notation this gives

$$
\begin{bmatrix}
f(0) & \cdots & f(m) \\
f(1) & \cdots & f(m+1) \\
\vdots & \ddots & \vdots \\
f(N-m) & \cdots & f(N)
\end{bmatrix}
\begin{bmatrix} p_0 \\ \vdots \\ p_m \end{bmatrix}
$$

$$
=
\begin{bmatrix}
1 & \cdots & 1 \\
e^{\omega_1} & \cdots & e^{\omega_s} \\
\vdots & \ddots & \vdots \\
e^{(N-m)\omega_1} & \cdots & e^{(N-m)\omega_s}
\end{bmatrix}
\begin{bmatrix} f_1 & & \\ & \ddots & \\ & & f_s \end{bmatrix}
\begin{bmatrix} p\,(e^{\omega_1}) \\ \vdots \\ p\,(e^{\omega_s}) \end{bmatrix},
\tag{5.6}
$$

and shows the appearance of yet another Vandermonde matrix. Taking into account the above remark, we get the following result that is at the heart of Prony's method.

Lemma 5.1 *If f is of the form (5.1) and $N \geq 2m - 1$, $m \geq s$, then*

$$\sum_{k=0}^{m} f(j+k)\, p_k = 0 \qquad j = 0, \ldots, N - m, \tag{5.7}$$

if and only if

$$p\,(e^{\omega_j}) = 0, \qquad j = 1, \ldots, s, \tag{5.8}$$

where $p = p_0 + p_1 x + \cdots + p_m x^m$.

Proof The requirement (5.7) means that the product in (5.6) equals zero, and since the Vandermonde matrix is nonsingular because $N - m \geq m - 1 \geq s - 1$, while the diagonal matrix is nonsingular as $f_j \neq 0$, $j = 1, \ldots, s$, it follows that (5.7) is equivalent to

$$
\begin{bmatrix} p\,(e^{\omega_1}) \\ \vdots \\ p\,(e^{\omega_s}) \end{bmatrix} = 0,
$$

which is (5.8).   □

Definition 5.2 The least degree polynomial p with $p(e^{\omega_j}) = 0$, $j = 1, \ldots, s$, is called the *Prony polynomial* for the function f from (5.1).

Clearly the Prony polynomial is always a polynomial of degree s of the form

$$p = p_s \prod_{j=1}^{s} (\cdot - e^{\omega_j}), \qquad p_s \neq 0,$$

and unique up to normalization. There are two convenient ways to normalize the Prony polynomial: namely, fixing the leading coefficient as $p_s = 1$, which gives a *monic* polynomial or choosing $p_s = \prod_j e^{-\omega_j}$ which then normalizes the constant coefficient to be $p_0 = 1$. Also observe that 0 is never a valid zero for a Prony polynomial, since there is no ω such that $e^\omega = 0$. At least no sensible one.

Lemma 5.1 already gives us a way to solve Prony's problem, i.e., to recover (5.1), provided that the number s of exponentials in the sum is known: determine the *kernel*

$$\{p \in \mathbb{C}^{s+1} : F_s p = 0\} \subseteq \mathbb{C}^{s+1}$$

of the *Hankel matrix*

$$F_s := \begin{bmatrix} f(0) & \ldots & f(s) \\ \vdots & \ddots & \vdots \\ f(s) & \ldots & f(2s) \end{bmatrix} \in \mathbb{C}^{s+1 \times s+1},$$

identify the nontrivial solution $0 \neq p \in \mathbb{C}^{s+1}$ of $F_s p = 0$ with a polynomial $p(x)$ and compute its zeros which are exactly the desired numbers e^{ω_j}, $j = 1, \ldots, s$. This procedure was already proposed by Prony[2] in his original paper [81], see also [90]; much later this was refined and extended, leading to the algorithms MUSIC [93] and ESPRIT [84], both in the context of multi-source radar data analysis.

There is, however, also an interpretation using continued fractions which we are going to explore here. To that end we first note that $f(k)$ can be interpreted as a *moment sequence* itself.

Definition 5.3 The *Dirac distribution* δ_x for $x \in \mathbb{R}$ is defined by the requirement

$$\int_{\mathbb{R}} f(t) \delta_x(t) \, dt = f(x), \qquad f \in C_{00}(\mathbb{R}),$$

where $C_{00}(\mathbb{R})$ denotes the (real- or complex-valued) continued functions on $\mathbb{R}$ with *compact support*. Alternatively, one could use the *point-measure*

---

[2] Essentially he says "and then compute the zeros of the polynomial" without caring for the numerical intricacies that only become apparent much later, cf. Wilkinson's excellent paper [111] on the "perfidious polynomial".

$$\int_{\mathbb{R}} f(t)\, d\mu_x(t) = f(x),\qquad\qquad (5.9)$$

whenever this is reasonably defined for the function f.

The Dirac distribution is a continuous *linear functional* on $C_{00}(\mathbb{R})$. Since any "reasonable" function φ defined on $\mathbb{R}$ defines the continuous linear functional $f \mapsto \int_{\mathbb{R}} f\,\varphi$, one can embed continuous functions into the context of distributions which is why they are sometimes also called *generalized functions*. For our purposes here the identity (5.9) is sufficient. The popular intuition for the Dirac distribution is that the "function" is spiky and zero everywhere except at x, where it is so infinite that the integral $\int f \delta_x$ is recovering the value of a function at this point. This intuitive concept is as helpful as it is incorrect and can be misleading.

We will not dwell on issues of Measure Theory about the point-measure here, as the issues are irrelevant in what follows; we only use point-measures formally to express point-evaluation functionals using a shorthand approach.

If we now define the measure

$$\mu := \sum_{j=1}^{s} f_j\, \mu_{e^{\omega_j}},$$

according to the distribution

$$\sum_{j=1}^{s} f_j\, \delta_{e^{\omega_j}},$$

then we obtain its moments as

$$\mu_k = \int_{\mathbb{R}} x^k d\left(\sum_{j=1}^{s} f_j\, \mu_{e_j^{\omega}}\right)(x) = \sum_{j=1}^{s} f_j \int_{\mathbb{R}} x^k\, d\mu_{e_j^{\omega}}(x) = \sum_{j=1}^{n} f_j\, (e^{\omega_j})^k$$

$$= \sum_{j=1}^{s} f_j e^{\omega_j k} = f(k);$$

hence $f(k)$ is indeed a moment sequence for the (possible signed) point-measure μ, and we can consider the Laurent series

$$\mu(x) := \sum_{j=0}^{\infty} \mu_j x^{-j}$$

it defines, or even better let us consider

$$\lambda(x) := x^{-1}\mu(x) = \sum_{j=1}^{\infty} \mu_{j-1} x^{-j}, \quad \text{i.e.,} \quad \lambda_j := \mu_{j-1},\ \lambda_0 = 0. \qquad (5.10)$$

The square Hankel matrices

$$M_n := \begin{bmatrix} \mu_0 & \cdots & \mu_n \\ \vdots & \ddots & \vdots \\ \mu_n & \cdots & \mu_{2n} \end{bmatrix} \in \mathbb{C}^{n+1 \times n+1} \tag{5.11}$$

can be considered as finite segments of the *Hankel operator*

$$M = \begin{bmatrix} \mu_0 & \mu_1 & \cdots \\ \mu_1 & \ddots & \ddots \\ \vdots & \ddots & \ddots \end{bmatrix}$$

that maps the *sequence space*

$$\ell(\mathbb{N}_0) := \{c = (c_k : k \in \mathbb{N}_0) : c_k \in \mathbb{C}\} \tag{5.12}$$

to itself by the *correlation*

$$(Mc)_k = \mu \star c := \sum_{j=0}^{\infty} \mu_{k+j} \, c_j, \quad k \in \mathbb{Z}.$$

The correlation is well defined whenever the infinite sum between μ and c is well defined. One way to obtain a well-defined operation is to restrict one of the two operands to finitely supported objects. Therefore in what follows c will always be the finitely supported object, since the intuition is that we will consider correlations between moment sequences and coefficient vectors of polynomials.

Definition 5.4 The *rank* of the *Hankel operator* M is defined as

$$\text{rank } M := \sup_{n \in \mathbb{N}_0} \text{rank } M_n = \sup_{n \in \mathbb{N}_0} \text{rank} \begin{bmatrix} \mu_0 & \cdots & \mu_n \\ \vdots & \ddots & \vdots \\ \mu_n & \cdots & \mu_{2n} \end{bmatrix}. \tag{5.13}$$

The sequence μ is called *nondegenerate* if it defines a finite rank Hankel operator with the property that with $n := \text{rank } M$,

$$1 = \text{rank } M_0 < \text{rank } M_1 < \cdots < \text{rank } M_{n-1} = \text{rank } M_n = \cdots = \text{rank } M \tag{5.14}$$

holds true.

We already know Hankel operators of finite rank. Indeed if we set

$$\mu_k = \sum_{j=1}^{s} f_j e^{\omega_j k}, \qquad k \in \mathbb{N}_0,$$

as in Prony's problem or moments of finite sums of point measures, then we continue
(5.6) to get for $k \in \mathbb{N}_0$ and $p \in \mathbb{C}^{k+1}$ the identity

$$M_k \, p = \begin{bmatrix} 1 & \cdots & 1 \\ e^{\omega_1} & \cdots & e^{\omega_s} \\ \vdots & \ddots & \vdots \\ e^{k\omega_1} & \cdots & e^{k\omega_s} \end{bmatrix} \begin{bmatrix} f_1 & & \\ & \ddots & \\ & & f_s \end{bmatrix} \begin{bmatrix} p\,(e^{\omega_1}) \\ \vdots \\ p\,(e^{\omega_s}) \end{bmatrix}$$

$$= \begin{bmatrix} 1 & \cdots & 1 \\ e^{\omega_1} & \cdots & e^{\omega_s} \\ \vdots & \ddots & \vdots \\ e^{k\omega_1} & \cdots & e^{k\omega_s} \end{bmatrix} \begin{bmatrix} f_1 & & \\ & \ddots & \\ & & f_s \end{bmatrix} \begin{bmatrix} 1 & e^{\omega_1} & \cdots & e^{k\omega_1} \\ \vdots & \vdots & \ddots & \vdots \\ 1 & e^{\omega_s} & \cdots & e^{k\omega_s} \end{bmatrix} \begin{bmatrix} p_0 \\ \vdots \\ p_k \end{bmatrix},$$

which can be summarized as

$$M_k = \underbrace{\begin{bmatrix} 1 & \cdots & 1 \\ e^{\omega_1} & \cdots & e^{\omega_s} \\ \vdots & \ddots & \vdots \\ e^{k\,\omega_1} & \cdots & e^{k\,\omega_s} \end{bmatrix}}_{k+1 \times s} \underbrace{\begin{bmatrix} f_1 & & \\ & \ddots & \\ & & f_s \end{bmatrix}}_{s \times s} \underbrace{\begin{bmatrix} 1 & e^{\omega_1} & \cdots & e^{k\,\omega_1} \\ \vdots & \vdots & \ddots & \vdots \\ 1 & e^{\omega_s} & \cdots & e^{k\,\omega_s} \end{bmatrix}}_{s \times k+1} =: V_{k,\Omega}^T \, F \, V_{k,\Omega}, \quad (5.15)$$

with the *Vandermonde matrix* $V_{k,\Omega} = V_k\left(e^{\Omega}\right)$ and the nonsingular diagonal matrix
$F := \mathrm{diag}\,(f_1, \ldots, f_s)$. Since the rank of $V_{k,\Omega}$ is $\min(k+1, s)$, we can record that

$$\mathrm{rank}\; M = s. \qquad (5.16)$$

Finite sums of exponentials define Hankel operators of finite rank, and this rank equals
the number of summands in (5.1). Moreover the coefficients of the Prony polynomial
are the components of the kernel vector of M_s, and since rank $M_s = s$, this vector and
therefore also the associated polynomial are unique up to normalization. This is often
naïvely turned into the algorithm of forming $M_k, k = 0, 1, \ldots,$ until rank $M_k \leq k$.
When this rank deficiency happens for the first time, that is when rank $M_k = k$, it
follows that the kernel of M_k is a one-dimensional subspace and defines a unique
monic polynomial of degree exactly k, see Problem 5.1. Is this already the Prony
polynomial for the sequence μ defined by sampling (5.1)? As tempting as it appears
at first, this conclusion is invalid.

Example 5.1 For the simplest counterexample, set $s = 2, \omega_1 = 0, \omega_2 = 1$ and $f_1 = -f_2 = 1$, hence $f(x) = 1 - e^x$. Then

$$M_0 = [0], \qquad M_1 = \begin{bmatrix} 0 & 1-e \\ 1-e & 1-e^2 \end{bmatrix},$$

and therefore rank $M_0 = 0$ while rank $M_1 = 2 = $ rank M_k, $k \geq 1$. Therefore M_0 gives the false expression that $f = 0$.

Example 5.1 may appear a bit too simplistic and "academic", but in fact the rank can even become stationary for an arbitrarily long time.

Example 5.2 We construct degenerate sequences directly with the exponential functions (5.1). The rank of the associated *Hankel operator* will then be s. For the purpose of the counterexample, we can even choose arbitrary frequencies ω_j, $j = 1, \ldots, s$, as well as[3] $0 \leq k < k' \leq s$ with $2k' - k < s - 1$, and any polynomial $p \in \Pi_k$ with $p(e^{\omega_j}) \neq 0$, $j = 1, \ldots, s$. Now we let the vector $f \in \mathbb{R}^s$ be any nonzero solution of the underdetermined $(2k' - k + 1) \times s$ system

$$
0 = Af = \begin{bmatrix} p(e^{\omega_1}) & \cdots & p(e^{\omega_s}) \\ e^{\omega_1} p(e^{\omega_1}) & \cdots & e^{\omega_s} p(e^{\omega_s}) \\ \vdots & \ddots & \vdots \\ (e^{\omega_1})^{2k'-k} p(e^{\omega_1}) & \cdots & (e^{\omega_s})^{2k'-k} p(e^{\omega_s}) \end{bmatrix} f \tag{5.17}
$$

$$
= \begin{bmatrix} 1 & \cdots & 1 \\ \vdots & \ddots & \vdots \\ (e^{\omega_1})^{2k'-k} & \cdots & (e^{\omega_s})^{2k'-k} \end{bmatrix} \begin{bmatrix} p_k(e^{\omega_1}) & & \\ & \ddots & \\ & & p_k(e^{\omega_s}) \end{bmatrix} f.
$$

Such a solution exists since the matrix $A \in \mathbb{C}^{(2k'-k+1) \times s}$ in (5.17) has rank $2k' - k + 1$. For $0 \leq \ell \leq k' - k$ we now have that

$$
\left(M_{k'} \begin{bmatrix} 0_{\ell \times 1} \\ p \\ 0_{(k'-k-\ell) \times 1} \end{bmatrix} \right)_r = \sum_{t=0}^{k} f(\ell + r + t) \, p_t = \sum_{t=0}^{k} \sum_{j=1}^{s} f_j \, e^{\omega_j(r+t+\ell)} \, p_t
$$

$$
= \sum_{j=1}^{s} f_j e^{\omega_j(\ell+r)} \sum_{t=0}^{k} e^{t\omega_j} \, p_t = \sum_{j=1}^{s} e^{\omega_j(\ell+r)} p(e^{\omega_j}) \, f_j
$$

$$
= e_{r+\ell}^T A f = 0, \qquad r = 0, \ldots, k',
$$

since then $r + \ell \leq 2k' - k < s - 1$, hence

$$
0 = M_{k+m} \begin{bmatrix} 0_{\ell \times 1} \\ p \\ 0_{(m-\ell) \times 1} \end{bmatrix}, \qquad \ell = 0, \ldots, m, \qquad m = 0, \ldots, k' - k,
$$

and since all these vectors are linearly independent, it follows immediately that

---

[3] That k and k' are limited by s is no restriction and only says that in order to get a "gap" of a certain size, the number s of terms has to be sufficiently large.

$$\text{rank } M_{k-1} = \text{rank } M_k = \cdots = \text{rank } M_{k'-1} = \text{rank } M_{k'} \leq \text{rank } M_{k'+1}$$

and all the Hankel matrices M_j, $j = k, \ldots, k'$, are automatically *singular*.

Remark 5.3 If $e^{\omega_j} \in \mathbb{R}$, the simple observation that

$$0 = \begin{bmatrix} 0_{1 \times \ell}, & p, & 0_{1 \times (k'-k-\ell)} \end{bmatrix} M_{k'} \begin{bmatrix} 0_{\ell \times 1} \\ p \\ 0_{(k'-k-\ell) \times 1} \end{bmatrix} = \sum_{j=1}^{s} \left(e^{\omega_j \ell} p(e^{\omega_j}) \right)^2 f_j \qquad (5.18)$$

shows that, as explicitly seen in Example 5.1, not all the coefficients f_j, $j = 1, \ldots, s$, can be positive. Indeed for positive weights we need a polynomial p that vanishes at all e^{ω_j}, which must be a multiple of the Prony polynomial.

Hankel operators of finite rank can be characterized in many equivalent ways, one of which we will give next, cf. [40, 77].

Theorem 5.1 (Kronecker's theorem[4]) *The Hankel operator M has finite rank if and only if $\mu(x)$ is a rational function, i.e.,*

$$\mu(x) = \frac{p(x)}{q(x)}, \qquad p, q \in \Pi, \qquad p(0) = 0. \qquad (5.19)$$

Remark 5.4 That the Hankel operator is of finite rank does *not* mean that the associated sequence μ is finitely supported, however it is quite the contrary. It can be shown that any finitely supported sequence μ always defines a Hankel operator of infinite rank—at least as long as it is nonzero, see Problem 5.2.

To prove the theorem we introduce the bilinear form

$$(\cdot, \cdot) : \ell(\mathbb{Z}) \times \Pi \to \ell(\mathbb{Z}), \qquad (\mu, p) := \mu \star p, \qquad (5.20)$$

and note that for the *shift operator* τ, $(\tau c)_k := c_{k+1}$, we have

$$(\tau \mu, p)_j = (\tau (\mu, p))_j = \sum_{k=0}^{\infty} \mu_{j+1+k} \, p_k = \sum_{k=1}^{\infty} \mu_{j+k} p_{k-1} = (\mu, (\cdot)p)_j. \qquad (5.21)$$

Although (5.21) is almost trivial to prove by a simple shift of the index, it has fundamental consequences.

---

[4] There are several quite different results attributed to LEOPOLD KRONECKER as *Kronecker's theorem*: for example also a number-theoretic one on lattices generated by real numbers that are linearly independent over $\mathbb{Q}$, see [45], which, by the way, also contains a nice chapter on continued fractions. So the lesson is that the name alone is not always helpful; one should look for the *meaning* of a result.

Lemma 5.2 *The set*

$$\ker(\mu, \cdot) = \{p \in \Pi : (\mu, p) = 0\} \tag{5.22}$$

is an ideal, *i.e., it is closed under addition and multiplication with arbitrary polynomials.*

Proof The shift-invariance of the zero sequence gives

$$0 = \tau 0 = \tau(\mu, p) = (\mu, (\cdot) p), \qquad p \in \ker(\mu, \cdot),$$

and closure under addition is trivial because of bilinearity.   □

Remark 5.5 The "ideal" property is a slight overkill in the univariate case which we consider here since all ideals in $\mathbb{K}[x]$ form a *principal ideal*, i.e., they are generated by a single polynomial. Univariate polynomials always form a *principal ideal ring*. The full value of the observation of Lemma 5.2 only occurs when one considers Prony's method in several variables, cf. [87, 91].

Proof (*of Theorem* 5.1) If M is of finite rank, then

$$0 \in \left\{ M \begin{bmatrix} p \\ 0 \end{bmatrix} : p \in \Pi \setminus \{0\} \right\},$$

as otherwise the rank would be infinite. Therefore there exists $0 \neq q \in \Pi$ of minimal degree such that $0 = (\mu, q) = (\mu, \Pi q)$, where $\Pi q = \{pq : p \in \Pi\}$ denotes the *principal ideal* generated by q. Therefore

$$0 = (\mu, q)(x) = \sum_{j=0}^{\infty} (\mu, q)_j \, x^{-j} = \sum_{j=0}^{\infty} \sum_{k=0}^{\infty} \mu_{j+k} q_k x^{-i} = \sum_{j,k=0}^{\infty} \mu_{j+k} x^{-j-k} q_k x^k$$

$$= \sum_{k=0}^{n} q_k x^k \sum_{j=k}^{\infty} \mu_j \, x^{-j} = \sum_{k=0}^{n} q_k x^k \left(\mu(x) - \sum_{j=0}^{k-1} \mu_j \, x^{-j} \right)$$

$$= q(x)\mu(x) - \sum_{k=0}^{n} q_k \sum_{j=0}^{k-1} \mu_j x^{k-j}$$

that is,

$$\mu(x) = \frac{p(x)}{q(x)}$$

with

$$p(x) = \sum_{k=0}^{n} q_k \sum_{j=0}^{k-1} \mu_j x^{k-j} = \sum_{k=0}^{n} q_k \sum_{j=0}^{k-1} \mu_{k-1-j} x^{j+1} = x \sum_{j=0}^{n-1} x^j \sum_{k=j+1}^{n} q_k \, \mu_{k-(j+1)} \tag{5.23}$$

as claimed, especially with $p(0) = 0$.

For the converse, we request that the denominator $q \in \Pi_n$ satisfies $\deg q = n$, that is $q_n \neq 0$. Furthermore we observe that the identity

$$\mu(x)\, q(x) = p(x), \qquad p \in \Pi_m,$$

can be extended using the convention $q_k = 0$ for $k < 0$, and then we have the following:

$$p(x) = \sum_{j=1}^{m} p_j x^j = \left(\sum_{j=0}^{\infty} \mu_j\, x^{-j}\right) \left(\sum_{k=0}^{n} q_k x^k\right) = \sum_{j=0}^{\infty} \sum_{k=0}^{n} \mu_j\, q_k x^{k-j}$$

$$= \sum_{k=0}^{n} \sum_{j=-k}^{\infty} x^{-j}\, q_k \mu_{j+k} = \sum_{j=-n}^{\infty} x^{-j} \sum_{k=-j}^{n} \mu_{j+k}\, q_k$$

$$= \sum_{j=-n}^{\infty} x^{-j} \sum_{k=\max(0,-j)}^{n} \mu_{j+k}\, q_k.$$

Since the left-hand side of this equation is a polynomial, it follows that all coefficients associated with the Laurent monomials $x^{-n}, \ldots, x^{-1}$ have to have a zero coefficient, i.e.,

$$0 = \sum_{k=0}^{n} \mu_{j+k}\, q_k = (\mu \star q)_j = \left(M \begin{bmatrix} q \\ 0 \end{bmatrix}\right)_j, \qquad j \in \mathbb{N}_0, \qquad (5.24)$$

so that by Lemma 5.2,

$$\ker\,(\mu, \cdot) \supseteq q\,\Pi \quad \Rightarrow \quad 0 = M \begin{bmatrix} (\cdot)^k\, q \\ 0 \end{bmatrix} = M \begin{bmatrix} 0_k \\ q \\ 0 \end{bmatrix}, \qquad k \in \mathbb{N}_0, \qquad (5.25)$$

hence rank $M \leq \deg q = n$. □

Remark 5.6 The assumption $p(0) = 0$ in (5.19) is equivalent to p being of the form $p(x) = x\,\tilde{p}(x)$; hence

$$\mu(x) = x\, \frac{\tilde{p}(x)}{q(x)}, \qquad \text{i.e.,} \qquad \lambda(x) = \frac{\tilde{p}(x)}{q(x)},$$

which indicates that the shifted sequence λ from (5.10) may be more appropriate to consider later.

Definition 5.5 A Hankel operator will be called *simple* if it has finite rank, and the denominator in the normalized representation (5.19) has only *simple zeros*.

The theory developed so far can be extended to the case of multiple zeros of q. The functions to be considered in that case then are still of the type (5.1), but now the coefficients are *polynomials* whose degree is one less than the multiplicity of the respective zero so that the number of coefficients of the polynomial coincides with the multiplicity. How the details can be worked out is described in Problems 5.7–5.11.

Indeed, this extension to exponential polynomials even works in several variables, cf. [75, 88, 89], although the concept of multiplicity of zeros then becomes more intricate and structural. Unfortunately these issues are beyond the scope of this book.

Remark 5.7 Prony himself was already aware of the situation of multiple zeros in [81]. Although he considers only the case of simple zeros, he remarks:

Je ne parle des cas où l'équation

$$A_0 + A_I\alpha + A_{II}\alpha^2 + \cdots\cdots\cdots + A_{(n)}\alpha^n = 0$$

a des racines égales ou imaginaires; j'ai donné dans le $n^o XX$ de mes leçons d'analyse les formules nécessaires pour les résoudre. On sait que les racines égales introduisent de coèfficiens variables et rationnels dans la valeur de z, et si ces racines sont égales à l'unité, z contiendra alors des termes entièrement rationnels …

Hence, Prony remarks in particular that if all zeros of q coincide, then the function f degenerates to a single polynomial.

A more careful inspection of the proof of Theorem 5.1 leads to the following observation.

Corollary 5.1 (Hankel & Prony)

1. *The polynomial q in the normalized representation $\mu(x) = \frac{p(x)}{q(x)}$ is the Prony polynomial for μ.*
2. *Any simple Hankel operator is generated by exponential functions, i.e., $\mu_j = f(j)$ for some f of the form (5.1).*
3. *Any simple Hankel operator factorizes as*

$$M = \underbrace{\begin{bmatrix} 1 & \cdots & 1 \\ e^{\omega_1} & \cdots & e^{\omega_n} \\ e^{2\omega_1} & \cdots & e^{2\omega_n} \\ \vdots & \ddots & \vdots \end{bmatrix}}_{=:V_\Omega} \begin{bmatrix} f_1 & & \\ & \ddots & \\ & & f_n \end{bmatrix} \underbrace{\begin{bmatrix} 1 & e^{\omega_1} & e^{2\omega_1} & \cdots \\ \vdots & \vdots & \vdots & \ddots \\ 1 & e^{\omega_n} & e^{2\omega_n} & \cdots \end{bmatrix}}_{=V_\Omega^T}. \tag{5.26}$$

Proof For (1), we note that q was defined by the property $(\mu, q) = 0$, which is in turn the definition of the Prony polynomial.

To verify (2), we divide q by any factor of the form $(\cdot)^k$, $k \in \mathbb{N}$, and if necessary, then normalize it into a monic polynomial and let $e^{\omega_1}, \ldots, e^{\omega_n}$ be the (remaining) zeros of q, i.e.,

$$q = (\cdot - e^{\omega_1}) \cdots (\cdot - e^{\omega_n}).$$

Then the proof of Theorem 5.1 shows that μ is a solution of the homogeneous *difference equation*

$$0 = \sum_{k=0}^{n} \mu_{j+k} q_k, \qquad j \in \mathbb{N}_0,$$

see also Sect. 6.5. The solution space has dimension n, cf. [55], or see Theorem 6.5, and since

$$\sum_{k=0}^{s} e^{\omega(j+k)} q_k = e^{\omega j} \sum_{k=0}^{s} q_k e^{\omega k} = e^{\omega j} q(e^{\omega}), \qquad j \in \mathbb{N}_0,$$

we see that the sequences $k \mapsto e^{\omega_j k}$, $j = 1, \ldots, n$, form a basis for this space. Consequently μ must be a linear combination of theses sequences, and therefore of the form (5.1).

For (3) we first record that according to (2), we can write

$$\mu_k = \sum_{j=1}^{s} f_j e^{\omega_j k}, \qquad k \in \mathbb{N}_0,$$

and therefore for $k, \ell \in \mathbb{N}_0$,

$$
\begin{aligned}
e_k^T M e_\ell = \mu_{k+\ell} &= \sum_{j=1}^{s} f_j e^{\omega_j (k+\ell)} = \sum_{j=1}^{s} f_j e^{\omega_j k} e^{\omega_j \ell} \\
&= \left[e^{\omega_1 k}, \ldots, e^{\omega_n k} \right] \begin{bmatrix} f_1 & & \\ & \ddots & \\ & & f_n \end{bmatrix} \begin{bmatrix} e^{\omega_1 j} \\ \vdots \\ e^{\omega_n j} \end{bmatrix} \\
&= e_k^T \begin{bmatrix} 1 & \cdots & 1 \\ e^{\omega_1} & \cdots & e^{\omega_n} \\ e^{2\omega_1} & \cdots & e^{2\omega_n} \\ \vdots & \ddots & \vdots \end{bmatrix} \begin{bmatrix} f_1 & & \\ & \ddots & \\ & & f_n \end{bmatrix} \begin{bmatrix} 1 & e^{\omega_1} & e^{2\omega_1} & \cdots \\ \vdots & \vdots & \vdots & \ddots \\ 1 & e^{\omega_n} & e^{2\omega_n} & \cdots \end{bmatrix} e_\ell,
\end{aligned}
$$

which is (5.26). Note that this is the "infinite version" of the argument that leads to the finite factorization (5.15). $\qquad \square$

Now we can combine our findings with Theorem 4.4. The shifted moment sequence λ from (5.10) has an associated continued fraction expansion if and only if $\det \Lambda_n \neq 0$, which is in turn equivalent to μ being *nondegenerate*. If this is satisfied, we can apply the full machinery of continued fractions to Prony's problem.

Corollary 5.2 *If for an exponential f of the form (5.1), the sequence $\lambda = (f(j - 1) : j \in \mathbb{N})$ and $\lambda_0 = 0$ is nondegenerate, then the continued fraction expansion of $\lambda(x)$ terminates after n steps, and the denominator of that convergent is the Prony polynomial for f.*

Proof Since the sequence is nondegenerate, there exists an associated continued fraction expansion. The recurrence for its convergents eventually computes a final fraction whose denominator is the Prony polynomial. □

5.1.1 Problems

5.1 Suppose that $k + 1 = \operatorname{rank} M_k = \operatorname{rank} M_{k+1}$. Show that there exists a unique $f = (f_0, \ldots, f_{k+1}) \in \mathbb{C}^{k+2}$ such that $M_{k+1} f = 0$ and $f_{k+1} = 1$.

5.2 (*Finite support, infinite rank*) Let μ be a finitely supported sequence, i.e., there exists k_0 such that $\mu_k = 0$, $k \geq k_0$. Show that the associated Hankel operator has infinite rank.

Hint: Look for rows of the bi-infinite matrix that are obviously linearly independent and find infinitely many of them.

5.3 Show that for any distinct $\omega_1, \ldots, \omega_n \in \mathbb{C}$ and $N \geq n - 1$, the matrix

$$\left[e^{j\omega_k} : \begin{array}{l} j = 0, \ldots, N \\ k = 1, \ldots, n \end{array} \right] := \begin{bmatrix} 1 & \cdots & 1 \\ e^{\omega_1} & \cdots & e^{\omega_n} \\ \vdots & \ddots & \vdots \\ e^{N\omega_1} & \cdots & e^{N\omega_n} \end{bmatrix} \in \mathbb{C}^{(N+1)\times n}$$

has rank n.

5.4 (*Determinant formula for square Vandermonde matrices*) Prove that

$$\det \begin{bmatrix} 1 & \cdots & 1 \\ x_0 & \cdots & x_n \\ \vdots & \ddots & \vdots \\ x_0^n & \cdots & x_n^n \end{bmatrix} = \prod_{0 \leq j < k \leq n} (x_k - x_j). \tag{5.27}$$

5.5 Show that the infinite vectors $\begin{bmatrix} 0_k \\ q \\ 0 \end{bmatrix}$, $k \in \mathbb{N}_0$ that appear in (5.25) are *linearly independent*.

5.6 Prove (5.18) from Remark 5.3.

5.1.1.1 Exponential Polynomials—the Full Truth

With these problems, the reader can extend the theory for Prony's problem to the general case of exponential polynomials.

Definition 5.6 (*Exponential polynomial*) An *exponential polynomial* is a finite sum of the form

$$f(x) = \sum_{j=1}^{n} f_j(x) e^{\omega_j x}, \qquad \omega_j \in \mathbb{R} + i\mathbb{T}, \qquad 0 \neq f_j \in \mathbb{C}[x]. \tag{5.28}$$

5.7 Show that Π_n is a *shift-invariant* space, i.e., $f \in \Pi_n$ implies that $f(\cdot + y) \in \Pi_n$ for any $y \in \mathbb{R}$. Show that for $f \in \Pi_n$, one has

$$f(x + y) = \sum_{j,k=0}^{n} \binom{k+j}{j} f_{k+j} \, x^j y^k, \qquad f(x) = \sum_{j=0}^{n} f_j x^j, \tag{5.29}$$

with the convention that $f_j = 0$ for $j > n$.

5.8 (*Stirling numbers, cf.* [39, 41]) The *Stirling numbers* of the second kind are defined as

$$\left\{ {n \atop k} \right\} := \frac{1}{k!} \left(\Delta^k (\cdot)^n \right)(0), \qquad k = 0, \ldots, n,$$

and 0 if $k \notin \{0, \ldots, n\}$. Verify the explicit representation

$$\left\{ {n \atop k} \right\} = \frac{1}{k!} \sum_{j=0}^{k} (-1)^{k-j} \binom{k}{j} j^n, \qquad k = 0, \ldots, n,$$

and the recurrence relation

$$\left\{ {n+1 \atop k} \right\} = k \left\{ {n \atop k} \right\} + \left\{ {n \atop k-1} \right\}. \tag{5.30}$$

5.9 Show that the θ *operator* defined as

$$(\theta f)(x) = x \, f'(x)$$

satisfies

$$\theta(\cdot)^n = n(\cdot^n) \quad \text{and} \quad \theta^n = \sum_{j=0}^{n} \left\{ {n \atop j} \right\} (\cdot)^j \frac{d^j}{dx^j}.$$

Hint: use the recurrence (5.30).

5.10 Show that for any $n \in \mathbb{N}$ and $x \in \mathbb{C} \setminus \{0\}$, there exists a nonsingular matrix $\Theta_n(x)$ such that

$$\begin{bmatrix} f(x) \\ \theta f(x) \\ \vdots \\ \theta^n f(x) \end{bmatrix} = \Theta_n(x) \begin{bmatrix} f(x) \\ f'(x) \\ \vdots \\ f^{(n)}(x) \end{bmatrix}.$$

Conclude from this identity that for any $x \in \mathbb{C} \setminus \{0\}$, we have

$$\theta^k f(x) = 0, \quad k = 0, \ldots, n, \qquad \Leftrightarrow \qquad f^{(k)}(x) = 0. \tag{5.31}$$

5.11 (*General factorization*) To generalize (5.15), assume that f is of the form (5.28) with

$$d_j := \deg f_j \quad \text{and} \quad f_j(x) = \sum_{\ell=0}^{d_j} f_{j,\ell} x^\ell, \quad j = 1, \ldots, n.$$

1. Show that the Hankel matrix

$$M_k = \begin{bmatrix} f(0) & \cdots & f(k) \\ \vdots & \ddots & \vdots \\ f(k) & \cdots & f(2k) \end{bmatrix}.$$

factorizes into

$$M_k = V_{k,\Omega} F V_{k,\Omega}^T$$

where

$$V_{k,\Omega} = \left[\left(\theta^r(\cdot)^j\right)(e^{\omega_\ell}) : \begin{matrix} j = 0, \ldots, k \\ r = 0, \ldots, d_\ell, \, \ell = 1, \ldots, n \end{matrix} \right]$$

$$= \begin{bmatrix} 1 & 0 & \cdots & 0 & \cdots & 1 & \cdots & 0 \\ e^{\omega_1} & (\theta(\cdot))(e^{\omega_1}) & \cdots & \left(\theta^{d_1}(\cdot)\right)(e^{\omega_1}) & \cdots & e^{\omega_n} & \cdots & \left(\theta^{d_n}(\cdot)\right)(e^{\omega_n}) \\ \vdots & \vdots & \ddots & \vdots & \ddots & \vdots & \ddots & \vdots \\ e^{k\omega_1} & \left(\theta(\cdot)^k\right)(e^{\omega_1}) & \cdots & \left(\theta^{d_1}(\cdot)^k\right)(e^{\omega_1}) & \cdots & e^{k\omega_n} & \cdots & \left(\theta^{d_n}(\cdot)^k\right)(e^{\omega_n}) \end{bmatrix}$$

and

$$F = \begin{bmatrix} F_1 & & \\ & \ddots & \\ & & F_n \end{bmatrix}, \quad F_\ell = \left[\binom{k+j}{j} f_{\ell,k+j} : \begin{matrix} j = 0, \ldots, d_\ell \\ k = 0, \ldots, d_\ell \end{matrix} \right]. \tag{5.32}$$

2. Show that F_ℓ from (5.32) is nonsingular.
3. Derive a factorization of M_k in terms of the Hermite–Vandermonde matrix from Definition (4.55).
   Hint: Use Problem 5.10.
4. Generalize Lemma 5.1 by showing that for any

$$\Pi_m \ni p = \sum_{j=0}^{m} p_j(\cdot)^j, \quad m \geq d := \sum_{j=1}^{n}(d_j + 1)$$

and $N \geq 2m - 1$, one has

$$\sum_{k=0}^{m} f(j+k)\, p_k = 0$$

if and only if

$$\theta^k p(e^{\omega_j}) = 0, \quad k = 0, \ldots, d_j - 1, \ j = 1, \ldots, n, \tag{5.33}$$

or

$$p^{(k)}(e^{\omega_j}) = 0, \quad k = 0, \ldots, d_j - 1, \ j = 1, \ldots, n, \tag{5.34}$$

respectively.

Hint: Prove (5.33), the equivalence to (5.34) is (5.31) from Problem 5.10.

5.2   Flat Extensions of Moment Sequences

Finally we touch on the issue of *truncated moment sequences*, i.e., the question how finite moment sequences can be extended into infinite ones in a consistent way. Here one usually restricts oneself to the case in which the initial segment $\mu_0, \ldots, \mu_{2n}$ of a moment sequence μ is known; and we assume that

$$M_n := \begin{bmatrix} \mu_0 & \cdots & \mu_n \\ \vdots & \ddots & \vdots \\ \mu_n & \cdots & \mu_{2n} \end{bmatrix}$$

is *symmetric* and strictly *positive definite* which implies the same for the Hankel submatrices $M_k, k = 0, \ldots, n - 1$, see Problem 5.12. Such a sequence can be interpreted as a sequence of moments coming from a linear functional that is at least *square positive* on Π_{2n}. Based on that knowledge, we define a particular type of extension of the *moment sequence* μ.

Definition 5.7 A sequence $\hat{\mu} \in \ell(\mathbb{N}_0)$ is called a *flat extension* of the moment sequence $\mu = (\mu_0, \ldots, \mu_{2n}, \ldots)$ if

1. $\hat{\mu}_j = \mu_j, j = 0, \ldots, 2n$,
2. rank $\hat{M}_k =$ rank $M_n = n + 1, k \geq n$.

In other words a flat extension leads to a moment sequence whose associated *Hankel operator* has rank $n + 1$; "flatness" means that the rank of the operator is fixed by rank $\hat{M}_n$ and the rank remains constant for all indices larger than n:

$$1 = \text{rank } \hat{M}_0 < \text{rank } \hat{M}_1 < \cdots < \text{rank } \hat{M}_n = \text{rank } \hat{M}_{n+1} = \cdots = n + 1. \quad (5.35)$$

Continued fractions help us to construct flat extensions.

Theorem 5.2 *Any moment sequence μ such that M_n is positive definite has a flat extension $\hat{\mu}$.*

Proof We construct the sequence of convergents for $\lambda(x)$ with $\Lambda_k = M_{k-1}, k = 1, \ldots, n + 1$. Since

$$0 < \det M_j, \qquad j = 0, \ldots, n,$$

Theorem 4.4 implies that

$$\lambda(x) = [0; r_1, \ldots, r_{n+1}, \ldots]$$

and

$$\frac{p_j(x)}{q_j(x)} - \sum_{k=1}^{2j-2} \mu_k x^{-k} = O\left(x^{-2j+1}\right), \qquad j = 1, \ldots, n + 1;$$

and at least the first $n + 1$ of the convergents p_j/q_j are well defined. Setting

$$\hat{\mu}(x) = \frac{p_{n+1}(x)}{q_{n+1}(x)}$$

gives the desired flat extension.   □

The proof of Theorem 5.2 relied only on the fact that there exists a continued fraction associated with the Laurent series $\lambda(x)$. For this, only the *nonsingularity* of the moment matrices is sufficient. In the case of square positive functionals, the nonsingularity and even the positive definiteness of the Hankel matrices $M_j, j < n$, already follows from the positive definiteness of the last one M_n, and Theorem 4.4 can be applied immediately. In the general case we can extend Theorem 5.2 in the following way by explicitly requesting that *all* Hankel matrices be nonsingular.

Corollary 5.3 *Any moment sequence μ such that $M_0, \ldots, M_n$ are nonsingular matrices admits a flat extension $\hat{\mu}$.*

In the following section in Theorem 5.3, we will see that for the existence of a flat extension, in fact the nonsingularity of M_n alone is already sufficient. This, however, will not make use of continued fractions any more, and we will lose the sequence of polynomials of the convergent that are connected by a three term recurrence relation. From this perspective, the above construction chose a special flat extension from, as we will see, a family of possible extensions.

Let us return to the case of positive definite moment matrices, i.e., moments coming from a square positive linear functional, where we can say even more. By Proposition 4.3, the zeros of q_{n+1} are real and simple, hence can be written as e^{ω_j}, $\omega_j \in \mathbb{R}$, $j = 1, \ldots, n+1$, as long as[5] $q_{n+1}(0) \neq 0$. In other words if we choose q_{n+1} as a monic polynomial,

$$q_{n+1}(x) = \prod_{j=1}^{n+1} (x - e^{\omega_j}), \tag{5.36}$$

which is possible since this does not affect the zeros and only corresponds to a renormalization of p_{n+1} by a unit in Π. Then Corollary 5.1 implies that in defining a finite rank Hankel operator, the flat extension $\hat{\mu}$ must be the samples of the exponential sum

$$f(x) = \sum_{j=1}^{n+1} f_j\, e^{\omega_j x}, \qquad f_j \in \mathbb{R}, \quad j = 1, \ldots, n+1, \tag{5.37}$$

in which the coefficients f_j are real since $\omega_j \in \mathbb{R}$ and $f : \mathbb{R} \to \mathbb{R}$ was assumed. Finally, define again the fundamental polynomials for Lagrange interpolation at e^{Ω}:

$$\Pi_n \ni \ell_j(x) = \prod_{k \neq j} \frac{x - e^{\omega_k}}{e^{\omega_j} - e^{\omega_k}} = \frac{q_{n+1}(x)}{x - e^{\omega_j}} \prod_{k \neq j} \frac{1}{e^{\omega_j} - e^{\omega_k}},$$

satisfying $\ell_j(e^{\omega_k}) = \delta_{j,k}$, $j, k = 1, \ldots, n+1$, and apply (5.6) to obtain that

$$M_n \ell_j = \begin{bmatrix} 1 & \ldots & 1 \\ e^{\omega_1} & \ldots & e^{\omega_{n+1}} \\ \vdots & \ddots & \vdots \\ e^{n\,\omega_1} & \ldots & e^{n\,\omega_{n+1}} \end{bmatrix} \begin{bmatrix} f_1 & & \\ & \ddots & \\ & & f_{n+1} \end{bmatrix} \begin{bmatrix} \ell_j(e^{\omega_1}) \\ \vdots \\ \ell_j(e^{\omega_{n+1}}) \end{bmatrix} = f_j \begin{bmatrix} 1 \\ e^{\omega_j} \\ \vdots \\ e^{n\,\omega_j} \end{bmatrix}. \tag{5.38}$$

This means that $f_j = \left(M_n \ell_j\right)_1$, which even gives a direct way to obtain the coefficient vector f of the exponential sum as

$$f^T = e_1^T M_n \left[\ell_1 \ldots \ell_{n+1}\right] = e_1^T M_n L, \qquad L := \left[\ell_1 \ldots \ell_{n+1}\right],$$

---

[5] A zero at the origin is a "spurious" zero when passing to Laurent polynomials and must be excluded in this theory.

which means that the weight vector f is simply the first row of the matrix $M_n L$. It has another nice property.

Lemma 5.3 *If M_n is positive definite, then the coefficients f_j from (5.37) are positive, i.e., $f_j > 0$, $j = 1, \ldots, n+1$.*

Proof Since the frequencies $\omega_j, j = 1, \ldots, n$, are real, the Vandermonde matrix $V_{n,\Omega}$ is real and the factorization $M_n = V_{n,\Omega}^T \operatorname{diag} f \, V_{n,\Omega}$ from (5.15) yields that

$$\operatorname{diag} f = V_{n,\Omega}^{-1} M_n V_{n,\Omega}^{-T}$$

is also positive definite, and thus $f_j > 0$, $j = 1, \ldots, n+1$. □

Moreover as in (5.38) we can show that

$$\ell_j^T M_n \ell_k = \left[\ell_j(e^{\omega_1}) \ldots \ell_j(e^{\omega_{n+1}}) \right] \begin{bmatrix} f_1 & & \\ & \ddots & \\ & & f_{n+1} \end{bmatrix} \begin{bmatrix} \ell_k(e^{\omega_1}) \\ \vdots \\ \ell_k(e^{\omega_{n+1}}) \end{bmatrix} = f_k \delta_{jk}$$

holds for $j, k = 1, \ldots, n+1$, and therefore we conclude that

$$L^T M_n L = \begin{bmatrix} f_1 & & \\ & \ddots & \\ & & f_{n+1} \end{bmatrix}. \tag{5.39}$$

Hence L and L^T diagonalize M_n, even if L is not a unitary matrix. Note that (5.39) gives yet another proof of Lemma 5.3. This is actually not very surprising in view of Problem 5.13.

Now let g be an arbitrary function, defined at least at the locations e^{ω_k}, $k = 1, \ldots, n+1$; then the interpolation polynomial $p \in \Pi_n$ at these sites is

$$p = \sum_{j=1}^{n+1} g(e^{\omega_j}) \ell_j,$$

and we get by (5.38) that the interpolatory quadrature formula at the knots e^{ω_j} takes the form

$$L(p) = \sum_{j=0}^{n+1} p_j L\left((\cdot)^j\right) = \sum_{j=0}^{n+1} p_j \mu_j = e_1^T M_n p = \sum_{j=1}^{n+1} g(e^{\omega_j}) e_1^T M_n \ell_j$$

$$= \sum_{j=1}^{n+1} f_j \, g(e^{\omega_j}). \tag{5.40}$$

In other words the coefficients f_j are the weights for the Gaussian quadrature formula

$$Q_n = \sum_{j=1}^{n+1} f_j \, \delta_{e^{\omega_j}}, \tag{5.41}$$

written in terms of the Dirac distributions at the nodes. Summarizing all of the above, we get the final small piece of insight.

Corollary 5.4 (Moments and Quadrature)

1. *A flat extension of a moment sequence is equivalent to a Gaussian quadrature formula.*
2. *The weights of a Gaussian quadrature formula are positive.*

Remark 5.8 Statement 2 of Corollary 5.4 is classical and can also be proved easily by other methods, see Problem 5.14.

Equation (5.41) relates Prony's problem and the moment problem, and allows us to formulate Prony's problem as a moment recovery problem, cf. [65]. The exponential function f can then be seen as a distributional *Fourier transform*

$$Q_n = \left(\sum_{j=1}^{n+1} f_j \, \delta_{\omega_j} \right)^{\wedge} (ix) = \sum_{j=1}^{n+1} f_j \int_{\mathbb{R}} \delta_{\omega_j}(t) \, e^{xt} \, dt = \sum_{j=1}^{n+1} f_j \, e^{\omega_j x}$$

of the Dirac distributions at ω_j, a point of view taken in the context of optical superresolution, cf. [14]. For an introductionary explanation, see also [90]; further details will be considered in Sect. 6.6.

5.2.1 Problems

5.12 Show that all principal submatrices of a symmetric and positive definite matrix are symmetric and positive definite again. Recall that the kth *principal submatrix* of $A \in \mathbb{R}^{n \times n}$ is the matrix

$$A_k = \begin{bmatrix} a_{11} & \dots & a_{1k} \\ \vdots & \ddots & \vdots \\ a_{k1} & \dots & a_{kk} \end{bmatrix} \in \mathbb{R}^{k \times k}.$$

5.13 Prove that

$$\begin{bmatrix} \ell_1 & \dots & \ell_{n+1} \end{bmatrix} =: L = \begin{bmatrix} 1 & \dots & 1 \\ e^{\omega_1} & \dots & e^{\omega_{n+1}} \\ \vdots & \ddots & \vdots \\ e^{n\,\omega_1} & \dots & e^{n\,\omega_{n+1}} \end{bmatrix}^{-1}.$$

5.14 Show that any quadrature formula of order n with exactness at least $2n$ must have positive weights.

Hint: The polynomials ℓ_j^2 are of degree $2n$, $j = 0, \ldots, n$.

5.3 Flat Extensions via Prony

The flat extension defined by the continued fractions, as in the proof of Theorem 5.2 is unique: the moments $\mu_0, \ldots, \mu_{2n}$ define the first $n + 1$ coefficients $r_1, \ldots, r_{n+1} \in \Pi_1 \setminus \Pi_0$ of the continued fraction expansion. These coefficients, in turn, yield the $(n + 1)$-convergent which expands into the Laurent series of the flat extension; its denominator is the orthogonal polynomial associated to the measure. Is this the only possible flat extension? The following example shows that this is not the case.

Example 5.3 We want to extend $M_0 = [\mu_0]$, $\mu_0 \neq 0$, into a Hankel matrix

$$M_1 = \begin{bmatrix} \mu_0 & b \\ b & c \end{bmatrix}$$

of rank 1. Clearly rank $M_1 = 1$ if and only if $0 = \det M_1 = \mu_0 c - b^2$ leads to $c = b^2/\mu_0$. The normalized kernel vector $[-b/\mu_0, 1]^T$ defines the Prony polynomial $p(x) = x - b/\mu_0$ with zero b/μ_0, which suggests that the underlying exponential function that generates μ is

$$f(x) = \mu_0 \left(\frac{b}{\mu_0} \right)^x.$$

Indeed,

$$f(0) = \mu_0, \qquad \hat{\mu}_1 := f(1) = b, \qquad \hat{\mu}_2 = f(2) = \frac{b^2}{\mu_0},$$

so that for *any* $b \in \mathbb{R}$, there exists a flat extension of M_0.

Of course this example is very simple, but nevertheless it already gives hints for a more general way to progress. To do so, assume that we are given a nonsingular $M_n \in \mathbb{C}^{(n+1)\times(n+1)}$ for which we want to obtain a flat extension

$$\hat{M}_{n+1} = \begin{bmatrix} M_n & v \\ v^T & c \end{bmatrix} \in \mathbb{C}^{(n+2)\times(n+2)}, \qquad v \in \mathbb{C}^{n+1}, \quad c \in \mathbb{C},$$

without further assumptions on $M_0, \ldots, M_{n-1}$. Keep in mind that as a Hankel matrix $\hat{M}_{n+1}$ must be symmetric and not Hermitian, even if the entries of the matrix may be complex, so that v^T in the lower left part of the matrix is justified.

The requirements that rank $\hat{M}_{n+1} = $ rank $M_n = n + 1$ and that $\hat{M}_{n+1}$ is a Hankel matrix can then be expressed equivalently as

$$c = v^T M_n^{-1} v \quad \text{and} \quad v = \begin{bmatrix} \mu_{n+1} \\ \vdots \\ \mu_{2n} \\ b \end{bmatrix}, \quad b \in \mathbb{R}; \tag{5.42}$$

see the *Schur formula* stated in Problem 5.15 for the first property. A vector $u = [q, 1]^T \in \mathbb{R}^{n+2}$ satisfies $\hat{M}_{n+1} u = 0$ if and only if

$$0 = M_n q + v = M_n \left(q + M_n^{-1} v \right) \tag{5.43}$$

$$0 = v^T q + c = v^T q + v^T M_n^{-1} v = v^T \left(q + M_n^{-1} v \right), \tag{5.44}$$

which is the case if and only if[6]

$$q = -M_n^{-1} v = -M_n^{-1} \begin{bmatrix} \mu_{n+1} \\ \vdots \\ \mu_{2n} \\ 0 \end{bmatrix} - b \, M_n^{-1} e_n. \tag{5.45}$$

In fact (5.45) implies (5.43) and (5.44) while, conversely, it follows already from (5.43) and the fact that M_n is nonsingular, so that (5.44) is even redundant. Consequently the vector q depends linearly on the number b, multiplied by the last column of the inverse M_n^{-1} which is a nonzero vector.

The associated Prony polynomial

$$q(x) = x^{n+1} + \sum_{j=0}^{n} q_j x^j = x^{n+1} - \sum_{j=0}^{n} \left(M_n^{-1} v \right)_j x^j = \prod_{j=1}^{n+1} (x - \zeta_j) \tag{5.46}$$

has $n + 1$ zeros which we call $\zeta_j \in \mathbb{C}$. We again assume for simplicity that these zeros are simple; how the case of multiple zeros will be treated is sketched in Problem 5.16. As shown in the preceding section, any function of the form

$$f(x) = \sum_{j=1}^{n+1} f_j \zeta_j^x, \quad f := \begin{bmatrix} f_1 \\ \vdots \\ f_{n+1} \end{bmatrix} \in \mathbb{C}^{n+1}, \tag{5.47}$$

defines a sequence

$$\mu_f : k \mapsto f(k)$$

with $\mu_f \star q = 0$. The sequences $v_j : k \mapsto \zeta_j^k$, $j = 1, \ldots, n + 1$, are linearly independent, and thus generate a linear space of dimension $n + 1$. However this property is already valid for their initial segments considered as vectors.

---

[6] Keep in mind that the the numbering of the rows and columns of M_n starts at 0; hence there also would be a unit vector e_0 and the vector appearing in (5.44) is $e_n = [\delta_{jn} : j = 0, \ldots, n] \in \mathbb{R}^{n+1}$.

Lemma 5.4 *If the zeros of q are simple, i.e., $\zeta_j \neq \zeta_k$, $1 \leq j < k \leq n+1$, then the vectors*

$$v_j := \begin{bmatrix} v_j(0) \\ \vdots \\ v_j(n) \end{bmatrix} = \begin{bmatrix} 1 \\ \zeta_j \\ \vdots \\ \zeta_j^n \end{bmatrix} \in \mathbb{C}^{n+1}, \qquad j = 1, \ldots, n+1,$$

are linearly independent.

Proof The matrix

$$\begin{bmatrix} v_1 & \ldots & v_{n+1} \end{bmatrix} = \begin{bmatrix} 1 & \ldots & 1 \\ \zeta_1 & \ldots & \zeta_{n+1} \\ \vdots & \ddots & \vdots \\ \zeta_1^n & \ldots & \zeta_{n+1}^n \end{bmatrix}$$

is a Vandermonde matrix for distinct points, hence invertible. This in turn is equivalent to the fact that the vectors are linearly independent. $\qquad\square$

Remark 5.9 If q has multiple zeros, one has to consider exponential polynomials which, however, are again linearly independent. This is left as an exercise for the reader, see Problem 5.19.

Any sequence μ_f satisfies[7] $\mu_f \star q = 0$ for the monic Prony polynomial q from (5.46); hence, in particular,

$$0 = (\mu_f \star q)_0 = \cdots = (\mu_f \star q)_n,$$

which can be written in matrix form as

$$0 = \begin{bmatrix} q_0 & \ldots & q_n & 1 & & \\ & q_0 & \ldots & q_n & 1 & \\ & & \ddots & \vdots & \ddots & \ddots \\ & & & q_0 & \ldots & q_n & 1 \end{bmatrix} \begin{bmatrix} \mu_{f,0} \\ \vdots \\ \mu_{f,2n+1} \end{bmatrix} =: Q \begin{bmatrix} \mu_{f,0} \\ \vdots \\ \mu_{f,2n+1} \end{bmatrix}, \tag{5.48}$$

where $Q \in \mathbb{C}^{(n+1) \times (2n+2)}$ has rank $n+1$, since its last $n+1$ columns form a lower triangular matrix with diagonal elements 1 and therefore they are linearly independent. Thus the kernel space of Q has dimension $n+1$ and contains the elements $(v_j(k) : k = 0, \ldots, 2n+2)$, which are linearly independent since their initial segments of length $n+1$ are already linearly independent as vectors in $\mathbb{R}^{n+1}$ by Lemma 5.4.

---

[7] Recall that this version of the correlation "$\star$" is defined on $\ell(\mathbb{Z}) \times \Pi$, so q stands for the polynomial and not for the vector of its coefficients of a lower degree. Not that it really makes a difference.

On the other hand, the property

$$\hat{M}_{n+1} \begin{bmatrix} q \\ 1 \end{bmatrix} = 0,$$

which defined q also yields that

$$0 = (\hat{\mu} \star q)_0 = \cdots = (\hat{\mu} \star q)_n,$$

where $\hat{\mu}$ is *any* sequence such that $\hat{\mu}_k = \mu_k$, $k = 0, \ldots, 2n$, and $\hat{\mu}_{2k+1} = b$. Therefore there exists an exponential sum f of the form (5.47) such that $\hat{\mu} = \mu_f$, i.e.,

$$\mu_k = \mu_{f,k}, \quad k = 0, \ldots, 2n, \qquad \mu_{f,2k+1} = b.$$

This is the required flat extension of μ which has the property that

$$\text{rank } M_{f,k} = n + 1, \qquad k \geq n.$$

Let us summarize the findings of this section.

Theorem 5.3 *For any sequence μ such that* $\det M_n \neq 0$, *there exists a one parameter family μ_b of flat extensions, namely*

$$\mu_{b,k} = \mu_k, \quad k = 0, \ldots, 2n, \qquad \text{rank } M_{b,k} = \text{rank } M_n = n + 1, \qquad k \geq n.$$

Proof The above reasoning shows that for each value b in (5.42), there exists an f such that μ_f is the flat extension for μ with $\mu_{f,2n+1} = b$, at least if the associated Prony polynomial q has simple zeros. In fact, if q^0 is the Prony polynomial for $b = 0$ and $w \in \mathbb{R}^{n+1}$ denotes the last column of M_n^{-1}, then the extension μ_b satisfies

$$0 = \mu_b \star q = \mu_b \star (q^0 - bw)$$

due to (5.45). Together with $\mu_{b,k} = \mu_k$, $k = 0, \ldots, 2n$, and $\mu_{b,2n+1} = b$, this implies that

$$\mu_{b,2n+2} + \sum_{j=0}^{n} \mu_{b,n+1+j} q_j^0 = (\mu_b \star q^0)_{n+1} = (\mu_b \star w)_{n+1} = b \sum_{j=0}^{n} \mu_{b,n+1+j} w_j,$$

yielding the difference equation

$$\mu_{b,2n+1+k} = \sum_{j=0}^{n} (b \, w_j - q_j^0) \, \mu_{b,j+n+k}, \qquad k \geq 1, \tag{5.49}$$

that successively computes the coefficients of the extension from the initial data $\mu_0, \ldots, \mu_{2k+1}$. $\square$

Remark 5.10 Equation (5.49) shows that the dependency of μ_b from b is a nonlinear one, but nevertheless each *element* of the sequence depends continuously on b. Since the growth of the elements of μ_b is usually exponential in b, we cannot expect that the complete sequence μ_b depends continuously on b in the sense that

$$\lim_{b' \to b} \|\mu_{b'} - \mu_b\|_p = 0;$$

if fact we even have $\|\mu_{b'} - \mu_b\|_p = \infty$ for $b' \neq b$ and any $1 \le p \le \infty$. The story is different once one considers only finite segments of the sequence.

Let us illustrate this behavior with a simple example.

Example 5.4 We consider the flat extension of $\mu = (0, 1, 0, \ldots)$ which requires us to consider

$$\hat{M}_2 = \begin{bmatrix} 0 & 1 & 0 \\ 1 & 0 & b \\ 0 & b & c \end{bmatrix}, \quad c = \begin{bmatrix} 0 & b \end{bmatrix} \begin{bmatrix} 0 & 1 \\ 1 & 0 \end{bmatrix}^{-1} \begin{bmatrix} 0 \\ b \end{bmatrix} = 0.$$

The coefficients of q are given as

$$-\begin{bmatrix} 0 & 1 \\ 1 & 0 \end{bmatrix} \begin{bmatrix} 0 \\ b \end{bmatrix} = \begin{bmatrix} -b \\ 0 \end{bmatrix};$$

that is

$$q(x) = x^2 - b = \left(x + \sqrt{b}\right)\left(x - \sqrt{b}\right)$$

with the zeros $\pm\sqrt{b}$. Therefore for $b \neq 0$,

$$f(x) = f_1 b^{x/2} + f_2 (-b)^{x/2}, \quad \mu_{b,k} = f_1 b^{k/2} + f_2 (-b)^{k/2}.$$

To determine f_1 and f_2, we interpolate the values of μ:

$$\begin{bmatrix} 0 \\ 1 \\ 0 \end{bmatrix} = \begin{bmatrix} \mu_0 \\ \mu_1 \\ \mu_2 \end{bmatrix} = \begin{bmatrix} f(0) \\ f(1) \\ f(2) \end{bmatrix} = \begin{bmatrix} 1 & 1 \\ \sqrt{b} & -\sqrt{b} \\ b & b \end{bmatrix} \begin{bmatrix} f_1 \\ f_2 \end{bmatrix} \quad \Rightarrow \quad f_1 = -f_2 = \frac{1}{2\sqrt{b}},$$

so that

$$f(x) = b^{(x-1)/2} \frac{1 - (-1)^{x/2}}{2}$$

and

$$\mu_{b,2k} = 0, \quad \mu_{b,2k+1} = b^k,$$

giving the flat extension

$$(0, 1, 0, b, 0, b^2, 0, b^3, \dots)$$

of μ.

Theorem 5.3 can also be interpreted in the sense of an interpolation problem with exponentials or, more precisely, exponential polynomials.

Corollary 5.5 *Given* $2m$ *values* $y_0, \dots, y_{2m-1} \in \mathbb{C}$, *there exists an exponential polynomial*

$$f(x) = \sum_{j=0}^{m} f_j(x) \, e^{\omega_j x}, \qquad \omega_j \in \mathbb{R} + i\mathbb{T}, \qquad \sum_{j=1}^{n} (d_j + 1) = m$$

such that

$$f(j) = y_j, \qquad j = 1, \dots, 2m,$$

if the Hankel matrix formed by $y_0, \dots, y_{2m-2}$ *is nonsingular.*

Proof The entries of the Hankel matrix M_{n-1} define the finite sequence $\mu_k = y_k$, $k = 0, \dots, 2m - 2$; and also set $b = y_{2m-1}$. The associated Prony polynomial q defines an exponential polynomial which gives a flat extension and in particular interpolates the data.   □

From an intuitive perspective, Corollary 5.5 seems very natural, as one fits $2m$ parameters, namely the frequencies ω_j and the coefficients of the polynomials f_j to $2m$ values $y_0, \dots, y_{2m-1}$. Even if this is a good indication that the problem may be qualified for a unique solution, it is not a proof of that of course, in particular since the problem is of a nonlinear nature, which would require the nonsingularity of a Jacobian to be uniquely solvable, at least locally.

5.3.1 Problems

5.15 (*Schur formula, see* [27], p. 47)] Let $A \in \mathbb{R}^{n \times n}$ be nonsingular, $B \in \mathbb{R}^{n \times m}$, $C \in \mathbb{R}^{m \times n}$ and $D \in \mathbb{R}^{m \times m}$. Then the block matrix

$$\begin{bmatrix} A & B \\ C & D \end{bmatrix} \in \mathbb{R}^{(n+m) \times (n+m)}$$

has rank n if and only if $D = CA^{-1}B$.

Hint: Gauss elimination.

5.16 Let $q \in \Pi_n$ be of the form

$$q = \prod_{j=1}^{s} (\cdot - \zeta_j)^{d_j}, \quad \zeta_j \in \mathbb{C}, \quad j = 1, \ldots, s, \quad \sum_{j=1}^{s} d_j = n, \quad (5.50)$$

Then $\mu \in \ell(\mathbb{Z})$ satisfies $\mu \star q = 0$ if and only if

$$\mu = (f(k) : k \in \mathbb{Z}), \quad f(x) = \sum_{j=1}^{s} f_j(x) \zeta_j^x, \quad f_j \in \Pi_{d_j-1}. \quad (5.51)$$

Hint: use the tools from Problem 5.7–5.11.

5.17 Show that the polynomial q from (5.50) is the unique monic solution of minimal degree of the θ *Hermite interpolation problem*

$$0 = \theta^k p(\zeta_j), \quad k = 0, \ldots, d_j - 1, \quad j = 1, \ldots, s.$$

5.18 With q from (5.50), show that

$$\{\mu : \mu \star q = 0\}$$

is a shift-invariant subspace of $\ell(\mathbb{Z})$ of dimension n.

5.19 Show that for $\zeta_j \neq \zeta_k$, $1 \leq j < k \leq s$, and

$$d_j \geq 1, \quad \sum_{j=1}^{s} d_j = n + 1,$$

the vectors

$$v_{j,k} = \left(\ell^k \zeta_j^\ell : \ell = 0, \ldots, n \right) = \begin{bmatrix} \delta_{k0} \\ \zeta_j \\ 2^k \zeta_j \\ \vdots \\ n^k \zeta_j^n \end{bmatrix}, \quad k = 0, \ldots, d_j - 1, \quad j = 1, \ldots, s$$

are linearly independent.

Hint: write the vectors in terms of θ operators.

5.20 Complete the proof of Theorem 5.3.

Chapter 6
Digital Signal Processing

When the epoch of analogue (which was to say also the richness of language, of analogy*) was giving way to the digital era, the final victory of the numerate over the literate.*

S. *Rushdie* , Fury

6.1   Signals and Filters

A *time discrete signal* is a doubly infinite sequence of the form

$$\sigma = \left(\sigma_j \; : \; j \in \mathbb{Z}\right).$$

We denote by

$$\ell(\mathbb{Z}) = \left\{\sigma = \left(\sigma_j \; : \; j \in \mathbb{Z}\right) : \sigma_j \in \mathbb{C}\right\}$$

the vector space of all such signals with possible complex coefficients.

Realistic signals normally have a beginning and an end, hence a *finite support* and at least *finite energy*. Nevertheless it is much more convenient to work with bi-infinite signals, so we do not have to worry about any boundary issues that are always very inconvenient to track.

The concepts of finite support and energy can be formalized.

Definition 6.1 (*p-norm*) The *p-norm* of a signal, $1 \leq p \leq \infty$, is defined as

$$\|\sigma\|_p := \left(\sum_{j \in \mathbb{Z}} |\sigma_j|^p\right)^{1/p}, \qquad 1 \leq p < \infty \qquad (6.1)$$

and

© The Author(s), under exclusive license to Springer Nature Switzerland AG 2021
T. Sauer, *Continued Fractions and Signal Processing*, Springer Undergraduate Texts
in Mathematics and Technology, https://doi.org/10.1007/978-3-030-84360-1_6

$$\|\sigma\|_\infty := \sup_{j\in\mathbb{Z}} |\sigma_j|.$$

Moreover, we denote by $\ell_p(\mathbb{Z})$ the space of all signals with a finite p-norm, i.e.,

$$\ell_p(\mathbb{Z}) = \left\{\sigma \in \ell(\mathbb{Z}) : \|\sigma\|_p < \infty\right\}, \qquad 1 \le p \le \infty. \tag{6.2}$$

The "0 norm"

$$\|\sigma\|_0 := \#\left\{j : \sigma_j \ne 0\right\}$$

recently gained a lot of popularity in connection with *Compressive Sensing* and is nowadays considered frequently. In this spirit, we use $\ell_0(\mathbb{Z})$ for the vector space of finitely supported sequences.

In Digital Signal Processing one usually considers only very special operators on the signal spaces.

Definition 6.2 (*Filter*) A *filter* $F : \ell(\mathbb{Z}) \to \ell(\mathbb{Z})$ is an operator defined on the discrete signal space. It is called an *LTI filter*, short for the **L**inear **T**ime **I**nvariant Filter, if F is a *linear operator* and *time invariant*, in which the latter means that

$$\sigma'_j = \sigma_{j+k}, \quad j \in \mathbb{Z} \quad \Rightarrow \quad \left(F\sigma'\right)_j = (F\sigma)_{j+k}, \quad j \in \mathbb{Z}. \tag{6.3}$$

Remark 6.1 It is common practice in signal processing to use the term *digital filter* or even just *filter* synonymously for "LTI filter", cf. [44].

With the *shift operator* $\tau : \ell(\mathbb{Z}) \to \ell(\mathbb{Z})$ defined as $(\tau\sigma)_j = \sigma_{j+1}$, there is a nice and simple way to describe LTI filters.

Lemma 6.1 *A filter F is an LTI filter if and only if it commutes with τ, i.e.,*

$$\tau F = F\tau. \tag{6.4}$$

Proof Writing σ' in (6.3) as $\sigma' = \tau^k\sigma$, the LTI property is equivalent to

$$F\tau^k\sigma = F\sigma' = \tau^k F\sigma, \qquad k \in \mathbb{Z},$$

hence (6.4) follows for any LTI filter by setting $k = 1$. Conversely we simply observe that (6.4) implies for $k > 0$ as follows:

$$\tau^k F = \tau^{k-1} F\tau = \cdots = F\tau^k,$$

and the argument for $k < 0$ is similar. $\square$

Any linear filter F can be written as a bi-infinite matrix $F = \left[F_{jk} : j, k \in \mathbb{Z}\right]$, such that

$$(Ff)_j = \sum_{k\in\mathbb{Z}} F_{jk} f_k, \qquad j \in \mathbb{Z}.$$

If F is an LTI filter, then

$$\left[F_{j+1,k} \ : \ j, k \in \mathbb{Z}\right] = \tau F = F \tau = \left[F_{j,k-1} \ : \ j, k \in \mathbb{Z}\right]. \tag{6.5}$$

To verify (6.5), recall that for $j \in \mathbb{Z}$,

$$((F\tau)\,f)_j = (F\,(\tau f))_j = \sum_{k \in \mathbb{Z}} F_{j,k}(\tau f)_k = \sum_{k \in \mathbb{Z}} F_{j,k} f_{k+1} = \sum_{k \in \mathbb{Z}} F_{j,k-1} f_k.$$

Since the two bi-infinite matrices in (6.5) define the same operator, they must coincide in all components; hence $F_{j+1,k} = F_{j,k-1}$, $j, k \in \mathbb{Z}$, or after iteration we have

$$F_{j+\ell,k} = F_{j,k-\ell}, \qquad \ell \in \mathbb{Z}.$$

This holds true whenever $F_{jk} = f_{j-k}$ for some $f \in \ell(\mathbb{Z})$, but conversely we also have that $j - k = \ell - m$ implies $j - \ell = k - m$, and therefore

$$F_{jk} = F_{\ell+(j-\ell),k} = F_{\ell,k-(k-m)} = F_{\ell,m},$$

so that F_{jk} depends only on $j - k$ which can be summarized as follows.

Proposition 6.1 *A filter* $F \ : \ \ell(\mathbb{Z}) \to \ell(\mathbb{Z})$ *is an LTI filter if there exists* $f \in \ell(\mathbb{Z})$ *such that* $F_{jk} = f_{j-k}$, $j, k \in \mathbb{Z}$. *In that case,*

$$(F\sigma)_j = \sum_{k \in \mathbb{Z}} f_{j-k}\,\sigma_k, \qquad j \in \mathbb{Z}. \tag{6.6}$$

Definition 6.3 The sum in (6.6) is called the *convolution* $f * \sigma$ between f and σ. An operator F of the form

$$F = \left[f_{j-k} : j, k \in \mathbb{Z}\right]$$

is called a *Toeplitz operator*.

If $\|f\|_0 = \|\sigma\|_0 = \infty$, the sum occurring in the definition of $f * \sigma$ in (6.6) is an infinite one and does not need to be well defined. One way to overcome this problem is to avoid it by requesting that at least one of the operands be finitely supported, usually it is the sequence f that represents the filter.

Remark 6.2 A Toeplitz operator for a sequence f is almost the same as a Hankel operator for f, just with the two differences that a Toeplitz operator is *bi*-infinitely supported and that its entries are of the form f_{j-k}. The Hankel operator on the other hand is usually only defined for nonnegative entries and the matrix elements are given as f_{j+k}.

Next we consider some fundamental terminology from signal processing.

Definition 6.4 *(Pulse, filter types and z-transform)*

1. The *pulse* $\delta \in \ell(\mathbb{Z})$ is defined as $\delta_j = \delta_{j0}, \ j \in \mathbb{Z}$.
2. The *impulse response* of a filter F is the signal $F\delta$.
3. The *support* of a signal $\sigma \in \ell(\mathbb{Z})$ is defined as

$$\mathrm{supp}\ \sigma = \{ j \in \mathbb{Z} \ : \ \sigma_j \neq 0 \} .$$

4. A filter is called *FIR Filter*, short for **F**inite **I**mpulse **R**esponse Filter, if it is an LTI filter with a finitely supported impulse response: $F\delta \in \ell_0(\mathbb{Z})$. Otherwise the filter is called an *IIR filter*, which abbreviates the name **I**nfinite **I**mpulse **R**esponse Filter.
5. The *z-transform* of $f \in \ell(\mathbb{Z})$ is the formal bi-infinite *Laurent series*

$$f(z) = \sum_{k \in \mathbb{Z}} f_k \, z^{-k}, \qquad z \in \mathbb{C}^\times = \mathbb{C} \setminus \{0\}.$$

It is a Laurent polynomial provided that $f \in \ell_0(\mathbb{Z})$.

One reason for introducing the z-transform is easily seen: for arbitrary signals $f, g \in \ell(\mathbb{Z})$, one has

$$(f * g)(z) = \sum_{j \in \mathbb{Z}} \left(\sum_{k \in \mathbb{Z}} f_{j-k} \, g_k \right) z^{-j}$$

$$= \sum_{j \in \mathbb{Z}} \sum_{k \in \mathbb{Z}} f_j \, g_k \, z^{-j-k} = \left(\sum_{j \in \mathbb{Z}} f_j z^{-j} \right) \left(\sum_{k \in \mathbb{Z}} g_k z^{-k} \right),$$

that is,

$$(f * g)(z) = f(z) \, g(z), \tag{6.7}$$

so that the z-transform turns convolutions into products. In particular any LTI filter F can be expressed as

$$(F\sigma)(z) = f(z) \, \sigma(z). \tag{6.8}$$

This property is not only provided by a convolution or filtering alone but also by a large class of operators.

Definition 6.5 Given $m \in \mathbb{Z}$, the *generalized convolution* with *dilation m* is defined as

$$(f *_m g)_j := \sum_{j \in \mathbb{Z}} f_{j-mk} g_k, \qquad j \in \mathbb{Z}. \tag{6.9}$$

In particular we get the convolution as $f * g = f *_1 g$ and the correlation as $f \star g = f *_{-1} g$. The associated *generalized Toeplitz operator* has the form

$$T_m(f) = \left[f_{j-mk} : j, k \in \mathbb{Z} \right].$$

The z-transform of a generalized convolution is easily computed as

$$(f *_m g)(z) = \sum_{j \in \mathbb{Z}} z^{-j} \sum_{k \in \mathbb{Z}} f_{j-mk} g_k = \sum_{\ell=0}^{m-1} \sum_{j \in \mathbb{Z}} z^{-mj-\ell} \sum_{k \in \mathbb{Z}} f_{mj+\ell-mk} g_k$$

$$= \sum_{\ell=0}^{m-1} \sum_{j \in \mathbb{Z}} z^{-mj-\ell} \sum_{k \in \mathbb{Z}} f_{mk+\ell} g_{j-k} = \sum_{\ell=0}^{m-1} \sum_{j \in \mathbb{Z}} \sum_{k \in \mathbb{Z}} f_{mk+\ell} z^{-mk-\ell} g_{j-k} z^{-m(j-k)}$$

$$= \left(\sum_{\ell=0}^{m-1} \sum_{k \in \mathbb{Z}} f_{mk+\ell} z^{-mk-\ell} \right) \left(\sum_{j \in \mathbb{Z}} g_{j-k} z^{-m(j-k)} \right),$$

hence

$$(f *_m g)(z) = f(z) g(z^m). \qquad (6.10)$$

With a different interpretation of refining sequences, generalized convolutions based on $m \geq 2$ are also known as *stationary subdivision operators*, cf. [15] but this would be a different story.

If $f \in \ell_0(\mathbb{Z})$ is finitely supported, then $f(z)$ is a Laurent polynomial that can be written as

$$f(z) = z^k p(z), \qquad p \in \Pi, \qquad p(0) \neq 0, \qquad (6.11)$$

for some appropriate index $k \in \mathbb{Z}$. In terms of filters this means that

$$F = \tau^k P, \qquad P\delta = p,$$

where P is the filter with p as its impulse response. This filter has the property that

$$(P\sigma)_j = \sum_{k \in \mathbb{Z}} p_{j-k} \sigma_k = \sum_{k \in \mathbb{Z}} p_k \sigma_{j-k} = \sum_{k \in \mathbb{N}_0} p_k \sigma_{j-k},$$

and the filtered signal at time j depends only on σ_k, $k \leq j$, which is knowing the value of σ from earlier steps in time. Since the future is usually slightly nontrivial to predict, these are the filters whose output depends only on data from the past and that can be implemented in reality.

Definition 6.6 An FIR filter F with as impulse response of the form (6.11) is called a *causal filter* if $f(z^{-1}) \in \Pi$, i.e., if $k \geq 0$ in (6.11).

6.1.1   Problems

6.1 Prove that the norms $\| \cdot \|_p$, $1 \leq p \leq \infty$ from Definition 6.1 are indeed norms.

6.2 Show that the "0-norm" is not a norm.

6.3 Show that the p-norms are antitone in p, i.e., for any $\sigma \in \ell_{p'}(\mathbb{Z})$ one has

$$\|\sigma\|_p \geq \|\sigma\|_{p'}, \qquad 1 \leq p \leq p' \leq \infty.$$

6.4 Show that the shift operator is a filter and has the z-transform $\tau(z) = z$.

6.5 Show that the convolution $f * g$ is symmetric for $f, g \in \ell(\mathbb{Z})$ and well defined in each of the following cases:

1. $f \in \ell_0(\mathbb{Z})$,
2. $f, g \in \ell_2(\mathbb{Z})$,
3. $f, g \in \ell_1(\mathbb{Z})$,
4. $f \in \ell_\infty(\mathbb{Z})$, $g \in \ell_1(\mathbb{Z})$.

6.6 The *upsampling operator* $\uparrow_m$ is defined as

$$(\uparrow_m \sigma)_{mj} = \sigma_j, \qquad j \in \mathbb{Z},$$

and $(\uparrow_m \sigma)_j = 0$ for $j \notin m\mathbb{Z}$. Show that

$$(\uparrow_m \sigma)(z) = \sigma(z^m).$$

6.2   Fourier and Sampling

One can restrict the z-transform of a signal to the complex unit circle

$$\partial \mathbb{D} = \{z \in \mathbb{C} : |z| = 1\} = \{e^{-i\theta} : \theta \in \mathbb{T}\},$$

by substituting $z = e^{-i\theta}$ into the z-transform which gives us

$$\hat{f}(\theta) := f\left(e^{-i\theta}\right) = \sum_{k \in \mathbb{Z}} f_k e^{ik\theta}, \qquad \theta \in \mathbb{T}.$$

This is the *Fourier series* associated with the sequence $f \in \ell(\mathbb{Z})$ which, continuing in the spirit of the preceding sections, we will treat this again as a *formal* series only. Convergence of the Fourier series is a nontrivial issue and the discovery (by Du Bois-Reymond in 1873) of a divergent Fourier series of a continuous function eventually triggered new directions in mathematical research, most notably in the field of *Approximation Theory* that started with Weierstrass' approximation result [108] in 1885.

Definition 6.7 If $f \in \ell_0(\mathbb{Z})$, then the function

$$\hat{f}(\theta) = \sum_{-n}^{n} f_k e^{ik\theta} \tag{6.12}$$

is called a *trigonometric polynomial*. Its *degree* is the smallest $n \geq 0$ such that (6.12) holds, i.e.,

$$\min \{n \in \mathbb{N}_0 : f_k = 0, \ |k| > n\}.$$

The important point is the fact that by switching from the z-transform to trigonometric polynomials, we have objects that are only defined on the unit circle $\partial \mathbb{D}$ instead of $\mathbb{C}^\times := \mathbb{C} \setminus \{0\}$ without losing information. So if the Fourier series associated to f converges (e.g., when $f \in \ell_0(\mathbb{Z})$), then the associated sequence f can be recovered as the *Fourier coefficients*

$$f_k = \frac{1}{2\pi} \int_{\mathbb{T}} \hat{f}(\theta) \, e^{-ik\theta} \, d\theta, \qquad k \in \mathbb{Z},$$

of the function $\hat{f} : \mathbb{T} \to \mathbb{C}$. This is the proper *Fourier transform* for functions defined on $\mathbb{T}$; the *dual group* for $\mathbb{T}$ is $\mathbb{Z}$, cf. [58, 66].

The trigonometric series then also satisfies

$$(f * \sigma)^\wedge (\theta) = (f * \sigma) \left(e^{-i\theta}\right) = f\left(e^{-i\theta}\right) \sigma \left(e^{-i\theta}\right) = \hat{f}(\theta) \, \hat{\sigma}(\theta), \tag{6.13}$$

and in general

$$(f *_m \sigma)^\wedge (\theta) = \hat{f}(\theta) \, \hat{\sigma}(m\theta), \qquad m \in \mathbb{Z} \setminus \{0\}, \tag{6.14}$$

because of (6.10).

Definition 6.8 For $f \in \ell(\mathbb{Z})$, the complex valued function $\hat{f}(\theta)$, provided that it exists, is called the *transfer function* of the filter with impulse response f.

In the signal processing literature the transfer function is the most common way to describe a filter, and it is usually given in the logarithmic *decibel*[1] scale and the unit *dB*. Instead of the value y, this scale uses the value $10 \log_{10} y$, and the unit *dB* is added to indicate this modification.

Since sine and cosine are odd and even functions, respectively, we have that

---

[1] Named after ALEXANDER GRAHAM BELL, despite the missing "'l'". The "'deci'" refers to the fact that a decimal logarithm with basis 10 is used.

$$\widehat{f}(\theta) = f_0 + \sum_{k=1}^{\infty} (f_k + f_{-k}) \cos k\theta + i \sum_{k=1}^{\infty} (f_k - f_{-k}) \sin k\theta;$$

hence the transfer function is real valued if and only if $f_k = f_{-k}$, i.e., if and only if the filter is a *symmetric filter*. On the other hand if f has real coefficients which is the usual assumption in signal processing, then

$$\hat{f}(-\theta) = f(e^{i\theta}) = \sum_{k\in\mathbb{Z}} f_k \, e^{ik\theta} = \overline{\hat{f}(\theta)}, \tag{6.15}$$

where $\overline{(\cdot)}$ denotes complex conjugation. This shows that for real filters it suffices to know the transfer function only on half of $\mathbb{T}$, and normally on $[0, \pi]$ since its values on $[-\pi, \pi]$ can then be determined by (6.15).

Definition 6.9 A filter F with the impulse response f is called a *lowpass* filter if supp $\hat{f} = [-a, a]$; it is called a *highpass* filter if supp $\hat{f} = [\pi - a, \pi + a]$ for some $0 < a < \pi$. It is called a *bandpass* filter if supp $f \cap [0, \pi] = [a, b], 0 < a < b < \pi$.

The name "lowpass filter" is due to the fact that a lowpass filter preserves the low frequency content and eliminates the high frequency content of a signal. In fact let σ defined by $\sigma_k = e^{i\omega k}, k \in \mathbb{Z}$ be the discretization or sampling of an oscillatory signal with frequency $\omega \in [-\pi, \pi]$; then

$$(F\sigma)_j = \sum_{k\in\mathbb{Z}} f_{j-k} \, e^{i\omega k} = \sum_{k\in\mathbb{Z}} f_k \, e^{i\omega(j-k)} = e^{i\omega j} \hat{f}(-\omega),$$

i.e.,

$$F\sigma = \hat{f}(-\omega) \, \sigma,$$

so that $\hat{f}(\omega)$ describes the behavior at the frequency ω. Then a lowpass filter cuts off frequencies that are away from $\omega = 0$ which corresponds with the least oscillatory constant signal, while a highpass filter deletes low frequency content. A bandpass filter just keeps something in-between.

Remark 6.3 It seems counterintuitive and contrary to our perception of reality that frequencies efficiently range between 0 and π, which seems to indicate that there is some highest frequency. However we will soon see that the "true" frequency also depends on the sampling density which we restricted to sampling at the integers here.

Usually a filter is designed by defining its transfer function as a trigonometric polynomial – the filter coefficients can be read from the coefficient vector. The filter itself is usually defined by the way it handles low and high frequencies.

Example 6.1 (*Filter design*) We look for a filter with a lowpass characteristic, determined by the conditions

$$\widehat{f}(0) = 1, \qquad \widehat{f}(\pi) = 0. \tag{6.16}$$

We try a simple filter $f_0 = a$, $f_1 = f_{-1} = b$, $f_2 = f_{-2} = c$, with five symmetric coefficients and thus we have the real transfer function

$$\widehat{f}(\theta) = a + b \left(e^{-i\theta} + e^{i\theta}\right) + c \left(e^{-2i\theta} + e^{2i\theta}\right) = a + 2b \cos \theta + 2c \cos 2\theta.$$

Then (6.16) means that

$$a + 2b + 2c = 1, \qquad a - 2b + 2c = 0,$$

hence $b = \frac{1}{4}$ and $a = \frac{1}{2} - 2c$, yielding

$$\widehat{f}(\theta) = \frac{1}{2} - 2c + \frac{1}{2} \cos \theta + 2c \cos 2\theta = \frac{1}{2} - 2c + \frac{1}{2} \cos \theta + 2c \left(2 \cos^2 \theta - 1\right)$$

$$= \frac{1}{2} - 4c + \frac{1}{2} \cos \theta + 4c \cos^2 \theta = 4c \left(\cos^2 \theta + \frac{1}{8c} \cos \theta + \frac{1}{8c} - 1\right)$$

$$= 4 \left(\cos \theta + 1\right) \left(c \cos \theta + \frac{1}{8} - c\right).$$

The first factor $\cos \theta + 1$ guarantees the zero at $\theta = \pi$; the second degenerates to a constant for $c = 0$ which is the case for the simplest or shortest filter with the required properties. Some choices for c are depicted in Fig. 6.1.

Given any continuous or at least an absolutely integrable function $g : \mathbb{T} \to \mathbb{C}$, one can construct the coefficients of the associated filter by computing the Fourier coefficients

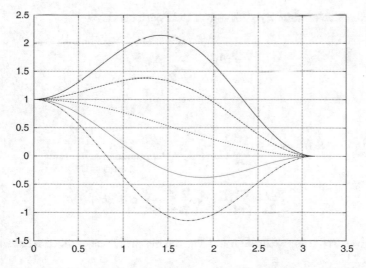

Fig. 6.1 Examples of the transfer function from Example 6.1 for $c = -0.4, -0.2, 0, 0.2, 0.4$ (top to bottom). In the simple case: $c = 0$ and the filter is even a *sigmoidal function*

$$f_k = \frac{1}{2\pi} \int_{\mathbb{T}} g(\theta)\, e^{-ik\theta}\, d\theta, \qquad k \in \mathbb{Z},$$

but this process will usually not lead to an FIR filter, especially not in practical relevant situations.

Example 6.2 (*Lowpass filter*) We construct an ideal lowpass filter that only preserves the lower half of the frequencies, and thus the lowpass filter has the transfer function $g = \chi_{[-\pi/2,\pi/2]}$ that cannot be a trigonometric polynomial since its support is a proper subset of $\mathbb{T}$; hence it cannot be represented by an FIR filter, see Problem 6.9. The Fourier coefficients of g and hence the filter coefficients are

$$f_0 = \frac{1}{2\pi} \int_{-\pi/2}^{\pi/2} dt = \frac{1}{2}$$

and for $k \neq 0$,

$$f_k = \frac{1}{2\pi} \int_{-\pi/2}^{\pi/2} e^{-ikt}\, dt = \frac{1}{2\pi} \left. \frac{e^{-ikt}}{-ik} \right|_{t=-\pi/2}^{\pi/2} = \frac{1}{2\pi} \frac{e^{-ik\pi/2} - e^{ik\pi/2}}{-ik} = \frac{1}{k\pi} \sin \frac{k}{2}\pi$$

$$= \begin{cases} 0, & k \in 2\mathbb{Z}, \\ \dfrac{(-1)^{(k-1)/2}}{k\pi}, & k \in 2\mathbb{Z}+1, \end{cases}$$

that is

$$f_{2k+1} = \frac{(-1)^k}{(2k+1)\pi}, \qquad f_{2k+2} = 0, \qquad k \in \mathbb{N}_0.$$

The *partial sum* of order $n = 2m+1$ of the Fourier series

$$\hat{f}(\theta) = \sum_{k=-\infty}^{\infty} f_k\, e^{ik\theta}$$

is

$$F_n(\theta) = \frac{1}{2} + \sum_{k=0}^{m} \frac{(-1)^k}{(2k+1)\pi} \underbrace{\left(e^{i(2k+1)\theta} + e^{-i(2k+1)x} \right)}_{2\cos(2k+1)\theta}$$

$$= \frac{1}{2} + \sum_{k=0}^{m} (-1)^k \frac{2}{(2k+1)\pi} \cos(2k+1)\theta,$$

which gives us the approximate filter. The quality problems of this filter can be seen in Fig. 6.2.

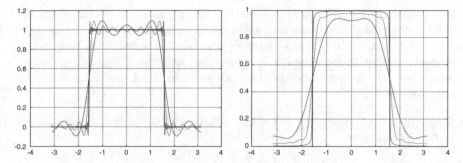

Fig. 6.2 *Left:* (Best) approximation of a bandpass by partial sums of the associated Fourier series for $n = 5, 15, 100$ to illustrate the Gibbs phenomenon. Observe that the overshooting effects only get more narrow, not smaller. *Right:* A *shape-preserving approximation* by so-called *Fejér means*. The overall quality is not so good, but the oscillations are gone

Remark 6.4 It is not really the partial sum that causes the trouble here. In fact the partial sum of the Fourier series is always the *best approximation* to the transfer function with respect to the norm

$$\|f\|_2 := \left(\int_{\mathbb{T}} |f(\theta)|^2 \, d\theta \right)^{1/2},$$

hence we cannot do any better and the Gibbs phenomenon is provably unavoidable. The shape-preserving approximations plotted on the right-hand side of Fig. 6.2 do not have this overshooting behavior and look more pleasant but they suffer from a lower rate of approximation, that is they are less accurate approximations.

The use of transfer functions is more intuitive since frequencies are now represented in quite a natural way: for example, a bandpass filter will now be really of the form $\widehat{f} = \chi_{[\omega_0, \omega_1]}$, where the precise frequencies associated to $0 \le \omega_1 < \omega_2 < \pi$ depend once more on the sampling rate. But since $\widehat{f}$ is always defined in $\mathbb{T}$, there has to be a relation between absolute frequencies and their representation in $\mathbb{T}$. This is done by the *sampling rate*. The point of view behind this is that the discrete signal σ is obtained from a continuous signal $s : \mathbb{R} \to \mathbb{R}$ by an *equidistant sampling*

$$\sigma_k = s\,(t_0 + kh), \qquad k \in \mathbb{Z}, \qquad t_0 \in \mathbb{R}, \, h > 0.$$

Here h is called the *sampling distance* and h^{-1} the *sampling rate*.

Intuitively, it should be quite clear that the frequency resolution of the sampling process will be related to the sampling rate: the finer the sampling, that is the higher the sampling rate, then the higher are the frequencies that can be detected. This is formalized in the famous *sampling theorem*, called the Shannon, Shannon–Whittaker or the Shannon–Whittaker–Kotelnikov sampling theorem. In fact Whittaker proved the recovery result in the context of infinite cardinal interpolation as early as 1915 in

[109], see also [110], but CLAUDE SHANNON not only rediscovered it in 1945, [97, 98], he also pointed out its meaning and importance in the context of digital signal processing. Kotelnikov [62] is historically somewhat in-between the two, and for obvious reasons was more popular in the Russian literature. The sampling theorem is based on a fundamental concept which identifies functions that can be recovered from samples.

Definition 6.10 A function $f \in L_1(\mathbb{R})$ is called *bandlimited* with *bandwidth T* if its *Fourier transform*

$$\widehat{f}(\xi) = \int_{\mathbb{R}} f(t)\, e^{-i\xi t}\, dt$$

vanishes outside $[-T, T]$:

$$\widehat{f}(\xi) = 0, \qquad \xi \notin [-T, T].$$

Note that Definition 6.10 uses the *continuous* Fourier transform of a function $f \in L_1(\mathbb{R})$, i.e., a function $f : \mathbb{R} \to \mathbb{C}$ such that

$$\|f\|_1 := \int_{\mathbb{R}} |f(x)|\, dx < \infty.$$

We will not dwell on *Fourier Analysis* here but simply recommend [30, 58, 105] and mention that the Fourier transform of a function defined on $\mathbb{R}$ is yet another function defined on $\mathbb{R}$. This fact can be phrased as the dual group of $\mathbb{R}$ being $\mathbb{R}$ again.

Being bandlimited means that the function f, seen as a signal defined on the continuum $\mathbb{R}$, contains only frequencies between $-T$ and T, hence the energy is localized in a compact subset of the spectrum of f. Bandlimited functions can be recovered *exactly* from discrete samples.

Theorem 6.1 (Sampling theorem) *If $f \in L_1(\mathbb{R})$ is a bandlimited function with bandwidth T and $\tau < \tau^* = \frac{\pi}{T}$, then*

$$f(x) = \sum_{k \in \mathbb{Z}} \sigma_k \frac{\sin \pi\, (x/\tau - k)}{\pi\, (x/\tau - k)}, \qquad \sigma_k = f(k\tau), \quad k \in \mathbb{Z}. \qquad (6.17)$$

The critical sampling rate $1/\tau^* = T/\pi$ or the half of it, depending on how precisely the Fourier transform and the bandwidth T are defined, is called the *Nyquist sampling rate* for the signal, and describes how finely the signal has to be sampled at least in order to recover it. The function

$$\frac{\sin \pi x}{\pi x} =: \operatorname{sinc} x, \qquad x \in \mathbb{R},$$

is called *sinus cardinalis* or *cardinal sine function*; the name is due to its behavior

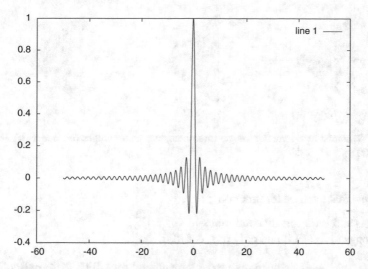

Fig. 6.3 The sinc function. It decays for $|x| \to \infty$, unfortunately only like $|x|^{-1}$ which makes it inconvenient for numerical applications

$$\text{sinc } k = \delta_{0k} = \begin{cases} 1, & k = 0, \\ 0, & \text{otherwise}, \end{cases}$$

at the *cardinal numbers* $\mathbb{Z}$, see Fig. 6.3. A proof of Theorem 6.1 can be found for example in [70].

Remark 6.5 (*Sampling rate & frequencies*) The sampling theorem answers the question regarding the frequency range of digital filters: the values $\theta \in [0, \pi]$ correspond to the frequencies $[0, h^{-1}]$, hence the frequency range is determined by the sampling rate.

6.2.1 Problems

6.7 Show that any trigonometric polynomial of the form (6.12) with $f_k \in \mathbb{C}$ can be written as

$$\hat{f}(\theta) = \sum_{k=0}^{n} a_k \cos k\theta + i \sum_{k=0}^{n} b_k \sin k\theta, \qquad a_k, b_k \in \mathbb{R},$$

$$= \sum_{k=0}^{n} c_k (\sin \theta)^k (\cos \theta)^{n-k}, \qquad c_k \in \mathbb{C},$$

which explains the name "trigonometric polynomial" because it is a polynomial in $\cos \theta$ and $\sin \theta$.

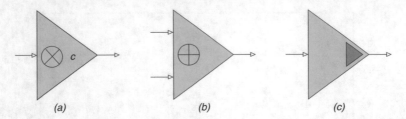

Fig. 6.4 Symbolic representation of the three components: multiplier (**a**), adder (**b**) and delay element (**c**)

6.8 Show that the transfer function $\hat{f}$:

1. exists and is uniformly continuous if $f \in \ell_1(\mathbb{Z})$,
2. belongs to $L_2(\mathbb{T})$ if and only if $f \in \ell_2(\mathbb{Z})$.

6.9 Let $f : \mathbb{T} \to \mathbb{C}$ be a trigonometric polynomial for which there exists an open set $\emptyset \neq I \subseteq \mathbb{T}$ such that $f(I) = 0$. Show that $f = 0$.

6.10 Show that if $f \in L_1(\mathbb{R})$ is a bandlimited function and there exists some non-degenerate interval $[a, b] \subset \mathbb{R}$, $a < b$, such that $f([a, b]) = 0$, then $f = 0$.
Hint: the proof requires the *inverse Fourier transform*

$$f^{\vee}(x) = \frac{1}{2\pi} \int_{\mathbb{R}} f(\xi) \, e^{i\xi x} \, d\xi,$$

that has the property that $(\hat{f})^{\vee} = f$ for any f such that $f, \hat{f} \in L_1(\mathbb{R})$.

6.3 Realization of Filters

The practical importance of FIR filters also results from the fact that they can be realized physically, at least if a certain *latency*, i.e., a delay of the output is accepted, and if one only considers a causal filter.

Example 6.3 Consider a causal filter F with zero components except for $f_0, \ldots, f_N$, and write its operation on a signal σ as

$$(F\sigma)_j = \sum_{k=0}^{n} f_k \sigma_{j-k} = \sum_{k=0}^{N} f_k \left(\tau^{-k}\sigma\right), \qquad j \in \mathbb{Z}. \tag{6.18}$$

The sum in (6.18) involves three operations: additions, multiplications and shift operations. More precisely for the output at index j, the kth coefficient of f has to be

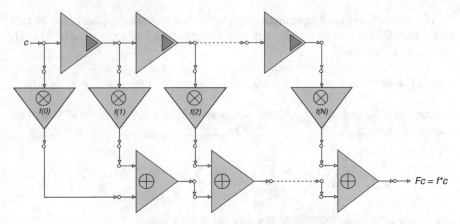

Fig. 6.5 Realization of a *FIR filter* as seen in the components from Fig. 6.4. The delay elements take care of the translations and the *latency* of the system is N clock tics

multiplied with σ_{j-k}, which is the state of σ just k tics earlier. In other words delayed information about σ is used at this point. This explains the third element in the filter, namely the *delay element*. The symbols for these three blocks: adder, multiplier and delay element are depicted in Fig. 6.4.

In this way any *FIR filter* can be built by cascading the three components from Fig. 6.4. Such a cascade for a causal FIR filter with coefficients $f_0, \ldots, f_N$ as in Example 6.3 is shown in Fig. 6.5.

We have already seen in the preceding chapters that FIR filters are somewhat limited in their capabilities and that the approximation they provide is often not very exciting, not the least due to the *Gibbs phenomenon*, see Fig. 6.2. And different approximation methods usually pay the price for loss in accuracy and localization.

To overcome this problem one extends the class of admissible filters by choosing a larger class of functions for z-transforms of the filter which are still controlled by a finite number of parameters. And in view of our experiences from the preceding chapters, it appears more than natural to switch from Laurent polynomials to rational functions.

Definition 6.11 A *rational filter* F is a filter that has a rational function as its z-transform, i.e.,

$$(F\sigma)(z) = f(z)\,\sigma(z) = \frac{p(z)}{q(z)}\,\sigma(z), \qquad p, q \in \Pi. \tag{6.19}$$

Keep in mind that it makes no difference whether we define numerator and denominator as Laurent polynomials or as polynomials, since we can always expand the fraction, multiplying numerator and denominator simultaneously by an arbitrary power of z and a nonzero constant.

We use this freedom of representation to normalize the rational filter in such a way that we can always assume p and q to be of the *causal filter* form $p(z^{-1}), q(z^{-1}) \in \Pi$, normalized such that

$$p(z) = p_0 + \cdots + p_m z^{-m}, \qquad q(z) = 1 + q_1 z^{-1} + \cdots + q_n z^{-n} = z^{-n} \tilde{q}(z),$$

for some $n \in \mathbb{N}_0$ and with $q_n \neq 0$. Hence $\tilde{q}(z) = q_n + q_{n-1} z + \cdots + z^n$ is a polynomial. By Lemma 4.1,

$$\frac{1}{q(z)} = z^n \frac{1}{\tilde{q}(z)} = z^n \sum_{j=n}^{\infty} \lambda_j z^{-j} = \sum_{j=0}^{\infty} \lambda_j z^{-j}, \qquad \lambda \in \ell(\mathbb{Z}),$$

is a purely Laurent series with no polynomial part so that

$$f(z) = \sum_{j=0}^{\infty} f_j z^{-j}, \qquad f \in \ell(\mathbb{Z}), \tag{6.20}$$

and normally the sequence f will not have finite support. Therefore we can neither hope nor expect that f is still an FIR filter. And why should we?

Remark 6.6 We could also choose $p, q \in \Pi$ as causal filters and that would give us an expansion of the form

$$f(z) = \sum_{j=0}^{\infty} f_j z^j$$

in terms of a formal power series. Both approaches are equivalent and correspond to the choice between the z-transform or the use of *symbol calculus* or *generating functions*. In the end it is a matter of taste, but since the z-transform is more common in signal processing, we decide for the approach in (6.20).

Although it is an IIR filter with infinitely many coefficients in the impulse response, a rational filter can still be implemented effectively with the physical building blocks from Fig. 6.4. To see this we use the above normalization to rephrase the definition of $(F\sigma)(z) = p(z)\sigma(z) q(z)^{-1}$ to be

$$p(z)\sigma(z) = (F\sigma)(z) q(z) = (F\sigma)(z) + (q(z) - 1)(F\sigma)(z),$$

that is

$$(F\sigma)(z) = p(z)\sigma(z) - \left(z^{-1}(F\sigma)(z)\right) \frac{q(z) - 1}{z^{-1}}$$

$$= p(z)\sigma(z) - \frac{q(z) - 1}{z^{-1}} \left(\tau^{-1} F\sigma\right)(z) \tag{6.21}$$

since

$$z^{-1}(F\sigma)(z) = \sum_{j\in\mathbb{Z}} (F\sigma)_j\, z^{-j-1} = \sum_{j\in\mathbb{Z}} (F\sigma)_{j-1}\, z^{-j} = \left(\tau^{-1} F\sigma\right)(z).$$

By our normalization $q(0) = 1$, it follows that

$$\frac{q(z)-1}{z^{-1}} = q_1 + q_2 z^{-1} + \cdots + q_n z^{1-n}$$

defines another FIR causal filter; therefore its outcome is determined at time step j only by the values of $\tau^{-1} F\sigma$ until time step j, which means the values of $F\sigma$ until time step $j-1$, and those values are known.

In other words: in (6.21) we compute $F\sigma$ by filtering σ with p and using a feedback loop filtered by the filter R with z-transform $r(z) = z^{-1}(1 - q(z))$. This is shown in Fig. 6.6; for details, see [43, 44, 95]. We record at this point that rational filters are definitely of real practical relevance since as we have seen above, they still can be implemented physically with a *finite number* of the same building blocks as FIR filters.

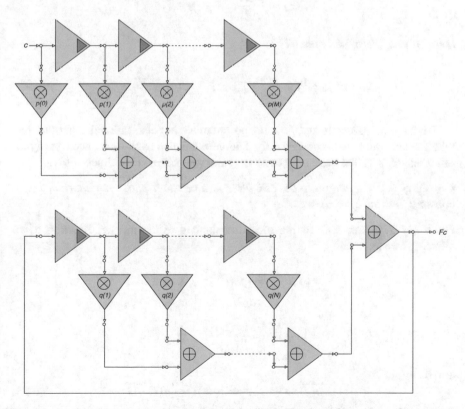

Fig. 6.6 A rational filter, realized by the *delayed feedback*: The signal filtered by p (filter on top) is sent into the filter q (filter below) and the results are added

6.4   Rational Filters and Stability

Besides the practical realization, rational filters already admit a much wider variety of possibilities; in particular this is due to an important class of filters that seem trivial at first, but can be applied for example very nicely for wavelet constructions, see [106].

Definition 6.12 A filter F is called an *allpass filter* if its transfer function satisfies $\left|\hat{f}(\theta)\right| = 1, \theta \in \mathbb{T}$.

An allpass filter does not change the *amplitude* of the frequency components of a signal, but just applies a *phase modulation* to them which however can depend on the respective frequency. The simplest allpass is of course the identity $F = I$ with $\hat{f} = 1$. But rational filters allow for a lot more allpasses.

Theorem 6.2 (Allpass filters) *A rational filter F with z-transform*

$$f(z) = \frac{p(z)}{q(z)}, \qquad p, q \in \Pi$$

is an allpass filter if and only if

$$q = z^n \, \overline{p}(z^{-1}) = \sum_{k=0}^{n} \overline{p_{n-k}} \, z^k, \qquad p(z) = \sum_{k=0}^{n} p_k \, z^k. \qquad (6.22)$$

Trivially, polynomials and Laurent polynomials are also rational functions just with the denominator 1; consequently Theorem 6.2 also applies to Laurent polynomials and tells us that there only exist very few and simple FIR allpass filters.

Corollary 6.1 *The only allpass FIR filter is a constant phase modulation, i.e., a constant function of modulus* 1.

Proof (*of Theorem* 6.2) Taking modulus squared of the transfer function of an allpass f, we get that

$$1 = \left|\widehat{f}(\theta)\right|^2 = \left|f\left(e^{i\theta}\right)\right|^2, \qquad \theta \in \mathbb{T},$$

hence

$$1 = |f(z)|^2 = \frac{|p(z)|^2}{|q(z)|^2}, \qquad z \in \partial\mathbb{D},$$

and therefore

$$p(z) \, \overline{p(z)} = |p(z)|^2 = |q(z)|^2 = q(z) \, \overline{q(z)}. \qquad (6.23)$$

In particular it follows that $\deg p = \deg q =: n$. Since for $z = e^{i\theta} \in \partial\mathbb{D}$, one has $\overline{z} = e^{-i\theta} = 1/z$, we then get

$$\overline{p(z)} = \sum_{j=0}^{n} \overline{p_j} z^j = \sum_{j=0}^{n} \overline{p_j} z^{-j} = z^{-n} \sum_{j=0}^{n} \overline{p_{n-j}} z^j =: z^{-n} p^\sharp(z), \qquad z \in \partial\mathbb{D}.$$

(6.24)

Substituting this into 6.23 and multiplying with z^n, it follows that

$$p(z)\, p^\sharp(z) = q(z)\, q^\sharp(z), \qquad z \in \partial\mathbb{C},$$

and since we can assume that p and q are coprime, it follows that $p = q^\sharp$ and $q = p^\sharp$ on $\partial\mathbb{D}$. This identity can be extended to all of $\mathbb{C}$, see Problem 6.11.

For the converse we only have to take into account that (6.22) yields that

$$\frac{|p(z)|^2}{|q(z)|^2} = \frac{p(z)\, z^{-n}\, p^\sharp(z)}{q(z)\, z^{-n}\, q^\sharp(z)} = \frac{p(z)}{q^\sharp(z)}\, \frac{p^\sharp(z)}{q(z)} = 1$$

which means that F is an allpass. $\qquad\square$

The polynomial defined in (6.24) is a special operation applied to the polynomial p that deserves a name of its own.

Definition 6.13 For $p \in \Pi_n$ with $\deg p = n$, we define the *reflection* $p^\sharp$ as

$$p^\sharp(z) = z^n \overline{p}(z^{-1}), \qquad z \in \mathbb{C} \setminus \{0\}.$$

(6.25)

The name "reflection" becomes clear when considering $p(z) = p_n z^n + \cdots + p_0$, $p_n \neq 0$. Then by (6.24), $p_j^\sharp = \overline{p_{n-j}}$, $j = 0, \ldots, n$. In other words the coefficients of the reflection are the complex conjugates of the polynomial read in reverse or *reflected* order. If we moreover write p in the factorized form

$$p = p_n (\cdot - \zeta_1) \cdots (\cdot - \zeta_n),$$

then

$$p^\sharp(z) = z^n \overline{p\left(\overline{z}^{-1}\right)} = z^n\, \overline{p_n} \prod_{j=1}^{n} \left(z^{-1} - \overline{\zeta_j}\right) = \overline{p_n} \prod_{j=1}^{n} \left(1 - \overline{\zeta_j}\, z\right).$$

(6.26)

Hence if p has the zeros ζ_j, $j = 1, \ldots, n$, then the zeros of $p^\sharp$ are the reflections $\overline{\zeta_j}^{-1}$, $j = 1, \ldots, n$, of those zeros at the unit circle.

Corollary 6.2 *A rational filter F is an allpass filter if and only if*

$$f(z) = c \prod_{j=1}^{n} \frac{z - \zeta_j}{1 - \overline{\zeta_j}\, z}, \qquad |c| = 1,$$

(6.27)

where $c = p_n/\overline{p_n}$.

The rational functions that define allpass filters are well-known objects from Function Theory, cf. [50].

Definition 6.14 A function of the form (6.27) is called a *Blaschke product* of order n.

Corollary 6.3 *A rational filter F is an allpass filter if and only if it is a Blaschke product.*

Let us return to the realization of rational filters and have a look at the simplest possible example.

Example 6.4 (*Summation filter*) The simplest example for a causal rational filter is the *summation filter S* defined by

$$(S\sigma)_j = \sum_{k=-\infty}^{j} \sigma_k, \qquad j \in \mathbb{Z}.$$

Since

$$(S\sigma)_j = \sum_{j=0}^{\infty} \sigma_{k-j} = \sum_{j \in \mathbb{Z}} s_j \, \sigma_{k-j} = s * \sigma, \qquad s_j = \begin{cases} 1, & j \geq 0, \\ 0, & j < 0, \end{cases}$$

we obtain the z-transform

$$s(z) = \sum_{k=0}^{\infty} z^{-k} = \frac{1}{1 - z^{-1}}, \qquad |z| > 1;$$

the condition on $|z|$ ensures convergence of the series, and obviously this is a rational IIR filter with an infinitely supported impulse response.

The realization is simple since

$$f(z) = \frac{p(z)}{q(z)}, \qquad p(z) = 1, \qquad q(z) = 1 - z^{-1}$$

implies that the feedback filter is $r(z) := 1 - q(z) = z^{-1}$. To summarize, the filter part itself is the identity filter, and the feedback is just a simple delay, see Fig. 6.7.

Unfortunately this feedback comes with a drawback that can lead to an unwanted behavior of rational filters. To understand what this means we expand $1/q$ as the Laurent series

$$\frac{1}{q(z)} = \sum_{j=0}^{\infty} \lambda_j z^{-j},$$

and once more we make the assumptions that supp $p \subseteq [0, m]$ and supp $\lambda \subseteq \mathbb{N}_0$, and obtain the identity

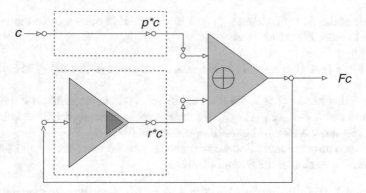

Fig. 6.7 Summation as a rational filter, numerator and denominator filter are framed

$$f(z) = \sum_{j=0}^{\infty} \sum_{k=0}^{m} p_k \lambda_j z^{-j-k} = \sum_{j=0}^{\infty} \left(\sum_{k=0}^{m} p_k \lambda_{j-k} \right) z^{-j}$$

$$= \sum_{j=0}^{\infty} \left(\sum_{k \in \mathbb{Z}} p_k \lambda_{j-k} \right) z^{-j} = \sum_{j=0}^{\infty} (\lambda * p)_j \, z^{-j} = (\lambda * p)(z).$$

Now we have a look at the behavior of λ_j and therefore also f_j for $j \to \infty$. This is the behavior of $F\delta$ for $j \to \infty$, which is the way in which the filter behaves when it is started with a single pulse at time 0 and then left to itself.

Indeed q can show a *damping* behavior if $\lambda_j \to 0$, $j \to \infty$, or it can be *exciting* in case $|\lambda_j| \to \infty$, $j \to \infty$. Since $f_j = (\lambda * p)_j$, this convergence or divergence pattern carries over to the impulse response f. Since the divergence phenomenon can lead to an uncontrollable behavior of the filter, the damping behavior is the more desirable one and leads to the following fundamental definition.

Definition 6.15 The filter F is called *stable* if

$$\lim_{j \to -\infty} f_j = \lim_{j \to \infty} f_j = 0.$$

Stability is a property of the denominator polynomial q of the impulse response f. To see what this means, let us look at the simplest nontrivial case of such a denominator first, namely

$$q(z) = 1 - \zeta z^{-1} = z^{-1}(z - \zeta), \qquad \zeta \in \mathbb{C} \setminus \{0\}.$$

Recalling (4.47) it follows that

$$\frac{1}{q(z)} = z \frac{1}{z - \zeta} = \sum_{j=0}^{\infty} \frac{\zeta^j}{z^j}, \qquad \text{that is } \lambda_j = \zeta^j,$$

and thus stability is equivalent to $|\zeta| < 1$ in this case. In other words: the *zero* ζ of $q(z)$ has to be *inside* the *unit disc*

$$z \in \mathbb{D}^0 = \{z \in \mathbb{C} \; : \; |z| < 1\} = \mathbb{D} \setminus \partial\mathbb{D}, \qquad \mathbb{D} := \{z \in \mathbb{C} \; : \; |z| \le 1\}. \qquad (6.28)$$

If, on the other hand, $|\zeta| > 1$, the filter coefficients will diverge, and if the zero lies *on* the *unit circle* $\partial\mathbb{D}$, i.e., $|\zeta| = 1$, we cannot make general statements about the impulse response: it may be bounded or divergent.

For an arbitrary rational filter, we will need a little bit more theory and start with a *quantitative* version of the Bézout identity.

Theorem 6.3 (Bézout identity) *Let* $f, g \in \Lambda$ *be two Laurent polynomials without common zeros. Then there exist* $p, q \in \Lambda$ *such that*

$$f\,p + g\,q = 1. \qquad (6.29)$$

If $f, g \in \Pi$, *then* $p, q \in \Pi$ *can be chosen such that* $\deg p = \deg g - 1$ *and* $\deg q = \deg f - 1$, *and these two polynomials are unique.*

Corollary 6.4 *Let* $f, g \in \Lambda$ *be two Laurent polynomials without common zeros. Then there exist for any* $h \in \Lambda$, *Laurent polynomials* $p, q \in \Lambda$ *such that*

$$f\,p + g\,q = h. \qquad (6.30)$$

Proof By Theorem 6.3 there are $\widetilde{p}, \widetilde{q} \Lambda$ such that $f\,\widetilde{p} + g\,\widetilde{q} = 1$ and

$$p = h\,\widetilde{p}, \qquad q = h\,\widetilde{q}$$

satisfy (6.30). □

The proof of Corollary 6.4 is not the way p and q are usually computed, since this is by the extended Euclidean algorithm which we know from Sect. 3.3. As shown there, it works on any Euclidean ring, especially on polynomials.

Proof (*of Theorem* 6.3) We begin with the case that $f, g \in \Pi$, set $m := \deg f$ as well as $n := \deg g$, and denote by $\zeta_1, \ldots, \zeta_m$ and $\eta_1, \ldots, \eta_n$ the zeros of f and g, i.e.,

$$f = f_m\,(x - \zeta_1) \cdots (x - \zeta_m), \qquad g = g_n\,(x - \eta_1) \cdots (x - \eta_n).$$

We choose $p \in \Pi_{n-1}$ as the unique solution of the interpolation problem

$$p\,(\eta_j) = \frac{1}{f\,(\eta_j)}, \qquad j = 1, \ldots, n, \qquad (6.31)$$

and $q \in \Pi_{m-1}$ as the solution of

$$q\left(\zeta_j\right) = \frac{1}{g\left(\zeta_j\right)}, \qquad j = 1, \ldots, m, \tag{6.32}$$

respectively, where in the case of multiple zeros, we once more consider the associated Hermite interpolation problem, see Problem 4.18. Since f and g have no common zeros, the right-hand sides are well defined in (6.31) and (6.32). The polynomial $h = f\, p + g\, q$ satisfies $\deg h \leq n + m - 1$ and

$$h\left(\zeta_j\right) = f\left(\zeta_j\right)\, p\left(\zeta_j\right) + g\left(\zeta_j\right)\, q\left(\zeta_j\right) = 1, \qquad j = 1, \ldots, m,$$

as well as

$$h\left(\eta_j\right) = f\left(\eta_j\right) + g\left(\eta_j\right)\, q\left(\eta_j\right) = 1, \qquad j = 1, \ldots, n,$$

and therefore $h - 1 \in \Pi_{n+m-1}$ vanishes at the $n + m$ points $\zeta_1, \ldots, \zeta_m, \eta_1, \ldots, \eta_n$, counting multiplicities if necessary, and hence $h - 1 = 0$.

If f, g are Laurent polynomials, we write them as $f = z^a\, \tilde f$, $g = z^b \tilde g$, $\tilde f, \tilde g \in \Pi$ and get p, q with the proper degree restrictions such that

$$1 = \tilde f\, p + \tilde g\, q = f\left((\cdot)^{-a} p\right) + g\left((\cdot)^{-b} q\right),$$

which is the desired representation, where we even have

$$\#\mathrm{supp}\, p < \#\mathrm{supp}\, g, \qquad \#\mathrm{supp}\, q < \#\mathrm{supp}\, f, \tag{6.33}$$

due to the degree restrictions for the polynomials p, q. $\square$

For the general case, we make use of a classical way to write rational functions as a sum of simple rational functions whose stability can be handled as in the case of $f(z) = (z - \zeta)^{-1}$.

Lemma 6.2 *If $f = p/q$, $p, q \in \Pi$, and $\zeta_1, \ldots \zeta_m$ are the zeros of q with multiplicities $\mu_1, \ldots, \mu_m$,*

$$q = \prod_{j=1}^{m} \left(\cdot - \zeta_j\right)^{\mu_j},$$

then there exist polynomials $p_j \in \Pi_{\mu_j - 1}$, $j = 1, \ldots, m$, and $p_0 \in \Pi$ such that

$$f(z) = p_0(z) + \sum_{j=1}^{m} \frac{p_j(z)}{\left(z - \zeta_j\right)^{\mu_j}}, \qquad z \in \mathbb{C} \setminus \{\zeta_1, \ldots, \zeta_m\}. \tag{6.34}$$

Definition 6.16 The representation (6.34) is called the *partial fraction decomposition* of f.

Lemma 6.2 can be easily carried over to Laurent polynomials p, q, just with two modifications in the details: first, zeros $\zeta_j = 0$ are not permitted any more as Laurent polynomials are undefined at the origin, and these zeros will be moved into the "integral" part g_0. Then the degree restriction becomes a support size restriction since there is no degree for Laurent polynomials, see Problem 3.18. The counterpiece of (6.34) is then

$$f(z) = g_0 + \sum_{j=1}^{m} \frac{p_j(z)}{\left(1 - \zeta_j z^{-1}\right)^{\mu_j}}, \qquad g_0, \dots, g_m \in \Lambda, \qquad z \in \mathbb{C} \setminus \{0, \zeta_1, \dots, \zeta_m\}$$

(6.35)

Proof (*of Lemma* 6.2) Induction on m, where $m = 1$ is clear. For the induction step, we assume the monic normalization $q(z) = z^{\mu_1 + \cdots + \mu_m} + \cdots$ and factor q to be

$$q(z) = \underbrace{(z - \zeta_1)^{\mu_1} \cdots (z - \zeta_{m-1})^{\mu_{m-1}}}_{=: q_0(z)} \underbrace{(z - \zeta_m)^{\mu_m}}_{=: q_m(z)},$$

so that q_0 and q_m are coprime. By Corollary 6.4 we have $p_0, p_m \in \Pi$ such that

$$p_m q_0 + p_0 q_m = p, \qquad \deg p_m < \deg q_m$$

and then we have that

$$\frac{p_0(z)}{q_0(z)} + \frac{p_m(z)}{(z - \zeta_m)^{\mu_m}} = \frac{p_0(z)}{q_0(z)} + \frac{p_m(z)}{q_m(z)} = \frac{p_0(z) q_m(z) + p_m(z) q_0(z)}{q_0(z) q_m(z)}$$

$$= \frac{p(z)}{q(z)} = f(z).$$

Applying the induction hypothesis to p_0/q_0, which has only $m - 1$ zeros $\zeta_1, \dots, \zeta_{m-1}$, we arrive at (6.2).   □

This leads to the final characterization of stability of a rational filter by using its poles. Recall that a *pole* of a rational function is a zero of its denominator.

Theorem 6.4 *A rational filter F with z-transform $f(z) = p(z)/q(z)$ is stable if and only if all the zeros of q belong to $\mathbb{D}^\circ$.*

Proof Let $\zeta_1, \dots, \zeta_m$ be the poles of $f(z)$, i.e., the zeros of q with associated multiplicities $\mu_1, \dots, \mu_m$. Using the partial fraction decomposition (6.35) and (6.36) from Problem 6.15, we get that

$$f(z) = \sum_{j=1}^{m} \frac{p_j(z)}{(1 - \zeta_j z^{-1})^{\mu_j}} = \sum_{j=1}^{m} \left(\sum_{k=-N}^{N} p_j(k) z^{-k} \right) \left(\sum_{\ell=0}^{\infty} q_j(\ell) \zeta_j^{\ell} z^{-\ell} \right)$$

$$= \sum_{j=1}^{m} \sum_{k=-N}^{N} \sum_{\ell=0}^{\infty} p_j(k) q_j(\ell) \zeta_j^{\ell} z^{-(\ell+k)}$$

$$= \sum_{j=1}^{m} \sum_{k=-N}^{N} \sum_{\ell \geq k} p_j(k) q_j(\ell - k) \zeta_j^{-k} \zeta_j^{\ell} z^{-\ell} = \sum_{j=1}^{m} \sum_{\ell=-N}^{\infty} r_j(\ell, \zeta_j) \zeta_j^{\ell} z^{-\ell}$$

where

$$r_j(\ell, \zeta_j) := \sum_{k=-N}^{\ell} p_j(k) q_j(\ell - k) \zeta_j^{-k}$$

is a polynomial in the variables ℓ and ζ_j. Hence,

$$f_k = \sum_{j=1}^{m} r_j(k, \zeta_j) \zeta_j^{k}, \qquad k \geq -N. \tag{6.36}$$

Whenever $|\zeta_j| < 1$, the respective summand in (6.36) converges to 0, so whenever $|\zeta_j| > 1$, the respective term diverges with the rate $|\zeta_j|^k$. For a zero on the unit circle, $|\zeta_j| = 1$ we either get a constant behavior and therefore boundedness if $\mu_j = 1$, or polynomial divergence for zeros of a higher order. And since the sequences $k \mapsto r_j(k, \zeta_j) \zeta_j^k$ are linearly independent as long as the ζ_j are all different, the divergence of one pole cannot be compensated by another one so that the ζ_j of maximal modulus determines the behavior of the filter. □

In view of Theorem 6.4, we will dedicate the next chapter on how to construct polynomials whose poles are inside the unit circle, since these will be the candidates for denominators of a stable rational filter.

6.4.1 Problems

6.11 Show that if two polynomials p, q coincide on $\partial \mathbb{D}$, then they are identical. *Hint:* Interpolation.

6.12 The *trapezoidal rule* is a quadrature formula that approximates the integral

$$\int_{k}^{k+1} f(x)\, dx \approx \frac{1}{2} \left(f(k+1) + f(k) \right).$$

Devise a rational filter that approximates $\int_{\mathbb{R}} f$ from integer samples $f(k)$ by the trapezoidal rule and compute its z-transform and transfer function.

6.13 (*Bézout identity*)

1. Define a Euclidean division for Laurent polynomials. What is the associated Euclidean function?
   Hint: use the degree of an appropriately defined polynomial.
2. Extend the extended Euclidean algorithm to arbitrary Euclidean rings and apply it to Laurent polynomials.
3. Write a `Matlab` program to solve the Bézout identity for (Laurent) polynomials.

6.14 Given $f, g \in \Pi$, consider the set of all solutions for

$$f\,p + g\,q = \gcd(f, g).$$

Show that given one solution p^*, q^*, all solutions are obtained as

$$p = p^* + r\,g, \qquad q = q^* - r\,f, \qquad r \in \Pi. \tag{6.37}$$

This is the mathematics behind the so-called *lifting scheme*, see [104].
Hint: first reduce the problem to the case that f, g are coprime and prove (6.37) in this case.

6.15 Show that for $\zeta \neq 0$ and $\mu \in \mathbb{N}$ there exists a polynomial $q = p_\mu \in \Pi_{\mu-1}$ depending only on μ, such that

$$\frac{1}{\left(1 - \zeta z^{-1}\right)^\mu} = \sum_{k=0}^\infty q_\mu(k)\, \zeta^k\, z^{-k}. \tag{6.38}$$

6.16 Show the following: for $\zeta, \eta \in \mathbb{C}, |\zeta| = |\eta| > 1$, the coefficients of the formal power series

$$\frac{p(z)}{1 - \zeta z^{-1}} + \frac{q(z)}{1 - \eta z^{-1}}, \qquad p \in \Pi,$$

are bounded if and only if they vanish identically.

6.5   Stability of Difference Equations

Presenting a different view of Signal Processing motivated by Systems Theory, we next look into the connection to the stability of difference equations with constant coefficients, which also bear a close relationship with the standard operations in signal processing since the action of a difference operator is nothing but filtering. Clearly we can only touch the surface of this classical topic here; for more information, see [35, 55].

Definition 6.17 (*Difference operator*) The *difference* $\Delta : \ell(\mathbb{N}_0) \to \ell(\mathbb{N}_0)$ is defined as $\Delta = \tau - I$. Higher order differences of order n are given as

$$\Delta^n = \Delta \, \Delta^{n-1} = (\tau - I)^n = \sum_{j=0}^{n} (-1)^{n-j} \binom{n}{j} \tau^j. \tag{6.39}$$

The difference can be written as a convolution and hence is a filter, so it is immediate that the associated z-transform is

$$\Delta(z) = \tau(z) - 1 = z - 1.$$

A weighted sum of differences

$$A = \sum_{k=0}^{n} a_k \Delta^k, \qquad a_k \in \mathbb{R},$$

is called a *difference operator* with constant coefficients. Substituting the right-hand side of (6.39), we then get that

$$A = \sum_{k=0}^{n} a_k \, (\tau - I)^k = \sum_{k=0}^{n} a_k \sum_{j=0}^{k} (-1)^{k-j} \binom{k}{j} \tau^j = \sum_{j=0}^{n} \left(\sum_{k=j}^{n} (-1)^{k-j} \binom{k}{j} a_k \right) \tau^j,$$

so that any difference operator can be seen as a polynomial in τ. This motivates the following definition.

Definition 6.18 (*Difference operator & difference equation*) Given a polynomial $q \in \Pi$, the associated *difference operator* is

$$q(\tau) := \sum_{j=0}^{\deg q} q_j \, \tau^j. \tag{6.40}$$

Given $g \in \ell(\mathbb{N})$, the *difference equation* is an equation of the form

$$q(\tau)u = g, \tag{6.41}$$

where u stands for the unknown from $\ell(\mathbb{N}_0)$.

The notion of $q(\tau)$ where q is a polynomial is in accordance with the standard way of writing a *differential operator* with constant coefficients as $q(D)$. In contrast to the case of derivatives, also τ^{-1} is well defined; therefore (6.40) does also allow q to be a Laurent polynomial without introducing any complications.

With the methods of the preceding sections, we can quite easily derive a result regarding the solvability of the difference equation (6.41) by noting that for $q \in \Pi$, $\deg q = n$, we have

$$(q(\tau)u)_k = \sum_{j=0}^{n} q_j \tau^j u_k = \sum_{j=0}^{n} q_j u_{k+j} = (u \star q)_k, \qquad k \in \mathbb{N}_0.$$

Therefore applying a difference operator is nothing but a correlation, and we can rewrite the difference equation (6.41) as the *decorrelation problem*

$$u \star q = g. \tag{6.42}$$

As in (5.48) this yields the linear system

$$g = \begin{bmatrix} q_0 \cdots q_n & & \\ & q_0 \cdots q_n & \\ & & \ddots \vdots \ddots \end{bmatrix} u =: Qu.$$

Beginning with the issue of uniqueness, let u, u' be two solutions and set $v := u - u'$, yielding $Qv = 0$. In particular note that the components $v_0, \ldots, v_{2n+1}$ satisfy

$$\begin{bmatrix} q_0 \cdots q_{n-1} \ q_n & & \\ \ddots \ \vdots \quad \vdots \ \ddots & \\ q_0 \ q_1 \cdots \ q_n & \\ q_0 \cdots q_{n-1} \ q_n \end{bmatrix} \begin{bmatrix} v_0 \\ \vdots \\ v_{2n+1} \end{bmatrix} = 0,$$

which yields by Lemma 5.4 that v is an exponential sequence

$$v_k = \sum_{j=1}^{n} c_j \zeta_j^k, \qquad k \in \mathbb{N}_0, \tag{6.43}$$

where $\zeta_1, \ldots, \zeta_n$ are the zeros of the polynomial $q(z)$.

Remark 6.7 Again we make the simplifying assumption here that the zeros of $\hat{q}$ are all simple. Multiple zeros of $\hat{q}$ in a normalized difference equation can be handled again by choosing v as samples of an exponential polynomial, where the degree of the polynomial connected to a certain zero coincides with its multiplicity. We also have to avoid the case that $q(0) = 0$, i.e., $q_0 = 0$, since then the respective ζ_j is not well defined. Therefore we assume that q is *normalized* in the sense that $q_0 \neq 0$, which does not really change the difference equation except maybe for a shift of data and the equation.

Under these assumptions we can solve the difference equation provided that its right-hand side is finitely supported.

Theorem 6.5 *Given $q \in \Pi_n$ with $q_0 q_n \neq 0$ and $g \in \ell_0(\mathbb{N}_0)$, there exists an n parameter family of solutions for the difference equation $q(\tau)u = g$.*

Proof Choose $m > n$ such that $g_j = 0$ for $j > m$, and let $u_0^*, \ldots, u_m^*$ denote the solution of the $(m+1) \times (m+1)$ system

$$\begin{bmatrix} q_0 & \cdots & q_{n-1} & q_n & & & \\ & \ddots & \vdots & \vdots & \ddots & & \\ & & q_0 & q_1 & \cdots & & q_n \\ & & & q_0 & \cdots & q_{n-1} & q_n \\ & & & & \ddots & \vdots & \vdots \\ & & & & & q_0 & q_1 \\ & & & & & & q_0 \end{bmatrix} \begin{bmatrix} u_0^* \\ \vdots \\ \\ u_m^* \end{bmatrix} = \begin{bmatrix} g_0 \\ \vdots \\ \\ g_m \end{bmatrix},$$

which has a unique solution since the matrix is upper triangular with a nonzero diagonal. Then we set

$$u^* = \left(u_0^*, \ldots, u_m^*, 0, \ldots\right)$$

for which

$$(Qu^*)_j = \begin{cases} g_j, & j = 0, \ldots, m, \\ 0, & j > m, \end{cases}$$

so that the set of all solutions is given by all $u = u^* + v$ with v of the form (6.43) and the n parameters are the coefficients c_j appearing there. $\square$

Example 6.5 (*Difference*) The simplest case is $\Delta u = g$, i.e., $q(z) = z - 1$. We then get the system

$$\begin{bmatrix} -1 & 1 & & \\ & \ddots & \ddots & \\ & & -1 & 1 \\ & & & -1 \end{bmatrix} \begin{bmatrix} u_0^* \\ \vdots \\ u_m^* \end{bmatrix} = \begin{bmatrix} g_0 \\ \vdots \\ g_m \end{bmatrix},$$

i.e.,

$$u_m^* = -g_m, \quad u_{m-1}^* = u_m^* - g_{m-1}, \ldots, \qquad u_j^* = -\sum_{k=j}^{m} g_k, \qquad j = 0, \ldots, m,$$

and indeed

$$(\Delta u^*)_j = u_{j+1}^* - u_j^* = -\sum_{k=j+1}^{m} g_k + \sum_{k=j}^{m} g_k = g_j, \qquad j = 0, \ldots, m,$$

as well as

$$(\Delta u^*)_j = 0, \qquad j > m.$$

Since the unique zero of q is 1, it follows that the general solution of the difference equations is $u^* + v$, where v is an arbitrary *constant sequence*. In particular choosing

$$v_k = \sum_{j=0}^{m} g_j, \qquad k \in \mathbb{N}_0$$

leads to the partial summation

$$u_k = \sum_{j=0}^{k-1} g_j, \qquad k \in \mathbb{N}_0,$$

which is often used as the "standard solution" for this difference equation.

The solution $u^* \in \ell_0(\mathbb{N}_0)$ from the proof of Theorem 6.5 is the unique finitely supported solution for a finitely supported right-hand side, while all the others are of an exponential type. In particular $u^* \in \ell_0(\mathbb{N}_0) \subset \ell_p(\mathbb{N}_0)$, see Problem 6.20, hence u^* is an ℓ_p sequence for any $1 \le p \le \infty$. On the other hand the difference equation in Example 6.5 is already a case where all solutions of the difference equation are at least *bounded*, i.e., belong to $\ell_\infty(\mathbb{N}_0)$. Of particular interest are difference equations where all solutions are summable or at least bounded.

Definition 6.19 (*Stable difference equation*) The *difference equation* $q(\tau)u = g$ is called *stable* if $u \in \ell_1(\mathbb{N}_0)$ for any $g \in \ell_0(\mathbb{N}_0)$. It is called *bounded* if $u \in \ell_\infty(\mathbb{N}_0)$ for any $g \in \ell_0(\mathbb{N}_0)$.

Theorem 6.6 *The difference equation $q(\tau)u = g$, $q \in \Pi$, $g \in \ell_0(\mathbb{N}_0)$ is stable if and only if all zeros of q lie in $\mathbb{D}^\circ$ and bounded if and only if all zeros lie in $\mathbb{D}$.*

Proof Let the zeros of q be ordered[2] as $|\zeta_1| \le \cdots \le |\zeta_n|$ and write $\zeta_j = |\zeta_j| e^{i\theta_j}$, $\theta_j \in \mathbb{T}$, $j = 1, \ldots, n$. Also assume that $c_j \ne 0$, which is not a restriction as otherwise one would simply ignore the zeros with zero weight in v. For k such that $g_j = 0$, $j \ge k$, we then have that

$$u_k = |\zeta_n|^k \sum_{j=1}^{n} c_j \left(\frac{\zeta_j}{|\zeta_n|} \right)^k = |\zeta_n|^n \left(\sum_{|\zeta_j| < |\zeta_n|} c_j \left(\frac{\zeta_j}{|\zeta_n|} \right)^k + \sum_{|\zeta_j| = |\zeta_n|} c_j e^{i\theta_j k} \right),$$

hence

$$\limsup_{k \to \infty} \frac{|u_k|}{|\zeta_n|^k} = \limsup_{k \to \infty} \left| \sum_{|\zeta_j| = |\zeta_n|} c_j e^{i\theta_j k} \right| \ne 0,$$

since the functions $e^{i\theta_j \cdot}$, $j = 1, \ldots, n$, are linearly independent because the θ_j were assumed to be distinct. Consequently the entries of u are exponentially increasing if

---

[2] Again we assume the zeros to be simple. This is even less of a restriction here, since polynomial factors are neglectable compared to exponential growth.

$|\zeta_n| > 1$, bounded if $|\zeta_n| = 1$, and exponentially decreasing if $|\zeta_n| < 1$. In the latter case we have that

$$\|u\|_1 \le \|u^*\|_1 + \sum_{k=1}^{\infty} \sum_{j=1}^{n} |c_j| \, |\zeta_j|^k = \|u^*\|_1 + \sum_{j=1}^{n} |c_j| \sum_{k=1}^{\infty} |\zeta_j|^k$$

$$= \|u^*\|_1 + \sum_{j-1}^{n} |c_j| \frac{1}{1 - |\zeta_j|},$$

while in the other two cases the ℓ_1 norm is ∞. $\qquad\qquad\square$

Remark 6.8 Theorem 6.6 could also be rephrased as the frequencies $\omega_j \in \mathbb{C}$ such that $\zeta_j = e^{\omega_j}$ have a negative real part, hence lie in the left half-plane $\mathbb{H}_-$. We will connect to the left half-plane in the next chapter.

In fact Theorem 6.6 should not even be too surprising if we consider it in terms of signal processing, and take the z-transform of the correlation $u \star q$ which gives

$$g(z) = (u \star q)(z) = (u \ast_{-1} q)(z) = u(z)\, q(z^{-1}) g(z^{-1}), \qquad g(z) = \sum_{j=0}^{m} g_j z^j.$$

Hence replacing z by z^{-1}, we find that

$$u(z) = \frac{g(z)}{q(z)}$$

is exactly the Laurent expansion of the rational function g/q, and we already know that its stability (in terms of a filter) depends on whether the zeros of q are inside the unit circle or not.

6.5.1 Problems

6.17 Can any difference operator be written in the form

$$q(\tau) = \sum_{k=0}^{n} a_k \Delta^k?$$

Prove it or give a counterexample.

6.18 Show that with the above definitions, the following statements are equivalent:

1. $v \star \hat{q} = 0$,
2. v is of the form (6.43),

3. v defines a finite rank Hankel operator M of rank at most n and $M_n q = 0$.

Can it happen that rank $M < n$?

6.19 Extend (6.43) to the case where q has multiple zeros.

6.20 Show that $\ell_1(\mathbb{N}_0) \subset \ell_p(\mathbb{N}_0)$ for any $p > 1$.

6.6   Superresolution via Continued Fractions and a Determinant Identity

Finally we consider a problem from sparse signal processing that connects Prony's original approach [81] with so-called *superresolution*, cf. [14]. The underlying model of this theory is that we have a signal $s : \mathbb{R} \to \mathbb{R}$ that consists of several *spikes*:

$$s = \sum_{j=1}^{m} f_j \, \delta_{x_j}, \qquad f_j > 0, \; x_j \in \mathbb{R}, \qquad j = 1, \ldots, m. \tag{6.44}$$

This signal is not sampled directly[3] but only after a lowpass filtering, which means that we can sample only a part of the coefficients of the Fourier transform

$$\hat{s}(\xi) = \int_{\mathbb{R}} s(x) \, e^{-i\xi x} \, dx = \sum_{j=1}^{m} f_j \, e^{-ix_j \xi},$$

namely its lowpass content

$$\sigma_k = \hat{s}(kh) = \sum_{j=1}^{m} f_j \, e^{-ix_j h k}, \qquad |k| \le N, \qquad h > 0,$$

where N stands for the number of lowpass coefficients that are available. Since s is real-valued, we again have $\sigma_{-k} = \overline{\sigma_k}$ and all information is already contained in $\sigma_0, \ldots, \sigma_N$. Of course, this is again Prony's problem, however with special constraints:

1. the weights f_j are positive, hence the underlying discrete measure is a *positive measure*. This assumption is indeed justified in many applications where f_j measures an intensity which is usually a nonnegative real number. For example, even in Prony's original application, the value f_j stood for the concentration of a certain component of fluid which can never be negative or imaginary.

---

[3] It would not even make sense to try to hit spikes by sampling! The result would be the zero function with probability 1.

2. The frequencies $\omega_j = -ihx_j$ are all *purely imaginary*. This is not only beneficial from the point of view of stability, since it restricts the problem to the torus $\mathbb{T}$, i.e., the values hx_j are only relevant modulo 2π, and have to be different on the torus.

The approach we present here is not intended as a numerical competitor to the well-established classical methods like MUSIC [93], ESPRIT [84] and their variations [80], or techniques like matrix pencil methods. The goal is just to show that the techniques developed in the preceding sections also give a way of solving the super-resolution problem in a numerically reasonable way entirely by using continued fractions.

We will reconstruct the frequencies by constructing the convergents of the Laurent series

$$\sigma(x) = \sum_{j=0}^{\infty} \sigma_j x^{-j},$$

where we assume that we have enough samples available, i.e., that N is chosen so large that

$$\text{rank} \begin{bmatrix} \sigma_0 & \cdots & \sigma_{N/2} \\ \vdots & \ddots & \vdots \\ \sigma_{N/2} & \cdots & \sigma_N \end{bmatrix} = m.$$

Then the remaining entries of σ can be obtained by the (positive) flat extension methods of Sect. 5.3 if necessary.

To compute the convergents, even the continued fraction expansion of the Laurent series $\sigma(x)$, we make use of the recurrence relation and the explicit formulas (4.37), (4.38) and (4.39) from Sect. 4.1.

Numerically the main catch is the quotient of determinants in (4.37) which defined the recurrence

$$\alpha_{n+1} = (-1)^n \left(\prod_{j=1}^{n} \alpha_j \right)^{-2} \frac{\det \Lambda_n}{\det \Lambda_{n+1}}, \quad \Lambda_n := \begin{bmatrix} \sigma_0 & \cdots & \sigma_{n-1} \\ \vdots & \ddots & \vdots \\ \sigma_{n-1} & \cdots & \sigma_{2n-2} \end{bmatrix}, \quad n \in \mathbb{N}.$$

Although the determinants are still all nonzero due to the positivity of the measure, i.e., $f_j > 0$, see Remark 5.3, their explicit computation is numerically unstable and therefore not recommended. To avoid this we note that[4]

---

[4] Note that here it is more convenient to start the indexing of α and Λ at 0 and not at 1 as we did in Sect. 4.1.

$$\det \Lambda_n = \det \begin{bmatrix} \sigma_0 & \cdots & \sigma_{n-1} & 0 \\ \vdots & \ddots & \vdots & \vdots \\ \sigma_{n-1} & \cdots & \sigma_{2n-2} & 0 \\ \sigma_n & \cdots & \sigma_{2n-1} & 1 \end{bmatrix}$$

so that by Cramer's rule for the last component of the solution for the linear system $\Lambda_{n+1}x = e_{n+1}$ and the Schur complement formula (6.53), see Problem 6.21, we get

$$\frac{\det \Lambda_n}{\det \Lambda_{n+1}} = e_{n+1}^T \Lambda_{n+1}^{-1} e_{n+1} = \left(\sigma_{2n} - v_n^T \Lambda_n^{-1} v_n\right)^{-1}, \qquad v_n := \begin{bmatrix} \sigma_n \\ \vdots \\ \sigma_{2n-1} \end{bmatrix}.$$

Hence we solve for $n = 1, 2, \ldots$, the square symmetric system[5]

$$\Lambda_n x_n = v_n, \qquad x_n \in \mathbb{C}^n, \tag{6.45}$$

e.g., by using a Cholesky factorization, cf. [36], and then compute the number

$$t_{n+1} := \sigma_{2n} - v_n^T x_n = \left(e_{n+1}^T \Lambda_{n+1}^{-1} e_{n+1}\right)^{-1} = \frac{\det \Lambda_{n+1}}{\det \Lambda_n}, \tag{6.46}$$

with the special case

$$t_1 = \left(e_0 \Lambda_1^{-1} e_0\right)^{-1} = \sigma_0.$$

This trick allows us to determine the quotient

$$t_{n+1} = \frac{\det \Lambda_{n+1}}{\det \Lambda_n} = (-1)^n \frac{1}{\alpha_{n+1}} \prod_{j=1}^n \alpha_j^{-2} = (-1)^n \alpha_{n+1} \prod_{j=1}^{n+1} \alpha_j^{-2} \tag{6.47}$$

of two successive determinants without having to really compute these determinants. Also these are the numbers needed for obtaining the recursion coefficients.

If this number is zero, then q_{n+1} is the Prony polynomial for the superresolution problem, otherwise we do the update to be

$$\alpha_{n+1} = \frac{(-1)^n}{t_{n+1}} \prod_{j=1}^n \alpha_j^{-2} = -\frac{1}{\alpha_n t_{n+1}} \left((-1)^{n-1} \alpha_n \prod_{j=1}^n \alpha_j^{-2}\right) = -\frac{t_n}{\alpha_n t_{n+1}} \tag{6.48}$$

by (6.46) and (6.47), initialized with $\alpha_1 = 1/\sigma_0$.

Since $m = \text{rank } \Lambda_m = \text{rank } \Lambda_{m+1} = \cdots$, the coefficients $\alpha_{m+1}, \alpha_{m+1}, \ldots$ have the value ∞ or $\alpha_n^{-1} = 0$ for $n > m$.

---

[5] It is tempting to call it a positive definite system, but due to the fact that the frequencies are purely imaginary, each Λ_n is a complex matrix that is symmetric but not necessarily hermitian.

Having obtained α_{n+1}, we can use (4.38) to compute

$$\beta_{n+1} = \alpha_{n+1} \left(\frac{\det \Lambda'_{n+2}}{\det \Lambda_{n+1}} + \frac{\det \Lambda'_{n+1}}{\det \Lambda_n} - 2 \sum_{j=1}^{n} \frac{\beta_j}{\alpha_j} \right), \qquad \beta_1 = -\frac{\sigma_1}{\sigma_0^2}.$$

To simplify this expression, we note again by Cramer's rule that

$$\frac{\det \Lambda'_{n+1}}{\det \Lambda_n} = -\frac{\det \begin{bmatrix} \sigma_0 & \cdots & \sigma_{n-2} & \sigma_n \\ \vdots & \ddots & \vdots & \vdots \\ \sigma_{n-1} & \cdots & \sigma_{2n-3} & \sigma_{2n-1} \end{bmatrix}}{\det \Lambda_n} = -x_{n,n}, \qquad n \geq 1,$$

which yields the simple and stable recurrence

$$\beta_{n+1} = \alpha_{n+1} \left(-x_{n+1,n+1} - x_{n,n} - 2 \sum_{j=1}^{n} \frac{\beta_j}{\alpha_j} \right) \tag{6.49}$$

for the coefficients β_n.

We now can turn our observations from above into an efficient algorithm to compute the recurrence coefficients: First, initialize

$$\alpha_1 = 1/\sigma_0, \qquad \tilde{\beta}_1 = 0, \qquad t_1 = \sigma_0, \tag{6.50}$$

and then, for $n = 1, 2, \ldots$, deal with the following steps:

1. Solve the linear system

$$\Lambda_n x_n = v_n = \begin{bmatrix} \sigma_n \\ \vdots \\ \sigma_{2n-1} \end{bmatrix},$$

and set

$$t_{n+1} = \sigma_{2n} - v_n^T x_n.$$

2. If $t_{n+1} \neq 0$, compute

$$\alpha_{n+1} = \frac{t_n}{-\alpha_n t_{n+1}},$$

and

$$\beta_n = \alpha_n \left(\tilde{\beta}_n - x_{n,n} \right), \qquad \tilde{\beta}_{n+1} = -x_{n,n} - 2 \sum_{j=0}^{n} \frac{\beta_j}{\alpha_j},$$

and continue with the next value of n.

3. If $t_{n+1} = 0$, then set $\beta_n = \alpha_n \left(\tilde{\beta}_n - x_{n,n} \right)$ and finish.

This way we construct the continued fraction expansion

$$\sigma(x) = \frac{1|}{|\alpha_1 x + \beta_1} + \cdots + \frac{1|}{|\alpha_m x + \beta_m}$$

for the power series associated with the recovered signal. This in fact has some consequences.

Example 6.6 (*Case $m = 1$*) In the special case $m = 1$, we immediately obtain $t_1 = 0$ at the first step of the iteration, and the update rule (6.48) yields $\alpha_2 = \infty$, or $1/\alpha_2 = 0$. Hence the resulting Prony polynomial is

$$q_1(x) = \alpha_1 x + \beta_1 = \frac{1}{\sigma_0} x - \frac{\sigma_1}{\sigma_0^2} = \frac{1}{\sigma_0} \left(x - \frac{\sigma_1}{\sigma_0} \right)$$

and has its zero at

$$\frac{\sigma_1}{\sigma_0} = \frac{s(1)}{s(0)} = \frac{f_1 e^{i x_1 h}}{f_1} = e^{i x_1 h}$$

from which the point x_1 can be reconstructed immediately. The tridiagonal matrix for the eigenvalue problem is the trivial 1×1 matrix

$$\left[-\frac{\beta_1}{\alpha_1} \right] = \left[\frac{\sigma_1}{\sigma_0} \right]$$

whose only eigenvalue is again $e^{i x_1 h}$.

Remark 6.9 Computing the continued fraction expansion can eventually be used to solve Prony's problem:

1. the final convergent p_m/q_m has the Prony polynomial $q_m \in \Pi_{m+1}$ as its denominator.
2. The recurrence coefficients can be used to set up the tridiagonal matrix

$$M = \begin{bmatrix} -\frac{\beta_1}{\alpha_1} & -\frac{1}{\alpha_2} & & & \\ \frac{1}{\alpha_1} & -\frac{\beta_2}{\alpha_2} & -\frac{1}{\alpha_3} & & \\ & \ddots & \ddots & \ddots & \\ & & \frac{1}{\alpha_{m-2}} & -\frac{\beta_{m-1}}{\alpha_{m-1}} & -\frac{1}{\alpha_m} \\ & & & \frac{1}{\alpha_{m-1}} & -\frac{\beta_m}{\alpha_m} \end{bmatrix} \in \mathbb{C}^{m \times m}, \qquad (6.51)$$

whose eigenvalues are the zeros of the Prony polynomial as pointed out in Sect. 4.4. There is no need to compute the polynomial itself; hence no need to explicitly deal with the recurrence for the convergents.

3. Since we computed the associated continued fraction, the other convergents p_n/q_n provide good approximations with a few spikes that capture $2n + 1$ coefficients from the measured sequence, $n = 0, \ldots, m - 1$.
4. The method can be used to detect m automatically, that is as the first n where the Schur complement

$$\sigma_{2n+2} - \left[\sigma_{n+1}, \ldots, \sigma_{2n+1}\right] \Lambda_n^{-1} \begin{bmatrix} \sigma_{n+1} \\ \vdots \\ \sigma_{2n+1} \end{bmatrix}$$

vanishes. This is only up to numerical precision, and depends on the f_j and the separation of the frequencies.

Although the nonsingularity of the Hankel matrices Λ_n can only be guaranteed for positive weights, the approach itself works as long as the sequence is non-degenerate, i.e., $\det \Lambda_n \neq 0, n = 0, \ldots, m$. This is the generic situation as degenerate sequences require a judicious choice of the coefficients, but nevertheless the condition is hard to check a priori.

Another side effect of the methods used in this section is that we can give an explicit formula for the Prony polynomial.

Theorem 6.7 *The monic Prony polynomial for the function*

$$f(x) = \sum_{j=1}^{m} f_j \zeta_j^x, \qquad \zeta_j \in \mathbb{R} + i\mathbb{T}, \qquad f_j \neq 0,$$

is given as

$$p(-x) = (-1)^m \frac{\det \begin{bmatrix} f(0) & f(1) & \cdots & f(m) \\ \vdots & \vdots & \ddots & \vdots \\ f(m-1) & f(m) & \cdots & f(2m-1) \\ 1 & x & \cdots & x^m \end{bmatrix}}{\det \begin{bmatrix} f(0) & \cdots & f(m-1) \\ \vdots & \ddots & \vdots \\ f(m-1) & \cdots & f(2m-2) \end{bmatrix}}. \tag{6.52}$$

Proof Since the Hankel matrices

$$\Lambda_n := \begin{bmatrix} f(0) & \cdots & f(n) \\ \vdots & \ddots & \vdots \\ f(n) & \cdots & f(2n) \end{bmatrix}, \qquad n \in \mathbb{N}_0,$$

satisfy $m = \text{rank } \Lambda_{m-1} = \text{rank } \Lambda_m = \cdots$, the polynomial in (6.52) is well defined; we have that $f(2m) = v^T \Lambda_{m-1}^{-1} v$ and the Schur determinant formula (6.54) from Problem 6.22 yields that

$$\det \tilde{\Lambda}_m := \det \begin{bmatrix} \Lambda_{m-1} & v \\ v^T & 1 + f(2m) \end{bmatrix} = \det \Lambda_{m-1} \left(1 + f(2m) - v^T \Lambda_{m-1}^{-1} v \right)$$
$$= \det \Lambda_{m-1}.$$

Expanding the determinant in the numerator of (6.52) with respect to the last row and incorporating Cramer's rule yields

$$q(x) := \det \begin{bmatrix} f(0) & f(1) & \cdots & f(m) \\ \vdots & \vdots & \ddots & \vdots \\ f(m-1) & f(m) & \cdots & f(2m-1) \\ 1 & x & \cdots & x^m \end{bmatrix}$$

$$= (-1)^{m+1} \sum_{j=0}^{m} (-1)^j x^j \det \begin{bmatrix} f(0) & \cdots & f(j-1) & 0 & f(j+1) & \cdots & f(m) \\ \vdots & \ddots & \vdots & \vdots & \vdots & \ddots & \vdots \\ f(m-1) & \cdots & f(m+j-2) & 0 & f(m+j) & \cdots & f(2m-1) \\ 0 & \cdots & 0 & 1 & 0 & \cdots & 0 \end{bmatrix}$$

$$= (-1)^{m+1} \sum_{j=0}^{m} (-x)^j \det \begin{bmatrix} f(0) & \cdots & f(j-1) & 0 & f(j+1) & \cdots & f(m) \\ \vdots & \ddots & \vdots & \vdots & \vdots & \ddots & \vdots \\ f(m-1) & \cdots & f(m+j-2) & 0 & f(m+j) & \cdots & f(2m-1) \\ f(m) & \cdots & f(m+j-1) & 1 & f(m+j+1) & \cdots & 1+f(2m) \end{bmatrix}$$

$$= (-1)^{m+1} \det \tilde{\Lambda}_m \sum_{j=0}^{m} (-x)^j e_j^T \tilde{\Lambda}_m^{-1} e_m = (-1)^{m+1} \det \Lambda_{m-1} \sum_{j=0}^{m} (-x)^j e_j^T \tilde{\Lambda}_m^{-1} e_m.$$

Now since the determinant for $j = m$ in the sum coincides with $\det \Lambda_{m-1} = \det \tilde{\Lambda}_m$, it follows that $q(x) = \det \Lambda_{m-1}(-x^m + \cdots)$, hence

$$p(x) := -\frac{q(-x)}{\det \Lambda_{m-1}} = (-1)^m \sum_{j=0}^{m} e_j^T \tilde{\Lambda}_m^{-1} e_m x^j$$

is monic. Its coefficient vector p satisfies

$$\Lambda_m p = \left(\tilde{\Lambda}_m - e_m e_m^T \right) (-1)^m \tilde{\Lambda}_m^{-1} e_m = (-1)^m \left(\tilde{\Lambda}_m \tilde{\Lambda}_m^{-1} e_m - e_m e_m^T p \right)$$
$$= (-1)^m (1 - p_m) e_m = 0$$

since $p_m = 1$ as $p(x)$ is monic. This completes the proof of (6.52). $\square$

The linear algebra in the proof of Theorem 6.7 works under much more general circumstances and with more general matrices, since all that is needed is Cramer's rule and the Schur complement formula for a rank preserving extension of a matrix.

The connection to Prony's problem is made by the fact that the Prony polynomial is the *unique* monic polynomial whose coefficient vector lies in the kernel of the Hankel matrix Λ_m.

6.6.1 Problems

6.21 (*Schur complement inversion*) Let X be given in the block matrix form

$$X = \begin{bmatrix} A & B \\ C & D \end{bmatrix} \in \mathbb{R}^{(m+n)\times(m+n)}, \qquad A \in \mathbb{R}^{m \times m}, \; D \in \mathbb{R}^{n \times n},$$

and the dimensions of the other matrices follow from the context. Also suppose that A is nonsingular.

1. Show that X has the *LU decomposition*

$$X = \begin{bmatrix} I & 0 \\ CA^{-1} & I \end{bmatrix} \begin{bmatrix} A & B \\ 0 & D - CA^{-1}B \end{bmatrix}.$$

2. Show that X is nonsingular if and only if the *Schur complement*

$$S := D - CA^{-1}B$$

of A in X is nonsingular.
3. Verify the formula

$$X^{-1} = \begin{bmatrix} A^{-1}\left(I + BS^{-1}CA^{-1}\right) & -A^{-1}BS^{-1} \\ -S^{-1}CA^{-1} & S^{-1} \end{bmatrix}. \tag{6.53}$$

6.22 ([27, p. 46]) Under the assumptions of Problem 6.21, prove the *Schur determinant formula*

$$\det \begin{bmatrix} A & B \\ C & D \end{bmatrix} = \det A \left(\det D - CA^{-1}B\right). \tag{6.54}$$

6.23 (*Cholesky factorization*) Show that any symmetric and positive definite matrix $A \in \mathbb{R}^{n \times n}$ can be written as

$$A = G^T G, \qquad G = \begin{bmatrix} * & \cdots & * \\ & \ddots & \vdots \\ & & * \end{bmatrix}, \tag{6.55}$$

where G has positive diagonal entries. Give an explicit algorithm for the computation of G.

Hint: Start with the representation of A as $G^T G$ and solve for the entries of G, starting with the equation for a_{11}.

6.24 Show that any symmetric complex valued matrix $A \in \mathbb{C}^{n \times n}$ with nonvanishing determinant has a Cholesky decomposition of the form (6.55).

6.25 Prove that the basis $q_0, \ldots, q_m$, defined by the recurrence

$$q_{n+1}(x) = (\alpha_{n+1}x + \beta_{n+1}) q_n(x) + q_{n-1}(x), \qquad q_0 = 1, \ q_{-1} = 0,$$

leads to the multiplication Table (6.51) for multiplication by x.
Hint: See how (4.66) was obtained.

6.26 Show that the inverse of the 2×2 matrix

$$A = \begin{bmatrix} a & b \\ c & d \end{bmatrix}$$

is

$$A^{-1} = \frac{1}{ad - bc} \begin{bmatrix} d & -b \\ -c & a \end{bmatrix}.$$

6.27 Show that the diagonal elements of the matrix

$$M = \begin{bmatrix} \eta_1 & -\frac{1}{\alpha_2} & & & \\ \frac{1}{\alpha_1} & \eta_2 & -\frac{1}{\alpha_2} & & \\ & \ddots & \ddots & \ddots & \\ & & \frac{1}{\alpha_{m-2}} & \eta_{m-1} & -\frac{1}{\alpha_m} \\ & & & \frac{1}{\alpha_{m-1}} & \eta_m \end{bmatrix}$$

from (6.51) can be computed recursively as

$$\eta_{n+1} = -x_{n+1,n+1} - x_{n,n} - 2 \sum_{j=1}^{n} \eta_j, \quad n = 1, \ldots, m-1, \qquad \eta_1 = \frac{\sigma_1}{\sigma_0}.$$

Chapter 7
Continued Fractions, Hurwitz and Stieltjes

This isn't shit; I've got a degree in mathematics, man.
R. Shea, The Illuminatus! Trilogy

7.1 The Problems

In the preceding chapter we learned that the stability of rational filters is determined by the location of its poles, and that stability is equivalent to all these poles being inside the unit circle. This leads us to two fundamental problems:

1. Given a polynomial $p \in \Pi$, can we determine or estimate the location of its zeros in a simple and efficient way. Or can we at least decide whether the zeros are inside the unit circle?
2. Can we *construct* polynomials whose zeros are located in a given region?

Clearly all of this concerns *complex* zeros now. There is a trivial way to answer the above questions: consider the polynomial in *factorized* form. However this is what we do *not* want to do here. We either assume that we are given the *coefficients* of a polynomial with respect to monomials and only want to work on those issues, being well aware of the fact that factorizing polynomials, i.e., computing their zeros is a difficult and numerically not a well-conditioned task.

7.2 Prelude: Zeros of Polynomials

We will have a general view of the problem when deciding whether or not a polynomial has all its zeros inside the unit circle. Probably the most general approach in estimating the locations of the zeros of a polynomial would be by analyzing *Gersh-*

© The Author(s), under exclusive license to Springer Nature Switzerland AG 2021
T. Sauer, *Continued Fractions and Signal Processing*, Springer Undergraduate Texts in Mathematics and Technology, https://doi.org/10.1007/978-3-030-84360-1_7

gorin circles; but since this method usually does not give very precise locations of the zeros, we only discuss this in the problem section.

Remark 7.1 We will see soon that the "good" location of poles can vary under simple and classical transformations used in Function Theory. Sometimes the desired locations are inside, sometimes outside of $\mathbb{D}$ and sometimes they have to lie in a certain half-plane. All this depends mainly in which way the result can be proved most easily.

A polynomial $q \in \Pi_n$ has all its zeros inside the disc $\mathbb{D}$ if the *mirror polynomial*

$$z^n q\left(z^{-1}\right) = \sum_{j=0}^{n} q_{n-j}\, z^j, \qquad z \in \mathbb{C} \setminus \{0\},$$

has all its zeros *outside* of $\mathbb{D}$. Therefore it is quite irrelevant for the theorems that follow whether they are targeted to polynomials with all zeros inside or outside the unit circle; it is only relevant that they all lie on the same side. Fortunately, the literature on complex analysis, for example [47] provides some quite general results that show under which circumstances a complex polynomial $f \in \mathbb{C}[z]$ has *all* or *no* zeros inside the unit disc. A classic in this respect which can also be found in [24] is the Eneström–Kakeya theorem that provides a *sufficient* condition for a polynomial to have no zeros *inside* the unit disc.

Theorem 7.1 (Eneström–Kakeya) *If $p_0 > p_1 > \cdots > p_n > 0$, then the polynomial $p(z) = p_0 + \cdots + p_n z^n$ has no zero in $\mathbb{D}$.*

Proof For $z \in \mathbb{C}$, we have

$$(1 - z)\, p(z) = p_0 + \sum_{j=1}^{n} \left(p_j - p_{j-1}\right) z^j - p_n z^{n+1},$$

and therefore for $|z| \le 1$, a double triangle inequality downwards yields

$$|1 - z|\, |p(z)| = |(1 - z)p(z)|$$

$$\ge p_0 - \sum_{j=1}^{n} |p_j - p_{j-1}|\, |z^j| - |p_n|\, |z^{n+1}| \ge p_0 + \sum_{j=1}^{n} \left(p_j - p_{j-1}\right) - p_n, = 0$$

with equality if and only if $|z| = 1$, i.e., $z = e^{i\theta}$ for some $\theta \in [0, 2\pi)$, and if all the powers $z^j = e^{i\theta j}$ have the same argument which is the case if and only if $\theta = 0$ or $z = 1$. Since $p(1) = p_0 + \cdots + p_n > 0$, the polynomial p cannot have a zero at $z = 1$, hence $0 \notin p(\mathbb{D})$. $\square$

While the Eneström–Kakeya is indeed a nice and interesting result, it is only a *sufficient* condition. The question is whether it is possible to characterize polynomials without zeros inside the unit circle *without* having to factorize. Recall that factorizing the polynomial would be an approach that is conceptionally cheap but computationally expensive, see Sect. 4.4. To derive a condition for all zeros being inside $\mathbb{D}$, we first modify the problem by means of a fractional linear *rational transform*

$$w = \frac{z+1}{z-1}, \qquad z = \frac{w+1}{w-1}. \tag{7.1}$$

These two transforms are inverses of each other which is easily verified by noting that both can be rewritten as $zw - z - w - 1 = 0$. Writing $w = u + iv$, we then get

$$|z|^2 = \left| \frac{w+1}{w-1} \right|^2 = \frac{(u+1)^2 + v^2}{(u-1)^2 + v^2} \qquad \Rightarrow \qquad \begin{cases} |z| > 1, & u > 0, \\ |z| = 1, & u = 0, \\ |z| < 1, & u < 0. \end{cases}$$

Consequently the transform $z \to w$ maps the complex plane $\mathbb{C}$ to itself and in such a way that $|z| < 1$ holds if and only if the associated w has negative real part: $\Re w < 0$. If now $p(z)$ is a Laurent polynomial, then

$$p(z) = \sum_{j=0}^{n} p_j z^{-j} = \sum_{j=0}^{n} p_j \left(\frac{w+1}{w-1} \right)^{-j} = \left(\frac{1}{w+1} \right)^n \sum_{j=0}^{n} p_j (w-1)^j (w+1)^{n-j}$$

$$= \left(\frac{1}{w+1} \right)^n \sum_{j=0}^{n} p_j^\dagger w^j = (1+w)^{-n} \, p^\dagger(w),$$

where

$$(1+w)^{-1} = \left(1 + \frac{z+1}{z-1} \right)^{-1} = \left(\frac{2z}{z-1} \right)^{-1} = \frac{z-1}{2z}.$$

If z is a zero of p such that $0 < |z| < 1$, then $w \neq 1$ and therefore $p^\dagger(w) = 0$ where w lies in the left half-plane. We record this fact in a formal way.

Theorem 7.2 *The Laurent polynomial $p(z)$ has all its zeros inside the unit circle if and only if $p^\dagger$ has all its zeros in the* left half-plane $\mathbb{H}_- := \{ z \in \mathbb{C} : \Re z < 0 \}$.

7.2.1 Problems

Definition 7.1 Given a matrix $A \in \mathbb{R}^{n \times n}$ the associated *Gershgorin circles* are defined as

$$\mathcal{D}_j := \mathcal{D}_j(A) := \left\{ z \in \mathbb{C} : |z - a_{jj}| \le \sum_{k \neq j} |a_{jk}| \right\}, \qquad j = 1, \ldots, n \tag{7.2}$$

Each Gershgorin circle is therefore defined in a single row of the matrix. We will now indicate how these circles can be used for estimating the locations of zeros of a given polynomial.

7.1 Show that if $\|A\| < 1$ for some norm $\|\cdot\|$, then $I - A$ is nonsingular and

$$(I - A)^{-1} = \sum_{j=0}^{\infty} A^j,$$

where the sum on the right-hand side converges absolutely (with respect to the norm $\|\cdot\|$.)

7.2 Show that any eigenvalue of λ of $A \in \mathbb{R}^{n \times n}$ belongs to the Gershgorin circles of A, i.e.,

$$\lambda \in \bigcup_{j=1}^{n} \mathcal{D}_j.$$

Hint: Write $A = D - B$ with $b_{jj} = 0$ and conclude from the fact that $D - \lambda I$ is nonsingular while $A - \lambda I$ is singular that $\left\|(D - \lambda I)^{-1} B\right\|_{\infty} \geq 1$ by Problem 7.1.

7.3 Use Gershgorin circles to give an estimate of the location of the zeros of a given polynomial. Can you give different methods?

Hint: Sect. 4.4.

7.3 Hurwitz Polynomials and Stieltjes' Theorem

Looking at the definition of the fractional linear transformation (7.1), we can immediately conclude that the coefficients of the modified polynomial $p^{\dagger}$ are real if and only if the coefficients of the original p are real. This leads us to a class of polynomials which will become the object of investigation for the rest of this chapter and to some extent follows the exposition in [28]. From now on we write polynomials as polynomials in the variable z to indicate that now we *explicitly* consider polynomials as functions $\mathbb{C} \to \mathbb{C}$. We also make that explicit by writing $\mathbb{C}[z]$ for the underlying ring of polynomials, even if the Hurwitz polynomials that we consider will have real coefficients by definition. This unfortunately is not shared by their factors which makes it necessary to consider $\mathbb{C}[z]$. Furthermore we are not so much interested in the unit circle as we are in the left half-plane.

Definition 7.2 A polynomial $f \in \mathbb{C}[z]$ is called a *Hurwitz polynomial* if it has real coefficients and all its zeros have a negative real part, i.e.,

$$Z(f) := \{z \in \mathbb{C} : f(z) = 0\} \subset \mathbb{H}_-. \tag{7.3}$$

Before we will collect further information on Hurwitz polynomials, we first address the question as to what justifies their appearance in the context of continued fractions. To achieve this we first mention a classical way of decomposing polynomials which is also used a lot in subdivision and wavelet theory, best known under the name *subsymbol decomposition*: we write $f(z)$ as

$$f(z) = \sum_{j=0}^{n} f_j z^j = \sum_{j \leq n/2} f_{2j} z^{2j} + \sum_{j < n/2} f_{2j+1} z^{2j+1} = h\left(z^2\right) + zg\left(z^2\right),$$

where h contains the coefficients of f with even indices, while g contains those with odd indices. Splitting a polynomial into such a pair can become useful and interesting if this pair has a special property.

Definition 7.3 Two *real* polynomials $p, q \in \mathbb{R}[z]$ with $\deg p = \deg q = n$ or $\deg p = n$ and $\deg q = n - 1$ are said to form a *positive pair* if their zeros $x_1, \ldots, x_n$ and $x'_1, \ldots, x'_n$ or $x'_1, \ldots, x'_{n-1}$, respectively, *interlace*, i.e., if

$$\begin{aligned} x'_1 < x_1 < x'_2 < \cdots < x'_n < x_n < 0, \quad q \in \Pi_n, \\ x_1 < x'_1 < x_2 < \cdots < x'_{n-1} < x_n < 0, \quad q \in \Pi_{n-1} \end{aligned} \tag{7.4}$$

and the leading coefficients of p and q have the same sign.

Remark 7.2 The requirement that two polynomials that form a positive pair have leading coefficients of the same sign is only a normalization issue; the zeros are obviously unaffected by that modification.

The punch line here is that positive pairs characterize Hurwitz polynomials and can in turn be characterized by continued fractions. In fact all positive pairs of polynomials as well as all Hurwitz polynomials can eventually be enumerated by continued fractions.

Theorem 7.3 (Stieltjes) *For a polynomial $f(z) = g\left(z^2\right) + zh\left(z^2\right)$, the following statements are equivalent:*

1. *f is a Hurwitz polynomial.*
2. *The polynomials g and h form a positive pair.*
3. *There exist $c_0 \geq 0$ and positive numbers $c_j, d_j > 0$, $j = 1, \ldots, m$, such that*

$$\frac{h(x)}{g(x)} = [c_0; d_1 x, c_1, d_2 x, c_2, \ldots, d_m x, c_m], \tag{7.5}$$

where $c_0 = 0$ if and only if $\deg f \in 2\mathbb{N}_0 + 1$, i.e., if f is of an odd degree.

Besides having positive coefficients, the continued fraction in (7.5) also offers a quite amazing structure; in the partial fraction expansion polynomials of degree 1 and degree 0 take turns so that the degrees in the convergents increase more slowly than those in Chap. 4 that had coefficients of degree 1 all the time.

To prove Theorem 7.3 we have to learn some more concepts and ideas, but the result is worth it and it is also a highlight. In fact Stieltjes' theorem will be a summary of two results as Theorems 7.5 and 7.6. In addition it allows us to construct and even to "enumerate" denominators of stable rational filters, hence this has a meaning in signal processing as well. But before we attack the steps of the proof of this theorem, we record another simple property of Hurwitz polynomials concerning the sign of their coefficients.

Lemma 7.1 *If $f \in \Pi_n$ is a Hurwitz polynomial of degree n with $f_n > 0$, then $f_j > 0$, $j = 0, \ldots, n$.*

Proof We factorize f as

$$f(z) = f_n \prod_{j=1}^{n} (z - \zeta_j), \qquad \zeta_j \in \mathbb{H}_-.$$

Since in a real polynomial, in particular in any Hurwitz polynomial, all zeros have to appear as complex conjugate pairs, f contains factors either of the form $(z + \alpha)$, $\alpha \in \mathbb{R}_+$ if the zero $-\alpha$ is real or of the form

$$(z - \zeta)(z - \overline{\zeta}) = z^2 - \underbrace{(\zeta + \overline{\zeta})}_{=2\Re\zeta\,<0} z + \underbrace{\zeta\,\overline{\zeta}}_{=|\zeta|^2>0} = z^2 + \beta z + \gamma, \qquad \beta, \gamma \in \mathbb{R}_+,$$

for a complex conjugate pair of zeros in $\mathbb{H}_-$. Hence

$$f(z) = f_n \left(\prod_{j=0}^{k} (z + \alpha_j) \right) \left(\prod_{j=0}^{k'} (z^2 + \beta_j z + \gamma_j) \right)$$

can only have positive coefficients.   □

Corollary 7.1 *In a Hurwitz polynomial all coefficients have to have strictly the same sign, i.e., they are either all positive or all negative.*

7.4 Cauchy Index and the Argument of the Argument

For the next concept we recall the Sturm chain from Sect. 4.3. So for an interval $I = [a, b]$ we counted the *weighted* sign changes $\Sigma_a^b f = \sigma(f, [a, b])$ of a function f, where "weighted" means that sign changes from $-$ to $+$ are counted positive and those from $+$ to $-$ negative. In the proof of Proposition 4.3, we then considered a rational function f defined as the quotient of two successive orthogonal polynomials or polynomials that satisfied a three term recurrence. However such a rational function does not only have zeros—which are zeros for the numerator—but it also included zeros of the denominator and this means poles. Each *pole* again provides a sign change, this time from $\pm\infty$ to $\mp\infty$ and the main tool of this section will be a device to count them as well.

Definition 7.4 *(Sign changes across poles & Cauchy index)* Let $f : \mathbb{C} \to \mathbb{C}$ be a rational function.

1. We say that f has a *singular sign change* or *sign change across a pole* at a point x if

$$\lim_{y \to x_-} f(y) = \pm\infty \quad \text{and} \quad \lim_{y \to x_+} f(y) = \mp\infty. \tag{7.6}$$

2. The *Cauchy index* $I_a^b f$ of a function f on the interval $[a, b]$ is the weighted sum of singular sign changes or sign changes across poles where the changes from $-\infty$ to $+\infty$ are counted as positive, and those from $+\infty$ to $-\infty$ as negative.

Recalling the approach from Sturm chains, the Cauchy index can also be formally defined by "normal" sign changes such as

$$I_a^b f := \Sigma_a^b \frac{1}{f}, \tag{7.7}$$

but keep in mind that also $1/f$ can have poles while one usually considers sign changes of a continuous function in the context of Sturm chains. It does not require much imagination to get the idea that also the Cauchy index will be strongly connected to Sturm chains. But to really follow the proof from [28], we need a little bit of Function Theory or complex analysis, cf. [24, Theorem 2, S. 175].

Definition 7.5 *(Argument)* The number $\theta =: \arg z \in \mathbb{R}$ is called the *argument* of the complex number $z = \Re z + i \Im z \in \mathbb{C}$ if it satisfies

$$\Re z + i \Im z = z = |z| e^{i\theta} = |z| (\cos\theta + i \sin\theta). \tag{7.8}$$

Clearly as is well known, the argument of a number is only defined modulo 2π, i.e., as an element of $\mathbb{T}$. Moreover it follows immediately from (7.8) that

$$\cos\theta = \Re z/|z|, \quad \sin\theta = \Im z/|z|. \tag{7.9}$$

The next result is a classic consequence of the residue theorem in Function Theory that can be found for example in [50, Theorem 9.2.2, p. 253], where it is formulated even for meromorphic functions, i.e., functions with possible poles. For our purposes, the following slightly simpler version is sufficient. We will not define all the details of this theorem such as positive orientation of a curve, since it is really only a tool in what follows, but we will refer to [50] once more.

Fig. 7.1 The domain of integration that certainly contains **no** zero of f if f is a Hurwitz polynomial, regardless of how large we choose R

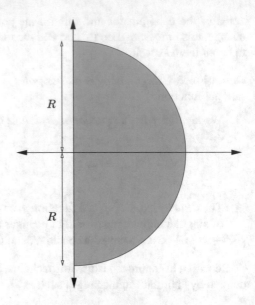

Theorem 7.4 (Principle of the argument for analytic functions) *If γ is a positively oriented piecewise smooth closed curve in D enclosing a domain $\Omega \subset D$, and f is analytic on a domain $D \subset \mathbb{C}$ with $f(\gamma) \neq 0$, then*

$$\frac{1}{2\pi} \Delta_\gamma \arg f(z) = \#\{z \in \Omega \;:\; f(z) = 0\}, \tag{7.10}$$

where Δ_γ stands for the number of changes in the argument modulo 2π along the curve γ.

Remark 7.3 The "Change of argument" in Theorem 7.4 means tracing the curve $\arg f(\gamma(t))$, $t \in [0, 1]$ for some parameterization of γ and to consider

$$\Delta_\gamma \arg f = \arg f\left(\gamma(1^-)\right) - \arg f\left(\gamma(0^+)\right) = \int_\gamma (\arg f)'(z)\,dz = \int_\gamma \frac{f'(z)}{f(z)}\,dz.$$

The expression on the right-hand side is then really well defined and corresponds to a continuous extension of $\arg f$ along the curve γ. For further details again consult [50].

Now let f be a Hurwitz polynomial and consider, for $R > 0$, the integral along the positively oriented curve γ that consists of the interval $[-Ri, Ri]$ and the semicircle of radius R in $\mathbb{H}_+$, see Fig. 7.1. Since f is a Hurwitz polynomial and has all zeros in the left half-plane, then no zero in the domain enclosed by the curve (7.10) yields that

$$0 = \Delta_\gamma f(z) = \Delta_{-R}^R \arg f(ix) - \Delta_{-\pi}^\pi f\left(R\,e^{ix}\right);$$

moreover for sufficiently large values of R, the change of argument along the semi-circle is determined by the leading term $f_n z^n$ of f, $n := \deg f$, and it has the value $n\pi$. In other words

$$\Delta_{-\infty}^{\infty} \arg f\,(ix) = \lim_{R \to \infty} \Delta_{-R}^{R} \arg f\,(ix) = \lim_{R \to \infty} \Delta_{-\pi}^{\pi} f\left(Re^{ix}\right) = n\pi. \qquad (7.11)$$

Writing f with respect to its odd and even coefficients as

$$f(z) = a_0 z^n + b_0 z^{n-1} + a_1 z^{n-2} + b_1 z^{n-3} + \cdots, \qquad a_0 \neq 0,$$

we get for $n = 2m$

$$f(ix) = (-1)^m a_0 x^n + i\,(-1)^{m-1} x^{n-1} + (-1)^{m-1} a_1 x^{n-2} + \cdots$$
$$= (-1)^m \left(a_0 x^n - a_1 x^{n-2} + a_2 x^{n-4} + \cdots\right) + i\,(-1)^{m-1}\left(b_0 x^{n-1} - b_1 x^{n-3} + \cdots\right)$$

and for $n = 2m + 1$

$$f(ix) = (-1)^m \left(b_0 x^{n-1} - b_1 x^{n-3} + \cdots\right) + i\,(-1)^m\left(a_0 x^n - a_1 x^{n-2} + \cdots\right),$$

respectively, which shows that in both cases

$$f(ix) = p(x) + i\,q(x), \qquad x \in \mathbb{R}, \qquad (7.12)$$

holds, where

$$p(x) = \begin{cases} (-1)^m \left(a_0 x^n - a_1 x^{n-2} + \cdots + (-1)^m a_m\right), & n = 2m, \\ (-1)^m \left(b_0 x^{n-1} - b_1 x^{n-3} + \cdots + (-1)^m b_m\right), & n = 2m + 1, \end{cases} \qquad (7.13)$$

and

$$q(x) = \begin{cases} (-1)^{m-1} \left(b_0 x^{n-1} - b_1 x^{n-3} + \cdots + (-1)^{m-1} b_{m-1} x\right), & n = 2m, \\ (-1)^m \left(a_0 x^n - a_1 x^{n-2} + \cdots + (-1)^m a_m x\right), & n = 2m + 1. \end{cases}$$
$$(7.14)$$

By (7.9), we have that

$$\tan\theta = \frac{\Im z}{\Re z}, \quad \cot\theta = \frac{\Re z}{\Im z} \quad \Rightarrow \quad \theta = \arctan\frac{\Im z}{\Re z} = \operatorname{arccot}\frac{\Re z}{\Im z}.$$

Applied to (7.12), this observation yields

$$\arg f\,(ix) = \arctan\frac{q(x)}{p(x)} = \operatorname{arccot}\frac{p(x)}{q(x)}.$$

Now any increment of the argument corresponds to a pole or singularity of the tangent of θ, and therefore

$$\frac{1}{\pi} \Delta_{-\infty}^{\infty} \arg f(ix) = \begin{cases} I_{-\infty}^{\infty} \dfrac{p(x)}{q(x)}, & n = 2m+1, \\[3mm] -I_{-\infty}^{\infty} \dfrac{q(x)}{p(x)}, & n = 2m, \end{cases}$$

so that we obtain for our Hurwitz polynomial by (7.11) the identity

$$n = I_{-\infty}^{\infty} \frac{b_0 \, x^{n-1} - b_1 \, x^{n-3} + \cdots}{a_0 \, x^n - a_1 x^{n-2} + \cdots} = -\Sigma_{-\infty}^{\infty} \frac{a_0 \, x^n - a_1 x^{n-2} + \cdots}{b_0 \, x^{n-1} - b_1 \, x^{n-3} + \cdots}. \qquad (7.15)$$

Now it is time to return to the decomposition $f(z) = g(z^2) + z \, h(z^2)$. Let us begin with the case $n = 2m$ where

$$\begin{aligned} g(x) &= f_n \, x^m + f_{n-2} \, x^{m-1} + \cdots + f_0, \\ h(x) &= f_{n-1} x^{m-1} + f_{n-3} x^{m-2} + \cdots + f_1. \end{aligned} \qquad (7.16)$$

Hence since $a_j = f_{n-2j}$ and $b_j = f_{n-1-2j}$, we have

$$\begin{aligned} g\left(-z^2\right) &= (-1)^m \left(a_0 \, z^n - a_1 \, z^{n-2} + \cdots\right), \\ h\left(-z^2\right) &= (-1)^m \left(b_0 \, z^{n-2} - b_1 \, z^{n-4} + \cdots\right), \end{aligned}$$

from which we conclude with the help of (7.15) that

$$n = -I_{-\infty}^{\infty} \frac{z \, h\left(-z^2\right)}{g\left(-z^2\right)}. \qquad (7.17)$$

The respective identities for the case $n = 2m + 1$ are

$$\begin{aligned} g(x) &= f_{n-1} x^m + f_{n-3} x^{m-1} + \cdots + f_0, \\ h(x) &= f_n \, x^m + f_{n-2} \, x^{m-1} + \cdots + f_1, \end{aligned} \qquad (7.18)$$

and

$$n = -I_{-\infty}^{\infty} \frac{g\left(-z^2\right)}{z \, h\left(-z^2\right)}. \qquad (7.19)$$

Now we derive a property of the Cauchy index by making use of its similarity to Sturm chains so that the following lemma is mainly a reformulation of Theorem 4.6.

Lemma 7.2 *Let $a < c < b$ and $\phi : [a, b] \to \mathbb{R}$ be a rational function. Then*

$$I_a^b \phi = I_a^c \phi + I_c^b \phi + \eta_c \phi,$$

where

$$\eta_c \phi := \begin{cases} 1 \\ -1 & \text{if} \quad \lim_{x \to c_\pm} \phi(x) \begin{cases} = \pm\infty \\ = \mp\infty \\ \in \mathbb{R}. \end{cases} \\ 0 \end{cases} \tag{7.20}$$

Proof Since the Cauchy index counts sign changes of ϕ^{-1}, we can proceed as in Theorem 4.6 by just taking into account the fact that any singular sign change of ϕ is a normal sign change of ϕ^{-1} and vice versa. If, on the other hand, such a sign change happens exactly at c, it is not recognized by the indices for the subintervals and has to be compensated explicitly by the quantity η_c from (7.20). $\qquad\square$

Taking into account that the factor z in the denominator is irrelevant for the Cauchy index since the denominator polynomial g satisfies $g(0) = f_0 \neq 0$, then there cannot be an η_0 term so we can expand (7.17) for $n = 2m$ in the following way:

$$n = -I_{-\infty}^\infty \frac{z\,h\,(-z^2)}{g\,(-z^2)} = -\left(I_{-\infty}^0 + I_0^\infty\right) \frac{z\,h\,(-z^2)}{g\,(-z^2)} = -2\,I_{-\infty}^0 \frac{z\,h\,(-z^2)}{g\,(-z^2)}$$

$$= 2\,I_{-\infty}^0 \frac{h\,(-z^2)}{g\,(-z^2)} = 2\,I_{-\infty}^0 \frac{h\,(x)}{g\,(x)} = I_{-\infty}^0 \frac{h\,(x)}{g\,(x)} - I_{-\infty}^0 \frac{x\,h\,(x)}{g\,(x)}$$

$$= I_{-\infty}^0 \frac{h\,(x)}{g\,(x)} - I_{-\infty}^0 \frac{x\,h\,(x)}{g\,(x)} + \underbrace{I_0^\infty \frac{h\,(x)}{g\,(x)} - I_0^\infty \frac{x\,h\,(x)}{g\,(x)}}_{=0} = I_{-\infty}^\infty \frac{h\,(x)}{g\,(x)} - I_{-\infty}^\infty \frac{x\,h\,(x)}{g\,(x)}.$$

For $n = 2m + 1$, we obtain the analogous

$$n = I_{-\infty}^\infty \frac{g\,(x)}{x\,h\,(x)} - I_{-\infty}^\infty \frac{g\,(x)}{h\,(x)},$$

and therefore

$$n = \begin{cases} I_{-\infty}^\infty \dfrac{h\,(x)}{g\,(x)} - I_{-\infty}^\infty \dfrac{x\,h\,(x)}{g\,(x)}, & n = 2m, \\[3mm] I_{-\infty}^\infty \dfrac{g\,(x)}{x\,h\,(x)} - I_{-\infty}^\infty \dfrac{g\,(x)}{h\,(x)}, & n = 2m + 1. \end{cases} \tag{7.21}$$

These technical computations already allow us to tackle one statement of Theorem 7.3 which even has a name of its own or to be precise: according to [28], this is a *special case* of the *Hermite–Biehler Theorem*.

Theorem 7.5 (Hermite–Biehler) *A polynomial $f(z) = g\left(z^2\right) + z\,h\left(z^2\right)$ is a Hurwitz polynomial if and only if g and h form a positive pair.*

Proof We have already shown that f is a Hurwitz polynomial if and only if (7.21) is satisfied. For the rest, we once more have to distinguish two cases.

$n = 2m$: the denominator polynomial g has degree m and therefore at most m zeros. Therefore with an argument as in the proof Proposition 4.3, we can draw the following conclusion:

$$2m = I_{-\infty}^{\infty} \frac{h(x)}{g(x)} - I_{-\infty}^{\infty} \frac{x\,h(x)}{g(x)} \quad\Rightarrow\quad I_{-\infty}^{\infty} \frac{h(x)}{g(x)} = -I_{-\infty}^{\infty} \frac{x\,h(x)}{g(x)} = m,$$

and the quotient $h(x)/g(x)$ can only have singular sign changes or sign changing poles from $-\infty$ to $+\infty$; the quotient $x\,h(x)/g(x)$ on the other hand has only those from $+\infty$ to $-\infty$. This in turn is possible if between any such pair of jumps there is a regular sign change, i.e., a zero of h. Since g has exactly the m zeros $x_1, \ldots, x_m$ and h has the $m-1$ zeros $x_1', \ldots, x_{m-1}'$, these zeros can therefore only be arranged to be

$$x_1 < x_1' < x_2 < x_2' < \cdots < x_{m-1}' < x_m < 0.$$

According to (7.16) and Lemma 7.1, f_n and f_{n-1} have to have the same sign and therefore we can assume that g and h both have leading coefficients of the same and even positive sign which makes g and h a positive pair. Since all arguments were equivalences, the converse is obtained by repeating the proof backwards.

$n = 2m + 1$: now the $n = 2m + 1$ sign changes across poles have to be obtained by $m + 1$ sign changes for $x\,h(x)$ and m sign changes for $h(x)$ with opposite parities. This just means that the $m + 1$ sign changes for $x\,h(x)$ occur at the positions $x_1' < \cdots < x_m' < 0$ and at 0; note that $x = 0$ is the only additional zero when passing from $h(x)$ to $xh(x)$ which has exactly one more zero. Between these sign changes, there must be the sign changes for g, and we then have

$$x_1' < x_1 < x_2' < \cdots < x_m' < x_m < 0$$

as claimed.   □

The identity (7.21) is equivalent to f being a Hurwitz polynomial, or equivalently to the fact that g and h form a positive pair. It allows us to draw yet another conclusion.

Proposition 7.1 *Two polynomials g and h, $\deg g = m$, form a positive pair if and only if*

$$m = I_{-\infty}^{\infty} \frac{h(x)}{g(x)} = -I_{-\infty}^{\infty} \frac{x\,h(x)}{g(x)} \tag{7.22}$$

and if in the case $\deg g = \deg h$, we additionally have

$$\epsilon_{\infty} = \lim_{x \to +\infty} \operatorname{sgn} \frac{h(x)}{g(x)} = 1. \tag{7.23}$$

Proof We have already seen that for $n = 2m$, Eq. (7.22) follows directly from (7.21). To also obtain a respective result for $n = 2m + 1$, which is to get from (7.21) to the statement of Proposition 7.1, we need a formula for the Cauchy index of a rational function wherein the degree of the numerator exceeds that of the denominator, namely

$$I_{-\infty}^{\infty} f(x) + I_{-\infty}^{\infty} f^{-1}(x) = \frac{\epsilon_{\infty} - \epsilon_{-\infty}}{2}, \qquad \epsilon_{\pm\infty} = \lim_{x \to \pm\infty} \operatorname{sgn} f(x). \qquad (7.24)$$

So we see that the expression on the left-hand side of (7.24) is exactly the number of all singular and normal sign changes for f which sum-up to the following:

1. 1, if $\epsilon_{\infty} = 1$ and $\epsilon_{-\infty} = -1$,
2. -1, if the limits have signs $-$ and $+$,
3. 0 whenever $\epsilon_{\infty} = \epsilon_{-\infty}$.

Using (7.24), we can now rewrite the second line of (7.21) to be

$$2m + 1 = n = I_{-\infty}^{\infty} \frac{g(x)}{x h(x)} - I_{-\infty}^{\infty} \frac{g(x)}{h(x)}$$

$$= I_{-\infty}^{\infty} \frac{h(x)}{g(x)} - \frac{1-1}{2} - I_{-\infty}^{\infty} \frac{x h(x)}{g(x)} + \frac{1+1}{2},$$

which gives (7.22) again. Furthermore the equal sign for the leading coefficients of g and h,—a necessary condition for being a positive pair—follows for $n = 2m$, and therefore $\deg h = \deg g - 1$ directly from (7.22); for $n = 2m + 1$, i.e., $\deg h = \deg g$, the additional assumption (7.23) becomes necessary. □

To prove the second equivalence in Theorem 7.3, we need the following auxiliary statement.

Lemma 7.3 *Suppose that the polynomials g and h, $\deg g = m$ form a positive pair and there are constants c, d as well as polynomials $g_1, h_1 \in \Pi_{m-1}$ such that*

$$\frac{h(x)}{g(x)} = c + \cfrac{1}{dx + \cfrac{g_1(x)}{h_1(x)}} = \left[c; dx, \frac{g_1(x)}{h_1(x)} \right]. \qquad (7.25)$$

Then c, d and g_1, h_1 are determined uniquely by g and h and the following holds true:

1. *$c \geq 0, d > 0$,*
2. *$\deg g_1 = \deg h_1 = m - 1$,*
3. *g_1 and h_1 form a positive pair.*

If, conversely, the numbers c, d and the polynomials g_1, h_1 satisfy the above three conditions and g, h are defined by (7.25), then g and h form a positive pair.

Proof If g, h are a positive pair with $\deg g = m$, the definition implies in particular that $\deg h \in \{m - 1, m\}$ and that g has m real zeros, so that we obtain by (7.25) and the fact that the Cauchy index is not affected by adding a constant, the conclusion is

$$m = I_{-\infty}^{\infty} \frac{h(x)}{g(x)} = I_{-\infty}^{\infty} \left[c + \cfrac{1}{dx + \cfrac{g_1(x)}{h_1(x)}} \right] = I_{-\infty}^{\infty} \frac{h_1(x)}{dx \, h_1(x) + g_1(x)}. \qquad (7.26)$$

This can only hold if the denominator is a polynomial of degree at least m, hence $d \neq 0$ and $\deg h_1 = m - 1$, as otherwise the degree of the denominator could not exceed $m - 1$. Without loss of generality, we can also assume that the leading term of h_1 is positive; otherwise we multiply both g_1 and h_1 by -1. Now (7.26) tells us that both rational functions $h(x)/g(x)$ as well as $h_1(x)/(dx\,h_1(x) + g_1(x))$ have a maximal number of singular sign changes from $-$ to $+$, and as a result both are *negative* for sufficiently small x and *positive* for sufficiently large x. Consequently

$$-1 = -\operatorname{sgn} d = \lim_{x \to -\infty} \frac{h_1(x)}{dx\,h_1(x) + g_1(x)}, \qquad 1 = \operatorname{sgn} d = \lim_{x \to -\infty} \frac{h_1(x)}{dx\,h_1(x) + g_1(x)},$$

implying $d > 0$. By (7.26) the function h/g has precisely m singular sign changes from $-\infty$ to $+\infty$ which interlace with $m - 1$ sign changes from $+$ to $-$, so that

$$- I_{-\infty}^{\infty} \left[dx + \frac{g_1(x)}{h_1(x)} \right] \geq m - 1. \tag{7.27}$$

Since $\deg h_1 = m - 1$ and therefore this Cauchy index is at most $m - 1$, equality has to hold in (7.27) and therefore

$$m - 1 = -I_{-\infty}^{\infty} \left[dx + \frac{g_1(x)}{h_1(x)} \right] = -I_{-\infty}^{\infty} \frac{g_1(x)}{h_1(x)}. \tag{7.28}$$

From the second identity in (7.22), we moreover conclude that

$$m = -I_{-\infty}^{\infty} \frac{x\,h(x)}{g(x)} = -I_{-\infty}^{\infty} \left[cx + \frac{x}{dx + \dfrac{g_1(x)}{h_1(x)}} \right] = -I_{-\infty}^{\infty} \left[cx + \frac{1}{d + \dfrac{g_1(x)}{x\,h_1(x)}} \right]$$

$$= -I_{-\infty}^{\infty} \left[\frac{1}{d + \dfrac{g_1(x)}{x\,h_1(x)}} \right] = I_{-\infty}^{\infty} \left[d + \frac{g_1(x)}{x\,h_1(x)} \right] = I_{-\infty}^{\infty} \frac{g_1(x)}{x\,h_1(x)} \tag{7.29}$$

so that also $\deg g = m - 1$ since there must be a sign change between any pair of singular sign changes. This completes the proof of claim (2).

Since the two polynomials g_1, h_1 have the same degree, it follows that

$$\lim_{x \to \pm\infty} \frac{g_1(x)}{h_1(x)} = \mu \neq 0,$$

hence

$$\lim_{x \to \pm\infty} dx + \frac{g_1(x)}{h_1(x)} = \pm\infty, \qquad \text{i.e.,} \qquad \lim_{x \to \pm\infty} \frac{1}{dx + \dfrac{g_1(x)}{h_1(x)}} = 0,$$

and therefore by (7.25),

$$c = \lim_{x \to \infty} \left[\frac{h(x)}{g(x)} - \frac{1}{d\,x + \dfrac{g_1(x)}{h_1(x)}} \right] = \lim_{x \to \infty} \frac{h(x)}{g(x)} \begin{cases} > 0, & \deg g = \deg h, \\ = 0, & \deg g > \deg h, \end{cases}$$

which also verifies claim (1).

It remains to show that g_1 and h_1 indeed form a positive pair. To achieve this we apply (7.24) to (7.29) and obtain that

$$I_{-\infty}^{\infty} \frac{x\,h_1(x)}{g_1(x)} = -m + \frac{\epsilon_\infty - \epsilon_{-\infty}}{2} = -m + \epsilon_\infty, \tag{7.30}$$

since

$$\lim_{x \to +\infty} \operatorname{sgn} \frac{h_1(x)}{g_1(x)} = \epsilon_\infty := \lim_{x \to +\infty} \operatorname{sgn} \frac{x\,h_1(x)}{g_1(x)} = -\lim_{x \to -\infty} \operatorname{sgn} \frac{x\,h_1(x)}{g_1(x)} = -\epsilon_{-\infty}.$$

If we normalize g_1 and h_1 in such a way that $\epsilon_\infty = 1$, then this identity together with (7.28) and (7.30) is exactly what is needed to apply Proposition 7.1, and therefore we conclude that g_1 and h_1 form a positive pair.

For the converse, we just note that all arguments used here were either identities or equivalences. $\square$

With this lemma at hand, the proof of Theorem 7.3 is not magic any more, since we are shown that positive pairs are mapped to positive pairs by an excellent "double step" of the continued fraction expansion. Indeed Theorem 7.3 eventually follows from assembling the Hermite–Biehler Theorem, Theorem 7.5, and the following result.

Theorem 7.6 *Two polynomials g and h, $\deg g = m$, form a positive pair if and only if there exist*

$$c_0 \begin{cases} > 0, & \deg g = \deg h, \\ = 0, & \deg g = \deg h + 1, \end{cases} \qquad c_j, d_j \in \mathbb{R}_+, \quad j = 1, \ldots, m,$$

such that

$$\frac{h(x)}{g(x)} = [c_0; d_1\,x, c_1, \ldots, d_m\,x, c_m]. \tag{7.31}$$

Proof By virtue of Lemma 7.3, we only have to show that to any positive pair g, h there exists a decomposition into g_1, h_1 as in (7.25). If $m = \deg g = \deg h$, then we can perform a division of h by g with the remainder h_1: this gives the decomposition $h = c_0\,g + h_1$, in which even $c_0 > 0$ since the leading coefficients of the positive pair g and h must have the same sign. Since by assumption $\deg h \in \{m-1, m\}$, it follows that $\deg h_1 = m - 1$. Therefore

$$\frac{h(x)}{g(x)} = \frac{c\,g(x) + h_1(x)}{g(x)} = c_0 + \frac{h_1(x)}{g(x)} = c_0 + \frac{1}{\dfrac{g(x)}{h_1(x)}}.$$

On the other hand $\deg g = m = \deg h_1 + 1$, hence $g(x) = d_1\,x\,h_1(x) + g_1(x)$, $\deg g_1 \le m - 1$, and therefore

$$\frac{h(x)}{g(x)} = c_0 + \frac{1}{\dfrac{d_1 x\,h_1(x) + g_1(x)}{h(x)}} = c_0 + \frac{1}{d_1\,x + \dfrac{g_1(x)}{h_1(x)}},$$

so that Lemma 7.3 implies $d_1 > 0$ and $\deg g_1 = \deg h_1 = m - 1$. For $\deg h = \deg g - 1$, the same holds only with $c = 0$, and therefore $h_1 = h$. To summarize, we have shown that in both cases

$$\frac{h(x)}{g(x)} = c_0 + \frac{1}{dx + \dfrac{1}{h_1(x)/g_1(x)}} = \left[c_0;\, d_1\,x,\, \frac{h_1(x)}{g_1(x)}\right], \qquad c_0 \ge 0, \qquad (7.32)$$

holds with

$$\deg g_1 = \deg h_1 = m - 1$$

and $c_0 > 0$ if and only if $\deg g = \deg h$. This allows us to write h_1/g_1 as $\left[c_1;\, d_2\,x,\, \frac{h_2(x)}{g_2(x)}\right]$ with $\deg g_2 = \deg h_2 = m - 2$, and so on. Iterating this decomposition in (7.32), we get for $j = 1, \ldots, m$ a representation

$$\frac{h(x)}{g(x)} = \left[c_0;\, d_1\,x,\, c_1, \ldots, d_j\,x,\, \frac{h_j(x)}{g_j(x)}\right], \qquad \deg g_j = \deg h_j = m - j, \quad (7.33)$$

and the case $j = m$, together with the observation that $g_m, h_m \ne 0$ gives $c_m \ne 0$ and therefore (7.31) as claimed. The converse follows directly from expanding the continued fraction. $\qquad\square$

Corollary 7.2 *Two polynomials g, h form a positive pair if and only if all even order convergents of the continued fraction expansion of h/g are positive pairs as well.*

Proof Assume that $\deg g = m$; then for $n = 1, \ldots, m$, the convergent of order $2n$ takes the form

$$\frac{p_{2n}(x)}{q_{2n}(x)} = [c_0;\, d_1 x,\, c_1, \ldots, d_n x,\, c_n]$$

and is a positive pair by Theorem 7.6. The converse is just the case $n = m$. $\qquad\square$

7.4.1 Problems

7.4 Prove for $m \in \mathbb{N}$ the *Fourier identity*

$$\frac{1}{m} \sum_{k=0}^{m-1} e^{2\pi ijk} = \delta_{j0} = \begin{cases} 1, & j = 0, \\ 0, & j \neq 0, \end{cases} \qquad j = 0, \ldots, m-1. \qquad (7.34)$$

Why does it suffice to consider $j = 0, \ldots, m-1$ in (7.34)?

7.5 Given $m \in \mathbb{N}$. Show that any polynomial $f \in \Pi$ can be uniquely written as

$$f(z) = \sum_{j=0}^{m-1} z^j f_j(z^m)$$

where

$$f_j(z^m) = \frac{1}{m} \sum_{j=0}^{m-1} e^{2\pi ijk/m} f\left(e^{2\pi ik/m} z\right). \qquad j = 0, \ldots, m-1. \qquad (7.35)$$

Give the explicit version of (7.35) for $m = 2$.

Hint: Use (7.34).

7.6 Show that if g and h with $\deg g = \deg h$ are a positive pair, then also $g + \alpha h$ and h are a positive pair, $\alpha > 0$. What about negative values of α?

7.5 The Routh–Hurwitz Theorem

The famous theorem by Routh–Hurwitz[1] provides another characterization for a Hurwitz polynomial, this time with respect to certain determinants. And since determinants cannot be imagined without (square) matrices, we start with another peculiar type of matrices.

Definition 7.6 Let $p \in \Pi$ be a polynomial of degree n. The *Hurwitz matrix* associated to p is the $n \times n$ matrix

---

[1] And this does not refer to the statement *"A PhD dissertation is a paper of the professor written under aggravating circumstances"* which is attributed in [1] to ADOLF HURWITZ, but also to OTTO TOEPLITZ. Since the matrices bearing their names are both representing generalized convolutions, this does not really make a difference anyway.

$$H_p := \begin{bmatrix} p_{n-1} & p_{n-3} & p_{n-5} & \cdots & 0 \\ p_n & p_{n-2} & p_{n-4} & \cdots & 0 \\ 0 & p_{n-1} & p_{n-3} & \cdots & 0 \\ 0 & p_n & p_{n-2} & \cdots & 0 \\ \vdots & \vdots & \vdots & \ddots & \vdots \\ 0 & 0 & 0 & \cdots & p_0 \end{bmatrix}. \tag{7.36}$$

Example 7.1 Let us consider some examples of such Hurwitz matrices for small values of n and a generic polynomial $p(x) = p_0 + \cdots + p_n x^n$ of that degree:

$n = 1$: the 1×1 matrix $H_p = [p_0]$.
$n = 2$: the Hurwitz matrix is

$$H_p = \begin{bmatrix} p_1 & 0 \\ p_2 & p_0 \end{bmatrix}$$

and contains a zero for the first time.
$n = 3$: some structure within the matrix becomes visible:

$$H_p = \begin{bmatrix} p_2 & p_0 & 0 \\ p_3 & p_1 & 0 \\ 0 & p_2 & p_0 \end{bmatrix}.$$

$n = 4$: we begin to see a pattern in the matrix:

$$H_p = \begin{bmatrix} p_3 & p_1 & 0 & 0 \\ p_4 & p_2 & p_0 & 0 \\ 0 & p_3 & p_1 & 0 \\ 0 & p_4 & p_2 & p_0 \end{bmatrix}.$$

Hurwitz matrices can be seen as initial $n \times n$ segments of a *Hurwitz operator* $H_p : \ell(\mathbb{N}_0) \to \ell(\mathbb{N}_0)$ defined as

$$H(p) = \begin{bmatrix} p_{n-1} & p_{n-3} & p_{n-5} & \cdots \\ p_n & p_{n-2} & p_{n-4} & \cdots \\ & p_{n-1} & p_{n-3} & \cdots \\ & p_n & p_{n-2} & \cdots \\ & \vdots & \vdots & \ddots \end{bmatrix} = \left[p_{n-1+j-2k} : \begin{matrix} j \in \mathbb{N}_0 \\ k \in \mathbb{N}_0 \end{matrix} \right], \tag{7.37}$$

so that for $\sigma \in \ell(\mathbb{N}_0)$,

$$(H(p)\sigma)_j = \sum_{k=0}^{\infty} p_{n-1+j-2k}\, \sigma_k = \sum_{k=0}^{\infty} f_{j-2k}\, \sigma_k, \qquad j \in \mathbb{Z}, \qquad f := \tau^{m-1} p;$$

and if we extend σ canonically to $\ell(\mathbb{Z})$ by setting $\sigma_j = 0$, $j < 0$, then it follows that

$$(H(p)\sigma)_j = \sum_{k\in\mathbb{Z}} f_{j-2k}\,\sigma_k = (f *_2 \sigma)_j = \left(T_2 f(\tau^{n-1}p)\sigma\right)_j, \qquad j \in \mathbb{Z}.$$

In other words the Hurwitz operator represents the positive part of the Toeplitz matrix for the *generalized convolution* as stated in (6.9) of Definition 6.5. The generalized convolution uses the shifted filter $\tau^{n-1}p$ and the dilation factor 2, but the shift of the filter is for technical reasons only. This points to the well-known fact that Hurwitz matrices and Hurwitz operators have a close relationship to stationary subdivision operators with dilation 2, cf. [15], which can be extended to general *slanted matrices* [12, 74].

Returning to the finite matrices H_p again, a careful look at Example 7.1 convinces us that once again we have to distinguish between odd and even values of n, namely

$$H_p = \begin{bmatrix} p_{n-1} & \cdots & p_3 & p_1 & 0 & 0 & \cdots & 0 & 0 \\ p_n & \cdots & p_4 & p_2 & p_0 & 0 & \cdots & 0 & 0 \\ \vdots & \ddots & \vdots & \vdots & \vdots & \vdots & \ddots & \vdots & \vdots \\ 0 & \cdots & 0 & p_{n-1} & p_{n-3} & p_{n-5} & \cdots & p_1 & 0 \\ 0 & \cdots & 0 & p_n & p_{n-2} & p_{n-4} & \cdots & p_2 & p_0 \end{bmatrix}, \qquad n = 2m, \tag{7.38}$$

and

$$H_p = \begin{bmatrix} p_{n-1} & \cdots & p_2 & p_0 & 0 & \cdots & 0 \\ p_n & \cdots & p_3 & p_1 & 0 & \cdots & 0 \\ \vdots & \ddots & \vdots & \vdots & \vdots & \ddots & \vdots \\ 0 & \cdots & 0 & p_{n-1} & p_{n-3} & \cdots & p_0 \end{bmatrix}, \qquad n = 2m+1, \tag{7.39}$$

respectively, where the difference lies in the nature of the *last* row. Now we need the fundamental concept of the minors of a matrix.

Definition 7.7 Let $A \in \mathbb{R}^{n\times n}$ and $I \subset \{1, \dots, n\}$. The *I–minor* of A is defined as

$$m_I(A) = \det A(I, I) = \det\big[a_{jk} : j, k \in I\big],$$

and the jth *principal minor* as

$$m_j(A) = m_{\{1,\dots,j\}}(A) = \det[a_{k\ell} : k, \ell = 1, \dots, j].$$

Theorem 7.7 (Routh–Hurwitz Theorem) *A polynomial $f \in \Pi$ with a positive leading coefficient is a Hurwitz polynomial if and only if*

$$m_k\left(H_f\right) > 0, \qquad k = 1, \dots, \deg f. \tag{7.40}$$

Given the Hurwitz criterion (see Problem 7.8) Theorem 7.7 can be rephrased as follows.

Corollary 7.3 *A polynomial $f \in \Pi$ with a positive leading coefficient is a Hurwitz polynomial if and only if H_f is strictly positive definite.*

Before we turn to the proof of this theorem, to which the next section will be devoted, we again have a look at the first special cases.

$n = 1$: According to Theorem 7.7, a polynomial $f(x) = f_1 x + f_0$, $f_1 > 0$ is a Hurwitz polynomial if and only if $0 < m_1 (H_f) = f_0$ which can be easily verified "manually":

$$f(x) = 0 \quad \Leftrightarrow \quad x = -\frac{f_0}{f_1};$$

Moreover the zero which is always real in this case is negative if and only if f_0 and f_1 have the same sign and hence are positive.

$n = 2$: Here, the positivity of the principal minors of

$$H_f = \begin{bmatrix} f_1 & 0 \\ f_2 & f_0 \end{bmatrix}$$

leads to

$$0 < f_1, \quad 0 < f_0 f_1 \quad \Leftrightarrow \quad 0 < f_0, f_1.$$

And indeed the zeros of f are the numbers

$$x = \frac{-f_1 \pm \sqrt{f_1^2 - 4 f_0 f_2}}{2 f_0} \quad \Rightarrow \quad \Re x < 0 \quad \text{for} \quad 0 < f_0, f_1, f_2,$$

since the value of $\sqrt{f_1^2 - 4 f_0 f_2}$ is either imaginary or less than f_1 as long as $f_0 f_2 > 0$ that is $f_2 > 0$. So we can verify the Routh–Hurwitz criterion directly again.

$n = 3$: Now all principal minors with the matrix

$$M_f = \begin{bmatrix} f_2 & f_0 & 0 \\ f_3 & f_1 & 0 \\ 0 & f_2 & f_0 \end{bmatrix}$$

have to be positive which in turn is equivalent to

$$f_0, \ f_2 > 0 \quad \text{and} \quad f_1 f_2 > f_0 f_3,$$

where the latter also implies that $f_1 > 0$.

7.5.1 Problems

7.7 Show that a symmetric matrix $A \in \mathbb{R}^{n \times n}$ is positive definite if and only if $m_k(A) > 0$, $k = 1, \ldots, n$. What about the case when the matrix is only positive *semi*definite?

7.8 Prove the *Hurwitz criterion*: a symmetric matrix $A \in \mathbb{R}^{n \times n}$ is (strictly) positive definite if and only if $m_k(A) > 0$, $k = 1, \ldots, n$.

7.9 (*LU factorization*) Show that a matrix $A \in \mathbb{R}^{n \times n}$ admits a factorization

$$A = LU, \qquad L = \begin{bmatrix} 1 & & \\ \vdots & \ddots & \\ * & \cdots & 1 \end{bmatrix}, \quad U = \begin{bmatrix} u_{11} & \cdots & * \\ & \ddots & \vdots \\ & & u_{nn} \end{bmatrix}, \qquad \prod_{j=1}^{n} u_{jj} \neq 0,$$

if and only if $m_k(A) \neq 0$, $k = 1, \ldots, n$.

7.10 ([27, p. 32]) Given $A \in \mathbb{R}^{n \times n}$. For fixed $1 \leq \ell \leq n$ and $j, k = \ell + 1, \ldots, n$ define

$$b_{jk} := \det \begin{bmatrix} a_{11} & \cdots & a_{1\ell} & a_{1k} \\ \vdots & \ddots & \vdots & \vdots \\ a_{\ell 1} & \cdots & a_{\ell\ell} & a_{\ell k} \\ a_{j1} & \cdots & a_{j\ell} & a_{jk} \end{bmatrix}, \qquad B := \left[b_{jk} : j, k = \ell + 1, \ldots, n \right],$$

and prove the *Sylvester determinant identity*

$$\det B = m_\ell(A)^{n-\ell-1} \det A. \tag{7.41}$$

7.6 The Routh Scheme or the Return of Sturm's Chains

Our starting point for the proof of Theorem 7.7 is the characterization (7.15) of a Hurwitz polynomial by the Cauchy index:

$$n = I_{-\infty}^{\infty} \frac{b_0 x^{n-1} - b_1 x^{n-3} + \cdots}{a_0 x^n - a_1 x^{n-2} + \cdots} =: I_{-\infty}^{\infty} \frac{f_1(x)}{f_0(x)}. \tag{7.42}$$

The two polynomials f_1 and f_0 cannot have a common zero since otherwise we could divide by the respective linear factor and the denominator would have only a degree at most $n - 1$ with at most $n - 1$ zeros leading to a Cauchy index of $n - 1$ that would contradict (7.42). Therefore we can construct a sequence of polynomials $f_2, \ldots, f_m$ by division with a remainder in the following way:

$$f_j(x) = q_j(x) f_{j+1}(x) - f_{j+2}, \qquad \deg f_{j+2} < \deg f_{j-1}. \qquad (7.43)$$

This is once again the Euclidean algorithm for which we have the following observation connected to Sturm chain; it is also the more classical way to obtain a Sturm chain.

Lemma 7.4 *If f_0, f_1 are two polynomials without a common zero and $f_m \in \Pi_0 \setminus \{0\}$ is the sequence from (7.43), then $f_0, \ldots, f_m$ form a Sturm chain.*

Remark 7.4 Note that the polynomials in the Sturm chain of Lemma 7.4 are now indexed in reverse order compared to Definition 4.11. Nevertheless this should provide no difficult problems.

Proof Since the two polynomials have no common zero, the Euclidean algorithm ends with the greatest common divisor $f_m \neq 0$ being a constant function. We have to show that at each zero of f_j, the two "neighboring" polynomials f_{j-1} and f_{j+1} have an opposite sign. If we replace j by $j-1$ in (7.43), then it follows for each zero x of f_j that

$$0 = q_j(x) f_j(x) = f_{j-1}(x) + f_{j+1}(x),$$

so that either $f_{j-1}(x) = f_{j+1}(x) = 0$ or the two polynomials indeed have opposite signs. If, on the other hand, $f_j(x) = f_{j+1}(x) = 0$, then[2] (7.43) implies $f_{j+2}(x) = 0$ and eventually $f_m(x) = 0$ which is a contradiction. $\qquad\qquad\square$

Dealing with the Euclidean algorithm explicitly, we obtain the following sequence of polynomials:

$$f_2(x) = \frac{a_0}{b_0} x \, f_1(x) - f_0(x) = c_0 \, x^{n-2} - c_1 \, x^{n-4} + \cdots$$

$$f_3(x) = \frac{b_0}{c_0} x \, f_2(x) - f_1(x) = d_0 \, x^{n-3} - d_1 \, x^{n-5} + \cdots$$

$$f_j(x) = a_0^j \, x^{n-j} - a_1^j \, x^{n-j-2} + \cdots = \frac{a_0^{j-2}}{a_0^{j-1}} x \, f_{j-1}(x) - f_{j-2}(x), \quad (7.44)$$

where

$$a_k^0 = a_k, \qquad a_k^1 = b_k, \qquad a_k^j = \frac{a_0^{j-1} a_{k+1}^{j-2} - a_0^{j-2} a_{k+1}^{j-1}}{a_0^{j-1}}, \quad j = 2, \ldots, m, \quad (7.45)$$

since

---

[2] Yes, this is precisely the argument that we already used in the proof of Proposition 4.3.

$$f_j(x) = \frac{a_0^{j-2}}{a_0^{j-1}}x\left(\sum_{k=0}^{(n-j+1)/2}(-1)^k a_k^{j-1}x^{n-j+1-2k}\right) - \left(\sum_{k=0}^{(n-j)/2+1}(-1)^k a_k^{j-2}x^{n-j+2-2k}\right)$$

$$= \sum_{k=1}^{(n-j)/2+1}(-1)^k \frac{a_0^{j-2}a_k^{j-1} - a_0^{j-1}a_k^{j-2}}{a_0^{j-1}}x^{n-j+2-2k}$$

$$= \sum_{k=0}^{(n-j)/2}(-1)^k \frac{a_0^{j-1}a_{k+1}^{j-2} - a_0^{j-2}a_{k+1}^{j-1}}{a_0^{j-1}}x^{n-j-2k} = \sum_{k=0}^{(n-j)/2}(-1)^k a_k^j x^{n-j-2k}.$$

In general it could happen that at some step in this process, we run into

$$0 = a_0^j = \frac{a_0^{j-2}a_1^{j-1} - a_0^{j-1}a_1^{j-2}}{a_0^{j-1}}, \qquad a_0^{j-1} \neq 0,$$

so that we would divide by zero in the next step. In that case we replace a_1^{j-2} by $a_1^{j-2} + \varepsilon$ with a number $\varepsilon \in \mathbb{R}$ that has a sufficiently small absolute value[3] leading to the following modification on level j:

$$\frac{a_0^{j-2}a_1^{j-1} - a_0^{j-1}(a_1^{j-2}+\varepsilon)}{a_0^{j-1}} = a_0^j + \varepsilon = \varepsilon;$$

we can now choose ε in such a way that the signs of a_0^j and a_0^{j-1} coincide, and also at the same time $a_1^{j-2} + \varepsilon$ still has the same sign as a_1^{j-2}. The first property can be obtained by a proper choice of the sign of ε; the second one by making its absolute value small enough. We might be forced to repeat that process at several levels, but for a sufficiently small ε all these modifications can become satisfied simultaneously, and in a sign preserving way as described above. In the end, the process terminates with a reasonable result. This result is a rational function in ε and we can eventually pass to the limit $\varepsilon \to 0$. The function is continuous with respect to ε as long as f has no zeros on the imaginary axis so that the limit is then well defined. For details see [28].

This reasoning allows us to restrict ourselves to the *regular* case that the Routh scheme (7.44) produces a Sturm chain of length n. Now all polynomials with even index $f_0, f_2, \ldots$, share the same *parity*, i.e., are all either an odd function or an even function, that is $f(-x) = -f(x)$ or $f(-x) = f(x)$, respectively, while those with odd indices, $f_1, f_3, \ldots$, have opposite parity. This implies that

$$V(-x) = V(f_0(-x), f_1(-x), \ldots, f_{n-1}(-x), f_n(-x))$$
$$= \begin{cases} V(f_0(x), -f_1(x), \ldots, -f_{n-1}(x), f_n(x)), & n = 2m, \\ V(-f_0(x), f_1(x), \ldots, f_{n-1}(x), -f_n(x)), & n = 2m+1. \end{cases}$$

and therefore

---

[3] The mathematicians' favorite joke "Let $\varepsilon < 0$" thus is **not** forbidden here.

$$n = V(-\infty) + V(\infty) := \lim_{x \to \infty} V(-x) + V(x), \tag{7.46}$$

as there is either a sign change from $f_j(\infty)$ to $f_{j+1}(\infty)$ or from $\pm f_j(\infty)$ to $\pm f_{j+1}(-\infty) = \mp f_{j+1}(\infty)$. On the other hand, (7.42), (7.7) and Theorem 4.6 imply that

$$n = I_{-\infty}^{\infty} \frac{f_1(x)}{f_0(x)} = -\Sigma_{-\infty}^{\infty} \frac{f_0(x)}{f_1(x)} = V(-\infty) - V(\infty),$$

hence f is a Hurwitz polynomial if and only if

$$0 = V(\infty) = V\left(a_0^j : j = 0, \dots, n\right), \qquad n = V(-\infty). \tag{7.47}$$

All together, this proves the following theorem.

Theorem 7.8 (Routh criterion) *The polynomial $f(z)$ is a Hurwitz polynomial if and only if all the numbers a_0^j, $j = 0, \dots, n$, are either strictly positive or strictly negative.*

Remark 7.5 According to (7.47) the vector with sign changes that define $V(\infty)$ has to have at least $n + 1$ entries for a Hurwitz polynomial—how else could one obtain n sign changes? This means that the Euclidean algorithm for a Hurwitz polynomial is not degenerate in the sense that it cannot have any degree jumps: all quotient polynomials q_j must be of degree 1 and no more. Or in other words: if we had divided by zero in (7.45) which forced us to apply the ε-modification, then the underlying polynomial cannot be a Hurwitz polynomial anyway, and would be of no interest to us. But obviously the Routh scheme works at a higher generality.

We can arrange all coefficients of the polynomials $f_0, f_1, \dots, f_n$ into a table which is called the *Routh scheme*:

$$\begin{matrix} a_0^0 & a_1^0 & \dots \\ a_0^1 & a_1^1 & \dots \\ \vdots & & \\ a_0^n. & & \end{matrix}$$

This table can be explicitly computed by (7.45). The Routh criterion of Theorem 7.8 can now be rephrased so that that we can recognize a Hurwitz polynomial from the property that all entries of the *first column* of the Routh scheme have the same strict sign which means that zero is forbidden and either everything is strictly positive or strictly negative. This is now really easy to check.

Example 7.2 Let us try to get an idea what the Routh criterion means.

1. For $n = 2$ and $f(z) = f_0 + f_1 z + f_2 z^2$, we get that $a_0^0 = f_2$, $a_1^0 = f_0$ and $a_0^1 = f_1$, hence

$$a_0^2 = \frac{a_0^1 \, a_1^0}{a_1^0},$$

leading to the scheme

$$\begin{array}{cc} f_2 & f_0 \\ f_1 & \\ \dfrac{f_1 f_2}{f_0} & \end{array}$$

and we see that this polynomial is a Hurwitz polynomial if and only if f_0, f_1, f_2 have the same strict sign.

2. A slightly more intricate example from [28], where one can also see the "ε–argument" applied is the polynomial $f(z) = z^4 + z^3 + 2z^2 + 2z + 1$ leading to the scheme

$$\begin{array}{ccc} 1 & 2 & 1 \\ 1 & 2 & \\ \varepsilon & 1 & \quad \leftarrow 0\ 1 \\ 2 - \dfrac{1}{\varepsilon} & & \\ 1 & & \end{array}$$

with length n. Here f is *not* a Hurwitz polynomial since any positive choice $0 < \varepsilon < \frac{1}{2}$ leads to a sign distribution $+, +, +, -, +$, while $\varepsilon < 0$ leads to $+, +, -, +, +$ and in both cases $V(\infty) = 2$. This shows by the way that f must have two zeros in $\mathbb{H}_+$ since f has real coefficients.

The way from the Routh scheme to Theorem 7.7 is now very short: we first observe that the Hurwitz matrix is

$$H_f = \begin{bmatrix} b_0 & -b_1 & b_2 & \cdots \\ a_0 & -a_1 & a_2 & \cdots \\ 0 & b_0 & -b_1 & \cdots \\ 0 & a_0 & -a_1 & \cdots \\ \vdots & \vdots & \vdots & \ddots \end{bmatrix}.$$

As in Gaussian elimination we multiply the first row by a_0/b_0 and subtract that from the third row; then the same with the second and fourth row and so on, leading to a matrix of the form

$$H_f' = \begin{bmatrix} b_0 & -b_1 & b_2 & \cdots \\ 0 & c_0 & -c_1 & \cdots \\ 0 & b_0 & -b_1 & \cdots \\ 0 & 0 & c_0 & \cdots \\ \vdots & \vdots & \vdots & \ddots \end{bmatrix}, \qquad c_k = \frac{b_0 \, a_{k+1} - a_0 \, b_{k+1}}{b_0}.$$

The formula for the c_k is already familiar to us as it is precisely (7.45), and consequently,

$$H_f^{(1)} := H_f' = \begin{bmatrix} a_0^1 & a_1^1 & a_2^1 & \dots \\ 0 & a_0^2 & a_1^2 & \dots \\ 0 & a_0^1 & a_1^1 & \dots \\ 0 & 0 & a_0^2 & \dots \\ \vdots & \vdots & \vdots & \ddots \end{bmatrix},$$

from where it starts to be fun. Now we multiply the second row by a_0^1/a_0^2, subtract that from the third row and apply similar operations to the fourth and fifth, the sixth and seventh row, and so on. Again we encounter the recurrence (7.45) and obtain the matrix

$$H_f^{(2)} = \begin{bmatrix} a_0^1 & a_1^1 & a_2^1 & \dots \\ 0 & a_0^2 & a_1^2 & \dots \\ 0 & 0 & a_0^3 & \dots \\ 0 & 0 & a_0^2 & \dots \\ \vdots & \vdots & \vdots & \ddots \end{bmatrix}.$$

Assuming that there was no division by zero during this process, which would once more request the aforementioned ε–modification that never happens for Hurwitz polynomials, this iteration ends with the upper triangular matrix

$$H_f^{(n)} = \begin{bmatrix} a_0^1 & \dots & * \\ & \ddots & \vdots \\ & & a_0^n \end{bmatrix},$$

and since we only subtracted multiples of the earlier rows $1, \dots, k-1$ from the kth row, $k = 1, \dots, n$, the principal minors of H_f and H_f^n coincide; therefore

$$m_k\left(H_f\right) = m_k\left(H_f^{(n)}\right) = \prod_{j=1}^{k} a_0^j, \qquad k = 1, \dots, n. \tag{7.48}$$

So we can finally complete all the proofs of this chapter.

Proof (*of Theorem* 7.7) According to Theorem 7.8, the polynomial $f(z)$ with $a_0^0 = f_n > 0$ is a Hurwitz polynomial if and only if $a_0^j > 0$, $j = 1, \dots, n$, which according to (7.48) is equivalent to all principal minors of $H_f^{(n)}$ being positive; this in turn is equivalent to all principal minors of H_f being positive. ◻

7.6.1 Problems

7.11 Show that any function $f : \mathbb{R} \to \mathbb{R}$ can be written as $f = g + h$ where g is even and h is odd. What does that mean for polynomials?

7.12 Compute the Routh scheme for $\dfrac{x+2}{x^2 + 3x + 1}$, see Example 7.3.

7.7 Markov Numbers and Back to Moments

There is yet another characterization of Hurwitz polynomials that perfectly fits into our context here since it is totally based on the continued fraction expansion

$$\frac{h(x)}{g(x)} = [c_0; d_1 x, c_1, \ldots, d_m x, c_m]$$

for the positive pair g, h that makes up the Hurwitz polynomial $f(z) = g(z^2) + z h(z^2)$. We assume that g and h are coprime and that $\deg g = m$. Recall that c_0 is positive if $\deg g = \deg h$ and zero otherwise as shown in Theorem 7.6. As in Sect. 4.1, we expand the rational function into the formal Laurent series

$$\frac{h(x)}{g(x)} = c_0 + \sum_{j=1}^{\infty} \mu_j x^{-j}. \tag{7.49}$$

According to Theorem 5.1, the Hankel operator M associated to the sequence μ has rank $M = \deg g = m$. Ignoring the constant part in (7.49), we consider the convergents of

$$\sum_{j=1}^{\infty} \mu_j x^{-j} = [0; d_1 x, c_1, \ldots, d_m x, c_m] = \frac{h(x)}{g(x)} - c_0 = \frac{h(x) - c_0 g(x)}{g(x)} = \frac{\tilde{h}(x)}{g(x)}$$

where by the definition of c_0 in the proof of Theorem 7.6, we now always have that $\deg g = \deg \tilde{h} + 1$ without a change in the coefficients μ_j.

Example 7.3 *(Positive pair)* We consider the positive pair defined by the continued fraction $[0; x, 1, x, 1]$. The convergents are

$$\frac{p_1(x)}{q_1(x)} = \frac{1}{x}$$

$$\frac{p_2(x)}{q_2(x)} = \frac{1}{x+1}$$

$$\frac{p_3(x)}{q_3(x)} = \cfrac{1}{x + \cfrac{1}{1 + \cfrac{1}{x}}} = \frac{x+1}{x^2 + 2x}$$

$$\frac{p_3(x)}{q_3(x)} = \cfrac{1}{x + \cfrac{1}{1 + \cfrac{1}{x+1}}} = \frac{x+2}{x^2 + 3x + 1} = \frac{h(x)}{g(x)},$$

and the two polynomials $g(x) = x^2 + 3x + 1$ and $x + 2$ are a positive pair as the zeros of g are $(-3 \pm \sqrt{5})/2$ and h vanishes at -2. This confirms the statement of Corollary 7.2 that all even order convergents are positive pairs, but the example also shows that the odd order convergents cannot be positive pairs as the denominator polynomial always has a zero at the origin. The approach of Lemma 4.1 yields the coefficients μ of the Laurent expansion as the unique solution to

$$\begin{bmatrix} 1 \\ 2 \\ 0 \\ \vdots \end{bmatrix} = \begin{bmatrix} 1 & & & \\ 3 & 1 & & \\ 1 & 3 & 1 & \\ & \ddots & \ddots & \ddots \end{bmatrix} \begin{bmatrix} \mu_1 \\ \mu_2 \\ \mu_3 \\ \vdots \end{bmatrix}$$

giving

$$\mu_1 = 1, \quad \mu_2 = -1, \quad \mu_n = -3\mu_{n-1} - \mu_{n-2}, \quad n \geq 3,$$

that is

$$\mu = (1, -1, 2, -5, 13, -34, 89, -233, 610, -1597, \dots).$$

In the general case we have from the definition of convergents,

$$\frac{p_2(x)}{q_2(x)} = \frac{1}{d_1 x + \frac{1}{c_1}}, \quad \text{i.e.,} \quad p_2(x) = 1, \quad q_2(x) = d_1 x + \frac{1}{c_1}. \qquad (7.50)$$

Note that q_2 and p_2 form a positive pair with the order of Definition 7.3.

A triple application of the recurrence formula for the convergents yields for $n = 2, \dots, m$ that

$$p_{2n}(x) = c_n \, p_{2n-1}(x) + p_{2n-2}(x) = c_n \, (d_n x \, p_{2n-2}(x) + p_{2n-3}(x)) + p_{2n-2}(x)$$
$$= (c_n d_n x + 1) \, p_{2n-2}(x) + c_n \, p_{2n-3}(x)$$
$$= (c_n d_n x + 1) \, p_{2n-2}(x) + \frac{c_n}{c_{n-1}} \, (p_{2n-2}(x) - p_{2n-4}(x)) \, ;$$

hence since the only differences between the recurrence relation for p_n and q_n are the respective initial conditions, and we get

$$\left.\begin{aligned} p_{2n} &= c_n \left(d_n x + \frac{c_n + c_{n-1}}{c_n c_{n-1}} \right) p_{2n-2} - \frac{c_n}{c_{n-1}} \, p_{2n-4}, \\ q_{2n} &= c_n \left(d_n x + \frac{c_n + c_{n-1}}{c_n c_{n-1}} \right) q_{2n-2} - \frac{c_n}{c_{n-1}} \, q_{2n-4}, \end{aligned}\right\} \qquad n = 2, \dots, m, \quad (7.51)$$

for the even order convergents. The renormalized polynomials

$$p_n^0 = b_n \, p_{2n}, \quad q_n^0 = b_n \, q_{2n}, \quad n = 2, \dots, m$$

with

$$b_{2k} := \frac{c_1 \cdots c_{2k-1}}{c_2 \cdots c_{2k}}, \quad b_{2k+1} := \frac{c_2 \cdots c_{2k}}{c_1 \cdots c_{2k+1}}, \quad k \le \left\lfloor \frac{n}{2} \right\rfloor, \qquad (7.52)$$

see (3.4), then satisfy[4] $p_n^0/q_n^0 = p_n/q_n$ and the recurrence

$$p_n^0(x) = \frac{b_n}{b_{n-1}} c_n \left(d_n x + \frac{c_n + c_{n-1}}{c_n c_{n-1}} \right) p_{n-1}^0(x) - \frac{b_n}{b_{n-2}} \frac{c_n}{c_{n-1}} \, p_{n-2}^0(x),$$
$$= \frac{b_n}{b_{n-1}} c_n \left(d_n x + \frac{c_n + c_{n-1}}{c_n c_{n-1}} \right) p_{n-1}^0(x) - p_{n-2}^0(x),$$

see (7.68) in Problem 7.14. Moreover for n even,

$$\frac{b_n}{b_{n-1}} c_n = \frac{c_1 \cdots c_{n-1}}{c_2 \cdots c_n} \frac{c_1 \cdots c_{n-1}}{c_2 \cdots c_{n-2}} c_n = \frac{c_1^2 \cdots c_{n-1}^2}{c_2^2 \cdots c_{n-2}^2} =: \alpha_n,$$

while for n odd we get

$$\frac{b_n}{b_{n-1}} c_n = \frac{c_2 \cdots c_{n-1}}{c_1 \cdots c_n} \frac{c_2 \dots c_{n-1}}{c_1 \cdots c_{n-2}} c_n = \frac{c_2^2 \cdots c_{n-1}^2}{c_1^2 \cdots c_{n-2}^2} =: \alpha_n.$$

As a result we finally end up with the recurrence relation

---

[4] It is definitely worthwhile to recall here that despite the fact that the numerator and denominator of a convergent are always coprime, they are always only defined up to a unit in the underlying ring. In the case of polynomials this means a nonzero number.

$$p_n^0(x) = \alpha_n \left(d_n x + \frac{1}{c_n} + \frac{1}{c_{n-1}} \right) p_{n-1}^0(x) - p_{n-2}^0(x), \quad p_0^0 = 0, \quad p_{-1}^0 = -1,$$

$$q_n^0(x) = \alpha_n \left(d_n x + \frac{1}{c_n} + \frac{1}{c_{n-1}} \right) q_{n-1}^0(x) - q_{n-2}^0(x), \quad q_0^0 = 1, \quad q_{-1}^0 = 0,$$

$$(7.53)$$

that now even works for $n = 1, \ldots, m$ provided we define $c_0 := \infty$ or $1/c_0 = 0$. Note the slightly different initialization of the numerator polynomials of the convergents due to $p_{-1} = -1$ which has to compensate for the "$-$" in the recurrence relation.

In particular since $\alpha_n > 0$, the denominator polynomials are orthogonal polynomials of degree n for some positive measure by Theorem 4.2. We collect some properties of these convergents.

Proposition 7.2 (Convergents for positive pairs) *The even order convergents of the continued fraction* $[0; d_1 x, c_1, \ldots, d_m x, c_m]$ *satisfy*

$$\deg p_{2n} = \deg p_n^0 = n - 1, \quad \deg q_{2n} = \deg q_n^0 = n, \quad n = 1, \ldots, m, \quad (7.54)$$

and

$$[0; d_1 x, c_1, \ldots, d_m x, c_m] - \frac{p_n^0(x)}{q_n^0(x)} = O\left(x^{-2n-1}\right), \quad n = 1, \ldots, m. \quad (7.55)$$

Proof The degree formula (7.54) follows by simple induction from (7.51) or (7.53), respectively, with the initialization of (7.50). For (7.55) we first note that

$$\frac{p_m^0(x)}{q_m^0(x)} = [0; d_1 x, c_1, \ldots, d_m x, c_m],$$

and thus the convergent and the Laurent series for the rational function even coincide up to an arbitrary order. On the other hand if (7.55) holds for some $n \leq m$, then

$$[0; d_1 x, c_1, \ldots, d_m x, c_m] - \frac{p_{n-1}^0(x)}{q_{n-1}^0(x)}$$

$$= [0; d_1 x, c_1, \ldots, d_m x, c_m] - \frac{p_n^0(x)}{q_n^0(x)} + \frac{p_n^0(x)}{q_n^0(x)} - \frac{p_{n-1}^0(x)}{q_{n-1}^0(x)}$$

$$= O\left(x^{-2n-1}\right) - \frac{(-1)^n}{q_n^0(x)q_{n-1}^0(x)} = O\left(x^{-2n-1}\right) + O\left(x^{-\deg q_n^0 - \deg q_{n-1}^0}\right)$$

$$= O\left(x^{-2n+1}\right),$$

which advances the claim from n to $n - 1$.   □

This result shows that we found a condensed way to write the relevant part of the continued fraction expansion of h/g.

Corollary 7.4 (Condensed representation) *We have that*

$$\sum_{j=1}^{m} \cfrac{1|}{\left|-\alpha_j \left(d_j x - \dfrac{1}{c_j} + \dfrac{1}{c_{j-1}}\right)\right.} = [0; d_1 x, c_1, \ldots, d_m x, c_m] = \sum_{j=1}^{\infty} \mu_j x^{-j},$$

$$(7.56)$$

and the continued fraction on the left is associated to the series.

Proof The identity (7.56) follows from multiplying the recurrence (7.53) by -1 and the compressed continued fraction on the left is associated with $\mu(x)$ due to its order of approximation shown in (7.55). □

The formula (7.56) is called a "condensed" representation since the continued fraction expansion

$$\left[0; -\alpha_1 \left(d_1 x - \frac{1}{c_1} + \frac{1}{c_0}\right), \ldots, -\alpha_m \left(d_m x - \frac{1}{c_m} + \frac{1}{c_{m-1}}\right)\right]$$

only has m coefficients instead of the $2m$ coefficients of the standard expansion we know from the characterization of positive pairs. On the other hand all coefficients in the condensed representation have degree 1 which finally gives the same approximation behavior with fewer coefficients. All these observations allow us to give a necessary criterion for positive pairs.

Corollary 7.5 *If the polynomials $g, h \in \Pi_m$ are a positive pair, then*

$$\det \begin{bmatrix} \mu_1 & \cdots & \mu_k \\ \vdots & \ddots & \vdots \\ \mu_k & \cdots & \mu_{2k-1} \end{bmatrix} \begin{cases} > 0, & k = 1, \ldots, m, \\ = 0, & k = m+1, m+2, \ldots \end{cases} \qquad (7.57)$$

Equivalently, the Hankel matrix

$$\begin{bmatrix} \mu_1 & \cdots & \mu_m \\ \vdots & \ddots & \vdots \\ \mu_m & \cdots & \mu_{2m-1} \end{bmatrix}$$

is positive definite.

Proof Since μ_1 is the quotient of positive leading terms, (7.57) holds automatically. On the other hand the explicit formula (4.37) for the coefficients in the recurrence (7.53) yields that

$$\alpha_n d_n \left(\prod_{j=1}^{n-1} \alpha_j d_j\right)^2 = \frac{\det M_{k-1}}{\det M_k}, \qquad M_k := \begin{bmatrix} \mu_1 & \cdots & \mu_k \\ \vdots & \ddots & \vdots \\ \mu_k & \cdots & \mu_{2k-1} \end{bmatrix}, \qquad (7.58)$$

from which the claim follows directly as all $\alpha_j, d_j > 0$. The other determinants vanish, since according to Kronecker's Theorem, Theorem 5.1, the rank of the associated Hankel operator is m. $\qquad\square$

Example 7.4 (*Example* 7.3 *continued*) The determinants are

$$\det M_1 = 1, \qquad \det M_2 = \det \begin{bmatrix} 1 & -1 \\ -1 & 2 \end{bmatrix} = 1, \qquad \det M_3 = \begin{bmatrix} 1 & -1 & 2 \\ -1 & 2 & -5 \\ 2 & -5 & 13 \end{bmatrix} = 0$$

since $m = 2$.

Corollary 7.5 provides only a necessary condition without characterization of positive pairs as yet. Intuitively this makes sense, since so far we only considered *half* of the convergents of the continued fraction $[0; d_1x, c_1, \ldots, d_mx, c_m]$ to get a characterization only of the elements d_j. In fact the following observation which follows immediately from (7.58) shows that positive definiteness of M_n and positivity of the coefficients d_j are mutually equivalent.

Corollary 7.6 *For*

$$\mu(x) = [0; d_1x, c_1, \ldots, d_mx, c_m]$$

we have that $\det M_n > 0$, $j = 1, \ldots, m$, *if and only if* $d_j > 0$, $j = 1, \ldots, m$.

For a complete characterization of positive pairs in terms of the moment sequence μ, we have to take into account the other half of the convergents as well. Looking at Example 7.3 shows that initially something is "wrong" with the convergents of odd order p_{2n+1}/q_{2n+1} for h/g: the denominator is always of the form

$$q_{2n+1}(x) = x\,\hat{q}_{2n+1}(x). \tag{7.59}$$

This is easily verified by induction, taking into account that $q_1(x) = d_1x$, and that for $n \geq 1$,

$$\begin{aligned} q_{2n+1}(x) &= d_{n-1}x\,q_{2n}(x) + q_{2n-1}(x) = d_{n-1}x\,q_{2n}(x) + x\,\hat{q}_{2n-1}(x) \\ &= x\left(d_{n-1}q_{2n}(x) + \hat{q}_{2n-1}(x)\right), \end{aligned}$$

which can be extended into yet another recurrence when taking into account that

$$q_{2n}(x) = c_n q_{2n-1}(x) + q_{2n-2}(x) = c_n x\,\hat{q}_{2n-1}(x) + q_{2n-2}(x),$$

where the polynomials

$$\hat{q}_n(x) = \begin{cases} q_n(x), & n \in 2\mathbb{N}, \\ x^{-1}q_n(x), & n \in 2\mathbb{N}_0 + 1 \end{cases}$$

are the denominators of the convergents of the continued fraction

$$[0; d_1, c_2 x, d_2, c_2 x, \ldots, d_m, c_m x].$$

But this continued fraction is closely related to the original one.

Lemma 7.5 *One has*

$$[0; d_1, c_1 x, d_2, c_2 x, \ldots, d_m, c_m x] = x \, [0; d_1 x, c_1, \ldots, d_m x, c_m] = \sum_{j=0}^{\infty} \mu_{j+1} x^{-j}.$$

$$(7.60)$$

Proof (7.60) follows by applying the observation that

$$x \, [0; d_1 x, c_1, \ldots, d_m x, c_m] = x \, \cfrac{1}{d_1 x + \cfrac{1}{c_1 + [0; d_2 x, c_2, \ldots, d_m x, c_m]}}$$

$$= \cfrac{1}{d_1 + \cfrac{1}{c_1 x + [0; d_2 x, c_2, \ldots, d_m x, c_m]}}$$

recursively. The second identity in (7.60) is obvious. $\qquad \square$

The modified odd order convergents of $[0; d_1, c_1 x, d_2, c_2 x, \ldots, d_m, c_m x]$ follow the same pattern as before just with some minor modifications. With the initializations $p_0 = 0$, $q_0 = 1$ and $p_1 = 1$, $q_1 = d_1$, this satisfies the recurrence

$$p_{2n+1}(x) = d_{n+1} p_{2n}(x) + p_{2n-1}(x)$$
$$= d_{n+1} \left(c_n x + \frac{d_n + d_{n+1}}{d_n d_{n+1}} \right) p_{2n-1} - \frac{d_{n+1}}{d_n} p_{2n-3},$$

for $n = 1, \ldots, m - 1$, and the same for q_{2n+1}. As before we renormalize these functions to be

$$p_n^1 = b_n \, p_{2n+1}, \quad q_n^1 = b_n \, q_{2n+1}, \quad n = 1, \ldots, m - 1,$$

with

$$b_{2k} = \frac{d_2 \cdots d_{2k}}{d_1 \cdots d_{2k+1}} \qquad b_{2k+1} = \frac{d_1 \cdots d_{2k+1}}{d_2 \cdots d_{2k+2}}, \qquad k \le \left\lfloor \frac{n}{2n} \right\rfloor$$

to obtain the recurrences

$$p_n^1(x) = \beta_n \left(c_n x + \frac{1}{d_n} + \frac{1}{d_{n+1}} \right) p_{n-1}^1(x) - p_{n-2}^1(x), \quad p_0^1 = \frac{1}{d_1}, \; p_{-1}^1 = 1,$$

$$q_n^1(x) = \beta_n \left(c_n x + \frac{1}{d_n} + \frac{1}{d_{n+1}} \right) q_{n-1}^1(x) - q_{n-2}^1(x), \quad q_0^1 = 1, \; q_{-1}^1 = 0,$$

$$\tag{7.61}$$

for $n = 1, \ldots, m - 1$ with

$$\beta_n = \begin{cases} \dfrac{d_2^2 \cdots d_n^2}{d_1^2 \cdots d_{n-1}^2}, & n \in 2\mathbb{N}, \\[3mm] \dfrac{d_1^2 \cdots d_n^2}{d_2^2 \cdots d_{n-1}^2}. & n \in 2\mathbb{N}_0 + 1. \end{cases}$$

Explicitly, we have the particular values

$$\begin{aligned} p_0^1(x) &= \tfrac{1}{d_1}, & q_0^1(x) &= 1, \\ p_1^1(x) &= c_1 d_1 x + \tfrac{d_1}{d_2}, & q_1^1(x) &= c_1 d_1^2 x + d_1 + \tfrac{d_1^2}{d_2}. \end{aligned} \tag{7.62}$$

It remains to get rid of the constant term μ_1 in the Laurent series; since

$$[0; d_1, c_1 x, d_2, c_2 x, \ldots, d_m, c_m x] = \frac{1}{[d_1; c_1 x, d_2, c_2 x, \ldots, d_m, c_m x]},$$

we have $\mu_1 = 1/d_1$ and we can consider

$$\frac{x h(x) - \mu_1 g(x)}{g(x)} = \sum_{j=1}^{\infty} \mu_{j+1} x^{-j}$$

and its convergents

$$\hat{p}_n = p_n^1 - \mu_1 q_n^1, \quad \hat{q}_n = q_n^1, \qquad n = 0, \ldots, m,$$

that still satisfy the recurrences (7.61) with the initialization

$$\hat{p}_0 = 0, \quad \hat{p}_{-1} = 1, \quad \hat{q}_0 = 1, \quad \hat{q}_{-1} = 0.$$

Finally setting $\hat{p}_n^1 = -\hat{p}_n$ and $\hat{q}_n^1 = \hat{q}_n$, we get the recurrence

$$\hat{p}_n^1(x) = \beta_n \left(c_n x + \frac{1}{d_n} + \frac{1}{d_{n+1}} \right) \hat{p}_{n-1}^1(x) - \hat{p}_{n-2}^1(x), \quad \hat{p}_0^1 = 0, \; \hat{p}_{-1}^1 = -1,$$

$$\hat{q}_n^1(x) = \beta_n \left(c_n x + \frac{1}{d_n} + \frac{1}{d_{n+1}} \right) \hat{q}_{n-1}^1(x) - q_{n-2}^1(x), \quad q_0^1 = 1, \; q_{-1}^1 = 0,$$

$$\tag{7.63}$$

that differs from (7.61) only by the initialization and it has the property that $\hat{p}_n^1 / \hat{q}_n^1$ are the odd order convergents for

$$- \left([0; d_1, c_1 x, d_2, c_2 x, \dots, d_m, c_m x] - \frac{1}{d_1} \right) = - \sum_{j=1}^{\infty} \mu_{j+1} x^{-j}.$$

Now as in the case of even order convergents, exactly the same arguments can be used to characterize the positivity of the c_j.

Proposition 7.3 *The coefficients $c_1, \dots, c_m$ in*

$$\mu(x) = [0; d_1 x, c_1, \dots, d_m x, c_m]$$

are all positive if and only if

$$0 < \det \begin{bmatrix} -\mu_2 & \cdots & -\mu_{k+1} \\ \vdots & \ddots & \vdots \\ -\mu_{k+1} & \cdots & -\mu_{2k} \end{bmatrix} = (-1)^k \det \begin{bmatrix} \mu_2 & \cdots & \mu_{k+1} \\ \vdots & \ddots & \vdots \\ \mu_{k+1} & \cdots & \mu_{2k} \end{bmatrix} \tag{7.64}$$

holds for $k = 1, \dots, m$.

Proof The proof can be copied literally from the preceding case, except for determining c_m. For this we use the standard ambiguity[5] that for any $d > 0$,

$$[0; d_1, c_1 x, d_2, c_2 x, \dots, d_m, c_m x] = \left[0; d_1, c_1 x, d_2, c_2 x, \dots, d_m, c_m x - \frac{1}{d}, d \right];$$

then letting $d \to \infty$ extends the validity of the recurrence (7.63) to $n = m$ by defining $d_{m+1} = \infty$, or as $\frac{1}{d_{m+1}} = 0$ which is the way it appears in the formula. The rest is a straightforward copy of the preceding case. $\qquad\square$

Example 7.5 *(Example 7.4 continued)* In our example, we obtain

$$\det[\mu_2] = -1, \quad \det \begin{bmatrix} -1 & 2 \\ 2 & -5 \end{bmatrix} = 4, \quad \det \begin{bmatrix} -1 & 2 & -5 \\ 2 & -5 & 13 \\ -5 & 13 & -34 \end{bmatrix} = 0$$

since still $m = 2$.

In other words: the positive pair is characterized by the fact that the nonzero Hankel matrices for the sequence μ are all positive, while those for the shifted sequence alternate in the sign. This can be formulated more elegantly if one compensates the alternating sign in an appropriate way. This approach even leads to objects with a name of their own.

---

[5] This is the ambiguity that we explicitly excluded in Chap. 2 to obtain a unique continued fraction expansion for real numbers.

Definition 7.8 *(Markov numbers)* For two polynomials $g, h \in \Pi$ with

$$\frac{h(x)}{g(x)} = p(x) + \sum_{j=1}^{\infty} \mu_j x^{-j}, \qquad p \in \Pi,$$

the *Markov numbers* are defined as

$$\gamma_j = (-1)^{j+1} \mu_j, \qquad j \in \mathbb{N}. \tag{7.65}$$

Example 7.6 *(Example* 7.3 *continued)* The Markov numbers for the sequence μ of Example 7.3 are

$$\gamma = (1, 1, 2, 5, 13, 34, 89, 233, 610, 1597, \dots).$$

Markov numbers give a very elegant way to characterize positive pairs.

Theorem 7.9 (Positive pairs and Markov numbers) *The two polynomials $g, h \in \Pi_m$ form a positive pair if and only if the Markov numbers γ for the series expansion*

$$\frac{h(x)}{g(x)} = \mu_0 + \sum_{j=1}^{\infty} \mu_j x^{-j} \tag{7.66}$$

have the property that

$$\begin{bmatrix} \gamma_1 & \cdots & \gamma_m \\ \vdots & \ddots & \vdots \\ \gamma_m & \cdots & \gamma_{2m-1} \end{bmatrix} \quad and \quad \begin{bmatrix} \gamma_2 & \cdots & \gamma_{m+1} \\ \vdots & \ddots & \vdots \\ \gamma_{m+1} & \cdots & \gamma_{2m} \end{bmatrix} \tag{7.67}$$

are strictly positive definite.

Proof The proof is based on the (simple) observation that the determinants of the Hankel matrix do not change when passing from μ to the associated Markov numbers γ, cf. Problem 7.18. Hence the property of the first matrix is exactly the positivity of the d_n by Corollary 7.6. The positivity of the second matrix in (7.67) is equivalent to the positivity of the c_n by Proposition 7.3, and the fact that the Markov numbers for $-(\mu_2, \mu_3, \dots)$ are $(\gamma_2, \gamma_3, \dots)$. $\qquad \square$

Remark 7.6 Markov numbers and Theorem 7.9 can be found in Gantmacher's book [28]. The proof there is based on the Cauchy index and its relation to the signature of the two matrices in (7.67). This relationship, known as *Hermite's Theorem*, is indeed powerful, interesting and also closely related to the problem of counting the *real* zeros of a polynomial with real coefficients; interesting enough, this approach can also be extended to the multivariate case as the trace method for multiplication tables; see the chapter [38] in the generally recommended book [17]. The approach here is different and much more in the spirit of this book, since it is based entirely on continued fraction techniques.

Example 7.7 (*Example* 7.3 *continued*) Besides the positivity of $\mu_1 = 1$ and $\mu_2 = 1$, the determinants to be considered are now

$$\det \begin{bmatrix} 1 & 1 \\ 1 & 2 \end{bmatrix} = 1, \quad \det \begin{bmatrix} 1 & 2 \\ 2 & 5 \end{bmatrix} = 1, \quad \text{while} \quad \det \begin{bmatrix} 1 & 1 & 2 \\ 1 & 2 & 5 \\ 2 & 5 & 13 \end{bmatrix} = \begin{bmatrix} 1 & 2 & 5 \\ 2 & 5 & 13 \\ 5 & 13 & 34 \end{bmatrix} = 0.$$

Summarizing the results from this section, the connection to moments is now clear to us in view of Chaps. 4 and 5: if g and h form a positive pair, i.e., if $f(z) = g(z^2) + z\,h(z^2)$ is a Hurwitz polynomial, then the rational function h/g defines an infinite sequence that is the moment sequence for a discrete positive measure. The locations of these point measures are the zeros of g.

A remarkable special property of positive pairs occurs when not only considering the coefficient sequence of the plain Laurent series expansion of the rational function, but also the associated Markov numbers: not only is the sequence $(\gamma_1, \gamma_2, \dots)$ of Markov numbers itself a moment sequence for a positive discrete measure, the same also holds true for the shifted sequence $(\gamma_2, \gamma_3, \dots)$.

7.7.1 Problems

7.13 Given $g, h \in \Pi$ with $n := \deg g = \deg h + 1$, show that the Laurent expansion

$$\frac{h(x)}{g(x)} = \sum_{j=1}^{\infty} \mu_j x^{-j}$$

is obtained by setting

$$\begin{bmatrix} \mu_1 \\ \vdots \\ \mu_n \end{bmatrix} = \begin{bmatrix} g_n & & & \\ g_{n-1} & g_n & & \\ \vdots & & \ddots & \\ g_1 & \cdots & g_{n-1} & g_n \end{bmatrix}^{-1} \begin{bmatrix} h_{n-1} \\ \vdots \\ h_0 \end{bmatrix}, \quad \mu_k = \frac{1}{g_n} \sum_{j=0}^{n-1} g_j\, \mu_{k-n+j}, \quad k \geq n+1.$$

Write a `Matlab` program to compute these expansions.

7.14 Prove that the numbers b_n from (7.52) satisfy

$$\frac{b_n}{b_{n-2}} = \frac{c_{n-1}}{c_n}, \quad n \geq 1. \tag{7.68}$$

7.15 Show that if $g, h \in \mathbb{Z}[x]$, $\deg g = \deg h + 1$ are monic polynomials with integer coefficients, then $\mu_j \in \mathbb{Z}$, $j \in \mathbb{N}$.

7.16 Show that the coefficients μ_j of the formal Laurent series for h/g belong to the same field as g, h.

7.17 Work out the proof of Proposition 7.3 in detail.

7.18 Let $\mu \in \ell(\mathbb{N}_0)$ be any sequence and set $\mu'_j := (-1)^{j+1} \mu_j$, $j \in \mathbb{N}$. Show that the determinants of the associated Hankel matrices coincide:

$$\det \begin{bmatrix} \mu_1 & \cdots & \mu_n \\ \vdots & \ddots & \vdots \\ \mu_n & \cdots & \mu_{2n-1} \end{bmatrix} = \det \begin{bmatrix} \mu'_1 & \cdots & \mu'_n \\ \vdots & \ddots & \vdots \\ \mu'_n & \cdots & \mu'_{2n-1} \end{bmatrix}, \qquad n \in \mathbb{N}.$$

Hint: Write $M'_n = X^{-1} M_n X$ for some appropriate matrix X.

7.8 Hurwitz Polynomials and Total Positivity

The results of Sect. 7.7 also connect to an important property of matrices, namely total positivity which plays a fundamental role in Numerical Linear Algebra; this is in the theory of splines and in CAGD in general, see [29, 56]. Total positivity means that *all* minors of a matrix are positive. We begin by generalizing Definition 7.7.

Definition 7.9 *(General minor & total positivity)* Given $A \in \mathbb{R}^{m \times n}$ and $J \subset \{1, \ldots, m\}$, $K \subset \{1, \ldots, n\}$ with $\#J = \#K$, the *minor* $m_{IJ}(A)$ is defined as

$$m_{JK}(A) := \det \left[a_{jk} : j \in J, k \in K \right],$$

where $m_I(A) = m_{II}(A)$, $I \subset \{1, \ldots, m\}$. The matrix A is called *totally positive* if

$$m_{JK}(A) > 0, \qquad \#J = \#K, \tag{7.69}$$

and *totally nonnegative* if

$$m_{JK}(A) \geq 0, \qquad \#J = \#K, \tag{7.70}$$

Keep in mind that the important thing about the minors in (7.69) or (7.70) is the fact that the elements of the submatrix are arranged in the same order as in the original matrix. In particular total positivity and total nonnegativity are very sensitive to perturbations of the rows and columns.

Example 7.8 *(Pascal matrix)* The *Pascal matrix* or order n,

$$\left[\binom{j+k-2}{j-1} : j, k = 1, \ldots, n \right], \qquad n \in \mathbb{N}, \tag{7.71}$$

is totally positive for any $n \in \mathbb{N}$.

Another example that will play a fundamental role in what follows will be given a name of its own.

Definition 7.10 *(Generalized Vandermonde matrix)* For $0 < x_1 < \cdots < x_n \in \mathbb{R}$ and $\alpha_1 < \cdots < \alpha_n \in \mathbb{R}$, the associated *generalized Vandermonde matrix* is defined as

$$V \begin{pmatrix} x_1, \ldots, x_n \\ \alpha_1, \ldots, \alpha_n \end{pmatrix} := \left[x_j^{\alpha_k} : j, k = 1, \ldots, n \right]. \tag{7.72}$$

Indeed the positivity and order of points and exponents ensures that these matrices are totally positive.

Proposition 7.4 *For $0 < x_1 < \cdots < x_n \in \mathbb{R}$ and $\alpha_1 < \cdots < \alpha_n \in \mathbb{R}$, the matrix* $V \begin{pmatrix} x_1, \ldots, x_n \\ \alpha_1, \ldots, \alpha_n \end{pmatrix}$ *is totally positive.*

Proof We first show that $\det V \begin{pmatrix} x_1, \ldots, x_n \\ \alpha_1, \ldots, \alpha_n \end{pmatrix} \neq 0$ which we will do by induction on n where the case $n = 1$ is obvious. To advance the induction hypothesis, we assume the contrary; then there exists a vector $c \in \mathbb{R}^n$ such that

$$0 = V \begin{pmatrix} x_1, \ldots, x_n \\ \alpha_1, \ldots, \alpha_n \end{pmatrix} = \left[\sum_{k=1}^n c_k x_j^{\alpha_k} : j = 1, \ldots, n \right],$$

hence

$$f(x) = \sum_{k=1}^n c_k x^{\alpha_k}, \qquad x \in \mathbb{R}_+,$$

vanishes at $x_1, \ldots, x_n$, as does the function $f_0 = (\cdot)^{-\alpha_1} f$. By Rolle's theorem, f_0' vanishes at points $0 < x_1^1 < \cdots < x_{n-1}^1$, hence

$$\det V \begin{pmatrix} x_1^1, \ldots, x_{n-1}^1 \\ \alpha_1 - \alpha_1, \ldots, \alpha_n - \alpha_1 \end{pmatrix} = 0,$$

which is a contradiction. Therefore $\det V \begin{pmatrix} x_1, \ldots, x_n \\ \alpha_1, \ldots, \alpha_n \end{pmatrix} \neq 0$. Moreover the explicit formula (5.27) for the Vandermonde determinant shows that

$$\det V \begin{pmatrix} x_1, \ldots, x_n \\ 0, \ldots, n-1 \end{pmatrix} = \prod_{0 \leq j \leq k \leq n} (x_k - x_j) > 0.$$

If we now set

$$\alpha_j(t) := (1 - t)(j - 1) + t \alpha_j, \qquad t \in [0, 1], \qquad j = 1, \ldots, n,$$

i.e., $\alpha_j(0) = j - 1$, $\alpha_j(1) = \alpha_j$ and $\alpha_1(t) < \cdots < \alpha_n(t)$, then

$$g(t) = \det V \begin{pmatrix} x_1, \ldots, x_n \\ \alpha_1(t), \ldots, \alpha_n(t) \end{pmatrix}, \qquad t \in [0, 1],$$

is a continuous function in t with $g(0) > 1$ and $g(t) \neq 0$, $t \in [0, 1]$, so that

$$0 < g(1) = \det V \begin{pmatrix} x_1, \ldots, x_n \\ \alpha_1, \ldots, \alpha_n \end{pmatrix}.$$

From this inequality, total positivity follows immediately since any submatrix of a generalized Vandermonde matrix is a generalized Vandermonde matrix as well. □

Now we return to a Hurwitz polynomial $f(z) = g(z^2) + z\,h(z^2)$ where $g, h \in \Pi_m$ form a positive pair. The series expansion

$$-\frac{h(-x)}{g(-x)} = -\mu_0 + \sum_{j=1}^{\infty}(-1)^{j+1}\mu_j\,x^{-j} = -\mu_0 + \sum_{j=1}^{\infty}\gamma_j\,x^{-j}$$

used for the Markov numbers defines a sequence of Hankel matrices

$$H_n := \begin{bmatrix} \gamma_1 & \cdots & \gamma_n \\ \vdots & \ddots & \vdots \\ \gamma_n & \cdots & \gamma_{2n-1} \end{bmatrix}, \qquad n \in \mathbb{N}$$

as finite segments of the Hankel operator H of rank m with $\det H_n > 0$ for $n = 1, \ldots, m$ as shown in Theorem 7.9. Since H is a Hankel operator of finite rank, it follows by Corollary 5.1 that

$$\gamma_k = \sum_{j=1}^{m} c_j\,\zeta_j^k, \qquad k \in \mathbb{N},$$

where $\zeta_1, \ldots, \zeta_m \in \mathbb{R}_+$ are the zeros of $g(-\cdot)$ and $c_j > 0$, $j = 1, \ldots, m$, since γ is a moment sequence for a positive discrete measure, again due to Theorem 7.9 that ensures that the associated Hankel matrices of order up to m are all positive definite. Therefore H can be factorized into

$$H = \begin{bmatrix} \gamma_1 & \gamma_2 & \cdots \\ \gamma_2 & \gamma_3 & \cdots \\ \vdots & \vdots & \ddots \end{bmatrix} = \begin{bmatrix} 1 & \cdots & 1 \\ \zeta_1 & \cdots & \zeta_m \\ \zeta_1^2 & \cdots & \zeta_m^2 \\ \vdots & \ddots & \vdots \end{bmatrix} \begin{bmatrix} c_1 & & \\ & \ddots & \\ & & c_m \end{bmatrix} \begin{bmatrix} 1 & \zeta_1 & \zeta_1^2 & \cdots \\ \vdots & \vdots & \vdots & \ddots \\ 1 & \zeta_m & \zeta_m^2 & \cdots \end{bmatrix}. \tag{7.73}$$

If we now consider *any* $m \times m$ minor of (7.73), i.e., $J, K \subset \mathbb{N}$, $\#J = \#K = m$, then

$$m_{JK}(H) = \det\left(\begin{bmatrix} \zeta_1^{j_1} & \cdots & \zeta_m^{j_1} \\ \vdots & \ddots & \vdots \\ \zeta_1^{j_m} & \cdots & \zeta_m^{j_m} \end{bmatrix}\begin{bmatrix} c_1 & & \\ & \ddots & \\ & & c_m \end{bmatrix}\begin{bmatrix} \zeta_1^{k_1} & \cdots & \zeta_1^{k_m} \\ \vdots & \ddots & \vdots \\ \zeta_m^{k_1} & \cdots & \zeta_m^{k_m} \end{bmatrix}\right)$$

$$= \left(\prod_{j=1}^{m} c_j\right)\det\begin{bmatrix} \zeta_1^{j_1} & \cdots & \zeta_m^{j_1} \\ \vdots & \ddots & \vdots \\ \zeta_1^{j_m} & \cdots & \zeta_m^{j_m} \end{bmatrix}\det\begin{bmatrix} \zeta_1^{k_1} & \cdots & \zeta_1^{k_m} \\ \vdots & \ddots & \vdots \\ \zeta_m^{k_1} & \cdots & \zeta_m^{k_m} \end{bmatrix} > 0,$$

by Proposition 7.4, hence all $m \times m$ minors are positive. For minors of order $p < m$ we make use of the Cauchy–Binet formula (7.76), see Problem 7.22, and obtain that

$$m_{JK}(H) = \det\left(\begin{bmatrix} c_1\zeta_1^{j_1} & \cdots & c_m\zeta_m^{j_1} \\ \vdots & \ddots & \vdots \\ c_1\zeta_1^{j_p} & \cdots & c_m\zeta_m^{j_p} \end{bmatrix}\begin{bmatrix} \zeta_1^{k_1} & \cdots & \zeta_1^{k_p} \\ \vdots & \ddots & \vdots \\ \zeta_m^{k_1} & \cdots & \zeta_m^{k_p} \end{bmatrix}\right)$$

$$= \sum_{1 \le \ell_1 < \cdots < \ell_p \le m} \det\begin{bmatrix} c_{\ell_1}\zeta_{\ell_1}^{j_1} & \cdots & c_{\ell_p}\zeta_{\ell_p}^{j_1} \\ \vdots & \ddots & \vdots \\ c_{\ell_1}\zeta_{\ell_1}^{j_p} & \cdots & c_{\ell_p}\zeta_{\ell_p}^{j_p} \end{bmatrix}\det\begin{bmatrix} \zeta_{\ell_1}^{k_1} & \cdots & \zeta_{\ell_1}^{k_p} \\ \vdots & \ddots & \vdots \\ \zeta_{\ell_m}^{k_1} & \cdots & \zeta_{\ell_m}^{k_p} \end{bmatrix}$$

$$= \sum_{1 \le \ell_1 < \cdots < \ell_p \le m} c_{\ell_1}\cdots c_{\ell_m}\det\begin{bmatrix} \zeta_{\ell_1}^{j_1} & \cdots & \zeta_{\ell_p}^{j_1} \\ \vdots & \ddots & \vdots \\ \zeta_{\ell_1}^{j_p} & \cdots & \zeta_{\ell_p}^{j_p} \end{bmatrix}\det\begin{bmatrix} \zeta_{\ell_1}^{k_1} & \cdots & \zeta_{\ell_1}^{k_p} \\ \vdots & \ddots & \vdots \\ \zeta_{\ell_m}^{k_1} & \cdots & \zeta_{\ell_m}^{k_p} \end{bmatrix} > 0.$$

To summarize,

$$m_{JK}(H) > 0, \qquad J, K \subset \mathbb{N}, \quad \#J = \#K \le m. \tag{7.74}$$

This has yet another striking consequence.

Theorem 7.10 *A polynomial $f(z) = g(z^2) + z\,h(z^2)$ is a Hurwitz polynomial if and only if the Hankel operator H defined by the Markov numbers for h/g has totally positive submatrices of the size $m \times m$:*

$$H_{JK} := \begin{bmatrix} \gamma_{j_1,k_1} & \cdots & \gamma_{j_1,k_m} \\ \vdots & \ddots & \vdots \\ \gamma_{j_m,k_1} & \cdots & \gamma_{j_m,k_m} \end{bmatrix}, \qquad \begin{array}{l} 1 \le j_1 < \cdots < j_m, \\ 1 \le k_1 < \cdots < k_m, \end{array} \tag{7.75}$$

where $J = \{j_1, \ldots, j_m\}$, $K = \{k_1, \ldots, k_m\}$.

Proof If f is a Hurwitz polynomial, hence g, h is a positive pair, and then the above reasoning leading to (7.74) shows that all minors of order $\le m$ are positive. Therefore all matrices of the form (7.75) have positive minors and thus are totally positive as claimed.

Conversely the choices $J = \{1, \ldots, m\}$, $K = \{1, \ldots, m\}$ and $J = \{2, \ldots, m\}$, $K = \{1, \ldots, m\}$ extract the two submatrices in (7.67); and since especially all their principal minors are positive, they are positive definite. Therefore Theorem 7.9 allows us to conclude that g, h are a positive pair and by Theorem 7.3 the polynomial f is a Hurwitz polynomial. $\square$

A word on the comparison between Theorems 7.9 and 7.10 is in order here. Of course the conditions in Theorem 7.9 on the positive definiteness of two particular symmetric Hankel matrices appear much weaker than the total positivity of all minors up to a certain size. But as soon as the sequence γ consists of Markov numbers for a positive pair, these two matrices determine everything.

This is not surprising of course: the infinite series connected to h/g or to $-h(-\cdot)/g(-\cdot)$ determines a Hankel operator of rank deg g and a flat extension of the initial segment of this Hankel operator uniquely determines the rest of the infinite series.

7.8.1 Problems

7.19 Show that

1. any totally positive matrix A satisfies $a_{jk} > 0$,
2. any symmetric totally positive matrix is strictly positive definite.

7.20 Prove the total positivity of the Pascal matrix from (7.71).

7.21 Let $A \in \mathbb{R}^{n \times n}$ be a nonsingular totally nonnegative matrix. Show that there exist a lower triangular matrix $L \in \mathbb{R}^{n \times n}$ and an upper triangular matrix $U \in \mathbb{R}^{n \times n}$, such that

$$A = LU, \quad \ell_{jj} = 1, \quad L \geq 0, U \geq 0,$$

i.e., L and U have *nonnegative entries*.

Hint: Gauss elimination and induction.

7.22 (*Cauchy–Binet formula*) Let $A \in \mathbb{R}^{m \times n}, B \in \mathbb{R}^{n \times m}$, $m \leq n$. Prove the *Cauchy–Binet formula*

$$\det(AB) = \sum_{1 \leq k_1 < \cdots < k_m \leq n} \det \begin{bmatrix} a_{1,k_1} & \cdots & a_{1,k_m} \\ \vdots & \ddots & \vdots \\ a_{m,k_1} & \cdots & a_{m,k_m} \end{bmatrix} \det \begin{bmatrix} b_{k_1,1} & \cdots & b_{k_1,m} \\ \vdots & \ddots & \vdots \\ b_{k_m,1} & \cdots & b_{k_m,m} \end{bmatrix}. \quad (7.76)$$

Hint: use the formula $(AB)_{jk} = \sum_\ell a_{j\ell} b_{\ell k}$ and the multilinearity of determinants.

References

1. The MacTutor History of Mathematics Archive. University of St. Andrews (2003). http://mathshistory.st-andrews.ac.uk
2. Barbour, J.M.: Tuning and Temperament. A Historical Survey. Michigan State Press (1951). Dover reprint (2004)
3. Becker, O.: Quellen und Studien zur Geschichte. Math. Astron. Physik **B2**, 311–333 (1933)
4. Ben-Or, M., Tiwari, P.: A deterministic algorithm for sparse multivariate polynomial interpolation. In: Proceedings of the Twentieth Annual ACM Symposium on Theory of Computing, pp. 301–309. ACM Press, New York (1988)
5. Benson, D.J.: Music: A Mathematical Offering. Cambridge University Press, Cambridge (2007)
6. Bernoulli, D.: Adversaria analytica miscellanes de fractionibus continuis. N. C. Petropol **20** (1775)
7. Bernoulli, D.: Disquisitiones ulteriores de idole fractionum continuarum. N. C. Petropol. **20** (1775)
8. Boor, C.D., Ron, A. Math. Z.: The least solution for the polynomial interpolation problem. **210**, 347–378 (1992)
9. Borwein, J., van der Poorten, A., Shallit, J., Zudillin, W.: Neverending Fractions: An Introduction to Continued Fractions. Cambridge University Press, Cambridge (2014)
10. Bowman, K.O., Shenton, L.R.: Continued Fractions in Statistical Applications. Marcel Dekker, Inc. (1989)
11. Brezinski, C.: Padé-Type Approximations and General Orthogonal Polynomials. In: International Series of Numerical Mathematics, vol. 50. Birkhäuser (1980)
12. Buhmann, M., Micchelli, C.A.: Using two slanted matrices for subdivision. Proc. Lond. Math. Soc. **69**, 428–448 (1994)
13. Bultheel, A.: Laurent Series and Their Padé Approximants. Birkhäuser (1987)
14. Candès, E.J., Fernandez-Granda, C.: Towards a mathematical theory of super-resolution. Commun. Pure Appl. Math. **67**, 906–956 (2012)
15. Cavaretta, A.S., Dahmen, W., Micchelli, C.A.: Stationary Subdivision. Memoirs of the AMS, vol. 93, p. 453. American Mathematical Society (1991)
16. Chihara, T.S.: An Introduction to Orthogonal Polynomials. Gordon and Breach (1978). Dover reprint (2011)
17. Cohen, A.M., Cuypers, H., Sterk, M. (eds.): Some Tapas of Computer Algebra. Algorithms and Computations in Mathematics, vol. 4. Springer (1999)

© The Editor(s) (if applicable) and The Author(s), under exclusive license to Springer Nature Switzerland AG 2021
T. Sauer, *Continued Fractions and Signal Processing*, Springer Undergraduate Texts in Mathematics and Technology, https://doi.org/10.1007/978-3-030-84360-1

18. Davis, P.J.: Interpolation and Approximation. Dover Books on Advanced Mathematics. Dover Publications, New York (1975)
19. Dunkl, C., Xu, Y.: Orthogonal Polynomials in Several Variables. Cambridge University Press (2001)
20. Dyson, F.J.: The approximation to algebraic numbers by rationals. Acta Math. (1947)
21. Eaton, J.W., Bateman, D., Hauberg, S.: GNU Octave version 3.0.1 manual: a high-level interactive language for numerical computations. CreateSpace Independent Publishing Platform (2009). http://www.gnu.org/software/octave/doc/interpreter. ISBN:1441413006
22. Eisenbud, D.: Commutative Algebra with a View Toward Algebraic Geometry. Graduate Texts in Mathematics, vol. 150. Springer (1994)
23. Fauvel, J., Flood, R., Wilson, R. (eds.): Music and Mathematics. From Pythagoras to Fractals. Oxford University Press (2003)
24. Fisher, S.D.: Complex Variables. Wadsworth & Brooks (1990). Dover Reprint (1999)
25. Freitag, E., Busam, R.: Complex Analysis. Springer (2005)
26. Frobenius, G.: Über Relationen zwischen den Näherungsbrüchen von Potenzreihen. J. Reine Angew. Math. **90**, 1–17 (1881)
27. Gantmacher, F.R.: Matrix Theory, vol. I. Chelsea Publishing Company (1959). Reprinted by AMS (2000)
28. Gantmacher, F.R.: Matrix Theory, vol. II. Chelsea Publishing Company (1959). Reprinted by AMS (2000)
29. Gasca, M., Micchelli, C.A. (eds.): Total Positivity and Its Applications. Mathematics and Its Applications, vol. 359. Springer (1996)
30. Gasquet, C., Witomski, P.: Fourier Analysis and Applications. Filtering, Numerical Computation, Wavelets. Texts in Applied Mathematics, vol. 30. Springer (1998)
31. Gathen, J.v.z., Gerhard, J.: Modern Computer Algebra. Cambridge University Press (1999)
32. Gauss, C.F.: Methodus nova integralium valores per approximationem inveniendi. Commentationes societate regiae scientiarum Gottingensis recentiores **III** (1816)
33. Gautschi, W.: Numerical Analysis. An Introduction. Birkhäuser, Boston (1997)
34. Gelfand, I.M., Shilov, G.E.: Generalized Functions. AMS Chelsea Publications, vol. I–VI (2016). Originally 1958–1966 in Russian
35. Goldberg, S.: Introduction to Difference Equations. Wiley (1958). Dover reprint (1986)
36. Golub, G., van Loan, C.F.: Matrix Computations, 3rd edn. The Johns Hopkins University Press (1996)
37. Golub, G.H., Welsch, J.A.: Calculation of Gauss quadrature rules. Math. Comput. **23**, 221–230 (1969)
38. González-Vega, L., Rouillier, F., Roy, M.F., Trujillo, G.: Symbolic recipes for polynomial system solving. In: A.M. Cohen, H. Cuypers, M. Sterk (eds.) Some Tapas in Computer Algebra, Algorithms and Computations in Mathematics, vol. 4, chap. 2, pp. 34–65. Springer (1999)
39. Gould, H.W.: Noch einmal die Stirlingschen Zahlen. Jber. Deutsch. Math.-Verein **73**, 149–152 (1971)
40. Gragg, W.B., Reichel, L.: On singular values of Hankel operators of finite rank. Linear Algebr. Appl. **121**, 53–70 (1989)
41. Graham, R.L., Knuth, D.E., Patashnik, O.: Concrete Mathematics, 2nd edn. Addison-Wesley (1998)
42. Gröbner, W.: Algebraische Geometrie I. No. 273 in B.I–Hochschultaschenbücher. Bibliographisches Institut Mannheim (1968)
43. Grüningen, D.C.v.: Digitale Signalverarbeitung. VDE Verlag, AT Verlag (1993)
44. Hamming, R.W.: Digital Filters. Prentice-Hall (1989). Republished by Dover Publications (1998)
45. Hardy, G.H., Wright, E.M.: An Introduction to the Theory of Numbers, 3rd edn. Oxford University Press (1954)
46. Helmholtz, H.: On the Sensations of Tone, p. 1954. Longmans & Co. Translated by A. J. Ellis, Dover reprint (1885)
47. Henrici, P.: Applied and Computational Complex Analysis, vol. 2. Wiley (1977)

48. Hensley, D.: Continued Fractions. World Scientific (2006)
49. Higham, N.J.: Accuracy and Stability of Numerical Algorithms, 2nd edn. SIAM (2002)
50. Hille, E.: Analytic Function Theory, 2nd edn. Chelsea Publishing Company (1982)
51. Hofstadter, D.R.: Gödel, Escher, Bach: ein Endloses Geflochtenes Band. Klett–Cotta (1985)
52. Horn, R.A., Johnson, C.R.: Matrix Analysis. Cambridge University Press (1985)
53. Ifrah, G.: Universalgeschichte der Zahlen. Campus Verlag, Frankfurt a. Main, New York (1987)
54. Isaacson, E., Keller, H.B.: Analysis of Numerical Methods. Wiley (1966)
55. Jordan, C.: Calculus of Finite Differences, 3rd edn. Chelsea (1965)
56. Karlin, S.: Total Positivity. Stanford University Press, Stanford CA (1968)
57. Karpenkov, O.: Geometry of Continued Fractions, Algorithms and Computations in Mathematics, vol. 26. Springer (2013)
58. Katznelson, Y.: An Introduction to Harmonic Analysis. Dover Books on advanced Mathematics, 2nd edn. Dover Publications (1976)
59. Khinchin, A.I.: Mathematical Foundations of Information Theory. Dover Publications (1957)
60. Khinchin, A.Y.: Continued Fractions, 3rd edn. University of Chicago Press (1964). Reprinted by Dover (1997)
61. Knuth, D.E.: The Art of Computer Programming. Seminumerical Algorithms, 3rd edn. Addison-Wesley (1998)
62. Kotelnikov: On the carrying capacity of the "ether" and wire in telecommunications. In: First All Union Conference of Communications I, zd. Red. Upr. Svyazi RKKA, Moscow (1933). In Russian
63. Kreyszig, E.: Introductionary Functional Analysis with Applications. Wiley (1978)
64. Krushchev, S.: Orthogonal Polynomials and Continued Fractions. From Euler's Point of View, Encyclopedia of Mathematics and Its Applications, vol. 122. Cambridge University Press (2008)
65. Kunis, S., Peter, T., Römer, T., von der Ohe, U.: A multivariate generalization of Prony's method. Linear Algebr. Appl. **490**, 31–47 (2016)
66. Loomis, L.H.: Introduction to Abstract Harmonic Analysis. Van Nostrand (1953). Dover reprint (2011)
67. Lorentz, G.G.: Approximation of Functions. Chelsea Publishing Company (1966)
68. Lorentzen, L., Waadeland, H.: Continued Fractions with Applications. Studies in Computational Mathematics. North Holland (1992)
69. Loy, G.: Musimathics. The Mathematical Foundations of Music. MIT Press (2011)
70. Mallat, S.: A Wavelet Tour of Signal Processing, 2nd edn. Academic Press (1999)
71. Mallat, S.: A Wavelet Tour of Signal Processing: The Sparse Way, 3rd edn. Academic Press (2009)
72. Markovsky, G.: Misconceptions about the golden ratio. Coll. Math. J. **23**, 2–19 (1992)
73. Mersenne, M.: Traité de l'harmonie universelle. Corpus des œvres de philosophie en langue française. Arthème Fayard (2003). Reprint, original Paris (1627)
74. Micchelli, C.A., Sauer, T.: Regularity of multiwavelets. Adv. Comput. Math. **7**(4), 455–545 (1997)
75. Mourrain, B.: Polynomial-exponential decomposition from moments (2016). arXiv:1609.05720v1
76. Padé, H.: Sur la représentation approchée d'une fonction par des fractions rationelles. Ann. éc. n. (1892)
77. Partington, J.R.: An Introduction to Hankel Operators, London Mathematical Society Student Texts, vol. 13. Cambridge University Press (2010)
78. Perron, O.: Die Lehre von den Kettenbrüchen I, 3rd edn. B. G. Teubner (1954)
79. Perron, O.: Die Lehre von den Kettenbrüchen II, 3rd edn. B. G. Teubner (1954)
80. Potts, D., Tasche, M.: Fast ESPRIT algorithms based on partial singular value decompositions. Appl. Numer. Math. **88**, 31–45 (2015)
81. Prony, C.: Essai expérimental et analytique sur les lois de la dilabilité des fluides élastiques, et sur celles de la force expansive de la vapeur de l'eau et de la vapeur de l'alkool, à différentes températures. J. de l'École polytechnique **2**, 24–77 (1795)

82. Roberts, G.E.: From Music to Mathematics. Exploring the Connections. Johns Hopkins University Press (2016)
83. Roth, K.F.: Rational approximations to algebraic numbers. Mathematika **2**, 1–20 (1955)
84. Roy, R., Kailath, T.: ESPRIT—estimation of signal parameters via rotational invariance techniques. IEEE Trans. Acoust. Speech Signal Process. **37**, 984–995 (1989)
85. Sagan, C.: Unser Kosmos. Droemersche Verlagsanstalt Th. Knaur Nachf. (1989). Deutsche Taschenbuchausgabe
86. Sauer, T.: Polynomial interpolation in several variables: Lattices, differences, and ideals. In: M. Buhmann, W. Hausmann, K. Jetter, W. Schaback, J. Stöckler (eds.) Multivariate Approximation and Interpolation, pp. 189–228. Elsevier (2006)
87. Sauer, T.: Prony's method in several variables. Numer. Math. **136**, 411–438 (2017). https://doi.org/10.1007/s00211-016-0844-8. arXiv:1602.02352
88. Sauer, T.: Reconstructing sparse exponential polynomials from samples: difference operators, Stirling numbers and Hermite interpolation. In: M. Floater, T. Lyche, M.L. Mazure, K. Moerken, L.L. Schumaker (eds.) Mathematical Methods for Curves and Surfaces. 9th International Conference, MMCS 2016, Tønsberg, Norway. Revised Selected Papers, Lecture Notes in Computer Science, vol. 10521, pp. 233–251. Springer (2017). arXiv:1610.02780
89. Sauer, T.: Hankel and Toeplitz operators of finite rank and Prony's problem in several variables (2018). Submitted for publication. arXiv:1805.08494
90. Sauer, T.: Prony's method: an old trick for new problems. Snapshots of modern mathematics from Oberwolfach (2018). https://doi.org/10.14760/SNAP-2018-004-EN
91. Sauer, T.: Prony's method in several variables: symbolic solutions by universal interpolation. J. Symbolic Comput. **84**, 95–112 (2018). https://doi.org/10.1016/j.jsc.2017.03.006. arXiv:1603.03944
92. Schelter, W.: Maxima—A Computer Algebra System. http://maxima.sourceforge.net (2001)
93. Schmidt, R.: Multiple emitter location and signal parameter estimation. IEEE Trans. Antennas Propag. **34**, 276–280 (1986)
94. Schmüdgen, K.: The Moment Problem. Graduate Texts in Mathematics. Springer (2017)
95. Schüßler, H.W.: Digitale Signalverarbeitung, 3rd edn. Springer (1992)
96. Seidel, L.: Bemerkungen über den Zusammenhang zwischen dem Bildungsgesetze eines Kettenbruchs und der Art des Fortgangs seiner Näherungsbrüche. Abh. München **7** (1855)
97. Shannon, C.E.: A mathematical theory of communication. Bell System Tech. J. **27**, 379–423 (1948)
98. Shannon, C.E.: Communications in the presence of noise. In: Proc. of the IRE, vol. 37, pp. 10–21 (1949)
99. Shilov, G.E.: Elementary Real and Complex Analysis. MIT Press (1973). Dover reprint (1996)
100. Silverman, R.A.: Introductionary Complex Analysis. Prentice Hall (1967). Dover reprint (1972)
101. Simoson, A.J.: Exploring Continued Fractions. From the Integers to Solar Eclipses. MAA Press (2019)
102. Stifel, M.: Arithmetica integra cum prefatione Philippi Melanchthonis. Johanne Petreius, Nürnberg (1544)
103. Stifel, M.: Vollständiger Lehrgang der Arithmetik. Königshausen & Neumann. German translation by E. Knobloch and O, Schönberger (2007)
104. Sweldens, W.: The lifting scheme: a custom-design construction of biorthogonal wavelets. ACHA **3**, 186–200 (1996)
105. Tolstov, G.P.: Fourier Series. Prentice-Hall (1962). Republished by Dover Publications (1972)
106. Vetterli, M., Kovačević, J.: Wavelets and Subband Coding. Prentice Hall (1995)
107. Wall, H.S.: Analytic Theory of Continued Fractions. D. Van Nostrand Company (1948). Dover reprint (2018)
108. Weierstraß, K.: Über die analytische Darstellbarkeit sogenannter willkührlicher Funktionen reller Argumente. Sitzungsber. Kgl. Preuss. Akad. Wiss. Berlin pp. 633–639, 789–805 (1885)
109. Whittaker, E.T.: On the functions which are represented by the expansions of the interpolation-theory. In: Edinburgh R. Soc. Proc., vol. 35, pp. 181–194 (1915)

110. Whittaker, J.: Interpolatory function theory, Cambridge Tracts in Mathematics and Mathematical Physics, vol. 33 (1935)
111. Wilkinson, J.H.: The perfidious polynomial. In: G.H. Golub (ed.) Studies in Numerical Analysis, MAA Studies in Mathematics, vol. 24, pp. 1–28. The Mathematical Association of America (1984)
112. Königliche Gesellschaft der Wissenschaften (ed.): Carl Friedrich Gauss. Werke. Dritter Band. Königliche Gesellschaft der Wissenschaften, Göttingen (1876)
113. Xu, Y.: Common Zeros of Polynomials in Several Variables and Higher Dimensional Quadrature. Pittman Research Monographs, Longman Scientific and Technical (1994)
114. Yosida, K.: Functional Analysis. Grundlehren der mathematischen Wissenschaften, Springer, New York (1965)

Index

Printed in the United States
by Baker & Taylor Publisher Services